International Housing Market Experience and Implications for China

Recent rapid housing market expansion in China is presenting new challenges for policy makers, planners, business people, and citizens. Now that housing in middle-income China is driven by consumer choices and is no longer dominated by state policy decisions, housing policy issues in Chinese cities are becoming increasingly similar to those encountered in other global housing markets. With soaring prices and imbalances in housing supply favoring high income groups and housing demand driven by rising inequality in household incomes, many middle and lower-income households face worsening choices in terms of the quality and location of their housing as well as greater financial difficulties, which together can have negative implications for standards of public health.

This book examines the impact of these changes on the general population, as well as on aspiring homeowners and developers. The contributors look at the effect on the widening of wealth gaps, slower economic growth, and threats to political and social stability.

Though focusing on China, the editors also present discussions of specific policy design challenges encountered in Australia, Japan, Korea, the Netherlands, the Nordic countries, Singapore, Taiwan, the UK, and the US. This book would be of interest to housing policy makers, as well as academics who are studying the social and political effects of the Chinese housing market.

Rebecca L. H. Chiu is a Professor and Head of the Department of Urban Planning and Design, and Director of the Centre of Urban Studies and Urban Planning and One Belt One Road Urban Observatory at the University of Hong Kong. She is an elected Fellow of the Academy of Social Sciences in the UK. Her current research interests include housing and urban sustainability in high-density Asian cities, housing in Hong Kong and China, comparative housing and planning studies, liveability in high-density cities and in ageing communities, and urban governance and urban management in China and the Belt and Road Region. She is the Founder Chairman of the Asia Pacific Network for Housing Research. She has been appointed to government boards and committees on housing, urban planning, land, urban renewal, and natural and heritage conservation in Hong Kong and elsewhere. She is co-author of *Politics, Planning and Housing Supply in Australia, England and Hong Kong* and chief editor of *Housing Policy and Social Development in Asia*, both published by Routledge.

Zhi Liu is a Senior Fellow and Director of the China Program at the Lincoln Institute of Land Policy, and Director of Peking University-Lincoln Institute Center for Urban Development and Land Policy in Beijing. Previously, he was a lead infrastructure specialist at the World Bank, where he gained many years of operational experience in the infrastructure and urban sectors. His research interests are infrastructure finance, municipal finance, land policy, and housing policy. He serves on several expert committees or advisory groups to the central and local governments in China on national social and economic planning, affordable housing policy, and fiscal policy reform.

Bertrand Renaud is an international consultant on urban development and financial markets development. Formerly, he was Adviser in the Financial Development Department of the World Bank, an institution where he has held various positions in finance and in urban affairs. He was the first Head of the Urban Affairs Division of the OECD in Paris. Previously, he was Professor of Economics at the University of Hawaii, where he specialized in Asian urban development. He has taught and done research in US and Asian universities, including MIT, Seoul National University, the University of Hong Kong, and the KDI School of Public Policy in Seoul, Korea. He has published extensively. His latest book, *The Dynamics of Housing in East Asia*, was co-authored with Kyung-Hwan Kim and Man Cho, and published by Wiley-Blackwell of Oxford in 2016. He holds MS and PhD degrees from the University of California at Berkeley, and an engineering degree from the Paris Institute of Technology for Life Sciences and the Environment, France. He is a Fellow of the Weimer School of the Homer Hoyt Advanced Studies Institute.

Routledge Studies in International Real Estate

The Routledge Studies in International Real Estate series presents a forum for the presentation of academic research into international real estate issues. Books in the series are broad in their conceptual scope and reflect an interdisciplinary approach to Real Estate as an academic discipline.

Econometric Analyses of International Housing Markets
Rita Li and Kwong Wing Chau

Sustainable Communities and Urban Housing
A Comparative European Perspective
Montserrat Pareja Eastaway and Nessa Winston

Regulating Information Asymmetry in the Residential Real Estate Market
The Hong Kong Experience
Devin Lin

Delhi's Changing Built Environment
Piyush Tiwari and Jyoti Rao

Housing Policy, Wellbeing and Social Development in Asia
Edited by Rebecca L. H. Chiu and Seong-Kyu Ha

Construction and Application of Property Price Indices
Anthony Owusu-Ansah

Residential Satisfaction and Housing Policy Evolution
Clinton Aigbavboa and Wellington Thwala

International Housing Market Experience and Implications for China
Edited by Rebecca L. H. Chiu, Zhi Liu and Bertrand Renaud

International Housing Market Experience and Implications for China

Edited by
Rebecca L. H. Chiu, Zhi Liu,
and Bertrand Renaud

Routledge
Taylor & Francis Group
LONDON AND NEW YORK

First published 2019
by Routledge
2 Park Square, Milton Park, Abingdon, Oxon OX14 4RN

and by Routledge
52 Vanderbilt Avenue, New York, NY 10017

Routledge is an imprint of the Taylor & Francis Group, an informa business

© 2019 selection and editorial matter, Rebecca Chiu, Zhi Liu, and Bertrand Renaud; individual chapters, the contributors

The right of Rebecca Chiu, Zhi Liu, and Bertrand Renaud to be identified as the authors of the editorial material, and of the authors for their individual chapters, has been asserted in accordance with sections 77 and 78 of the Copyright, Designs and Patents Act 1988.

All rights reserved. No part of this book may be reprinted or reproduced or utilised in any form or by any electronic, mechanical, or other means, now known or hereafter invented, including photocopying and recording, or in any information storage or retrieval system, without permission in writing from the publishers.

Trademark notice: Product or corporate names may be trademarks or registered trademarks, and are used only for identification and explanation without intent to infringe.

British Library Cataloguing-in-Publication Data
A catalogue record for this book is available from the British Library

Library of Congress Cataloging-in-Publication Data
Names: Chiu, Rebecca Lai-Har, editor. | Liu, Zhi, 1961– editor. | Renaud, Bertrand, 1939– editor.
Title: International housing market experience and implications for China / [edited by] Rebecca Chiu, Zhi Liu, and Bertrand Renaud.
Description: Abingdon, Oxon ; New York, NY : Routledge, 2019. | Series: Routledge studies in international real estate
Identifiers: LCCN 2018045490 |
Subjects: LCSH: Housing policy–China. | Housing development–China. | Residential real estate–China.
Classification: LCC HD7368.A3 I58 2019 | DDC 333.33/80951–dc23
LC record available at https://lccn.loc.gov/2018045490

ISBN: 978-1-138-34503-4 (hbk)
ISBN: 978-0-429-43814-1 (ebk)

Typeset in Goudy
by Newgen Publishing UK

Contents

List of figures x
List of tables xiii
List of contributors xv
Preface xxi

PART 1
Housing challenges, policies, and reforms 1

1 Global and local housing challenges 3
  BERTRAND RENAUD, REBECCA L. H. CHIU, AND ZHI LIU

PART 2
Housing challenges of developed economies 27

2 Taming real estate bubbles with technology 29
  EDWARD GLAESER

3 Thinking of the development of housing policy 44
  HANNU RUONAVAARA

4 House prices, land rents, and agglomeration benefits in the Netherlands 62
  COEN N. TEULINGS

5 Green Belts, edgelands, and urban sprawl: Reconsiderations of Green Belt concepts 88
  GUY M. ROBINSON

viii Contents

6 Land and housing market dynamics and housing policy in Japanese metropolitan cities 105
EIJI OIZUMI

7 The role of government and housing market dynamics in Korea 130
SEONG-KYU HA

8 The land and housing delivery system in Korea: Evolution, assessment, and lessons 150
MAN CHO, KYUNG-HWAN KIM, SOOJIN PARK, AND SEUNG DONG YOU

9 Housing price, housing mobility, and housing policy in Taiwan 177
CHIEN-WEN PENG, BOR-MING HSIEH, AND CHIN-OH CHANG

10 Housing challenges in Hong Kong's dualistic housing system: Implications for Chinese cities 201
REBECCA L. H. CHIU

11 Building an equitable and inclusive city through housing policies: Singapore's experience 230
SOCK-YONG PHANG

PART 3
Urban housing development and outcomes in China 255

12 The long-term dynamics of housing in China 257
BERTRAND RENAUD

13 Urban housing challenges in mainland China 288
ZHI LIU AND YAN LIU

14 Migration and urban housing market dynamics in China 307
YA PING WANG

15 Residential land supply and housing prices in China: An empirical analysis of large cities 330
HUINA GAO, ZHI LIU, AND YUE LONG

PART 4
Implications for China   353

16 A roadmap for housing policy: Lessons of international experience   355
EDWARD GLAESER, REBECCA L. H. CHIU, ZHI LIU, AND
BERTRAND RENAUD

Index   389

Figures

4.1	The CBD model of a city.	64
4.2	Land rents per square meter for each ZIP code.	68
4.3	Land rents in Amsterdam: a confirmation of the monocentric model.	69
4.4	Plot size negatively and population density positively related to land rent.	70
4.5	Land rent surplus due to production side externalities: private vs. public transport.	73
4.6	Decomposition of the value of land at various distances from the CBD.	78
4.7	Value of new construction per square meter for each ZIP code.	80
4.8	Value of new construction per square meter and actual new construction per square kilometer in the Amsterdam region.	81
4.9	New construction ranked by the value of new construction at the location.	82
4.10	Persistence in the growth of regions.	86
6.1	Changes in GDP and demand-side indices in Japan, 1990–2015.	106
6.2	Changes in average income of the bottom 90 percent, 1990–2010. Comparison of Japan with Korea, the UK, and the USA.	109
6.3	Distribution of relative frequency (percentage) of the household members with equivalent disposable income in Japan, 2012.	109
6.4	Changes in residential land prices of metropolitan Tokyo and Osaka and other cities, 1985–1994.	111
6.5	Cumulative changes in residential land prices in metropolitan Tokyo and Osaka and local areas, 1990–2015.	113
6.6	Average floor space per unit of owner-occupied and rental houses. Comparison of Japan with Britain, France, Germany, and the USA.	115
6.7	Construction of owner-occupied and rental houses in Japan, 1990–2015.	117
6.8	Changes in sales and average prices of condominiums in metropolitan Tokyo, 2006–2015.	118

Figures xi

6.9	Component ratio of the volume of private rental apartment houses classified by floor space, in 23 wards of metropolitan Tokyo, 2013.	119
6.10	Component ratio of each income group in homeowning households in Japan. Comparison between years of 1998 and 2013.	120
6.11	Component ratio of each income group in households residing in private rental houses in Japan. Comparison between years of 1998 and 2013.	121
7.1	Characteristics and dynamics of urban housing markets.	133
7.2	Percentage changes 1990–2015.	138
7.3	Housing purchase price index (June 2011=100).	138
8.1	New housing supply, 1962–2017.	160
8.2	Housing investment as a share of GDP and of total investment (TFCF).	161
8.3	Housing price, land price, and construction cost in real terms.	164
8.4	Number of unsold dwellings.	165
8.5	Real housing price cycles.	166
8.6	The process of land and housing development in Korea – a schematic view.	167
9.1	Annual house price (HP) growth and annual Gross Domestic Product (GDP) growth.	178
9.2	Annual house price (HP) growth and annual change of Consumer Price Index (CPI).	179
9.3	House price index and transaction volume.	179
9.4	House price-to-income ratio.	181
9.5	Mortgage payment-to-income ratio.	182
9.6	Homeownership rates.	183
9.7	Housing vacancy rates.	184
9.8	Map of Taiwan.	186
9.9	Net migration rate.	187
9.10	Migration rate.	187
9.11	Intra-city mobility rate.	187
9.12	Intra-district mobility rate.	188
10.1	Homeownership trend, 1982–2016.	202
10.2	The nominal price trend of private housing, 1981–2015.	203
10.3	Housing affordability trends, 1997–2016.	204
10.4	Trends of housing price and housing transaction volume, 2002–2016.	204
10.5	Rental trends and rental affordability, 1997–2016.	205
10.6	The trend of public rental housing applications, 2001–2016.	206
10.7	Housing ladder and housing tiers – by subsidy eligibility only.	209
10.8	Price changes of different housing market niches, 1997–2016.	213
10.9	Reclaimed land supply, 1985–2015.	216

xii *Figures*

10.10	Average best lending rates of Hong Kong and the U.S., mortgage rates, and the inflation rates of Hong Kong, 1990–2014.	217
10.11	Number of loans associated with co-financing schemes, 1999–2014.	219
10.12	Number of housing loans refinanced compared to price index of private housing, 2003–2014.	220
10.13	Changing housing vacancy ratio by unit size over time.	221
10.14	Housing price trends of four prime cities in China (2011–2016) with announcement dates of price-cooling measures.	223
11.1	Homeownership rate.	234
11.2	International comparison of income Gini coefficients.	235
12.1	The three stages of long-term economic development.	261
12.2	US regional specialization in manufacturing and other activities, 1860–1987.	264
12.3	Urban dimensions of long-term economic development: a restatement.	266
12.4	Housing investment and level of urbanization.	266
12.5	Growth transition and 50% level of urbanization in China, Japan, South Korea, and Taiwan.	269
13.1	Residential housing investment in China, 1998–2016.	290
13.2	Sales of residential housing in square meters, 2000–2015.	290
13.3	Average residential housing price index for 35 major cities, 2000–2016.	291
13.4	Average price-to-income ratio for 35 major cities, 2001–2015.	292
13.5	Residential housing prices for 100 cities since mid-2010.	293
13.6	Land concession revenue and local tax revenue, 2001–2016.	294
13.7	Population and urbanization rate in China, 2005–2015.	296
13.8	Average monthly rent in Beijing, 2012–2016.	298
14.1	Average annual salary and housing price in urban areas.	310
14.2	Aerial image of the county seat.	316
14.3	A new street paved for housing expansion in the town.	317
14.4	Housing estates under construction in 2012.	318
14.5	Rural labor migration and remittance flow.	321
14.6	Housing market catchment areas in the county seat.	322
14.7	Trends of migration and the buying-up and buying-down processes.	323
15.1	Residential land supply ratio in Beijing.	331
15.2	Residential land supply ratio in Hangzhou.	331
15.3	Residential land supply ratio in Taiyuan.	332
15.4	Residential land supply ratio in Zhengzhou.	332
15.5	Residential land supply ratio in Chengdu.	333
15.6	Residential land supply ratio in Huhehaote.	333

Tables

3.1	Types of housing tenure in 2014 in the Nordic countries (unit: percentages).	46
4.1	Which factors explain the wide variety in land rents per square meter across ZIP codes?	71
4.2	Decomposition of the average land rent surplus for a number of cities.	72
4.3	Population growth rate for various agglomerations in the Netherlands, 1985–2012.	77
4.4	Contribution of various characteristics of the log construction index h.	84
5.1	Land-use change in the edgelands: the example of Spelthorne Borough, South West London.	92
6.1	Average income and loan repayment of worker households borrowing housing loans in Japan. Comparison between years of 2000 and 2013.	120
6.2	Average income and rent payment of worker households residing in private rental houses in Japan. Comparison between years of 2000 and 2013.	121
6.3	Housing stock and component ratio of housing tenures in Japan, 1958–2013.	124
7.1	Housing conditions in Korea.	131
7.2	Dwelling stock by housing type in Korea (unit: no. of thousand dwelling units (%)).	131
7.3	Housing tenure in Korea, 1975–2015 (unit: %).	132
7.4	Public rental housing stock in 2014.	140
7.5	Percentage of public housing for 20-year or longer leases against total housing stock in Seoul.	141
8.1	Improvement in housing conditions in Korea.	160
8.2	The supply of public rental housing.	162
8.3	Supply of developable land by the public sector (unit: 1,000 square meters).	163
8.4	Supply of developable land by region (unit: 1,000 square meters).	163

xiv Tables

10.1	Changing composition of housing by unit size over time (%), selected years.	203
10.2	Living space by type of dwelling in Hong Kong, 2013/2014.	207
10.3	Housing tiers defined by housing subsidy eligibility.	210
10.4	Projected supply of subsidized owner-occupied housing and unmet demand, 2013.	213
11.1	Distribution of households by household income categories in 2015.	238
11.2	Proportion of families/households in low-, middle-, and high-income neighborhoods: US metropolitan areas > 500,000 and Singapore.	238
11.3	Singapore: relationship between average household income and house type, 2017.	239
11.4	Proportion of population residing in low-, middle- and high-income neighborhoods, 2017.	240
11.5	Socio-economic characteristics by ethnic group, 2015.	240
11.6	Ethnic limits under the HDB's Ethnic Integration Policy.	246
13.1	Housing for sale in China, 2011–2015.	292
13.2	The planned and actual provision of affordable housing in China, 2010–2015.	300
14.1	Average price of commercial housing in selected cities, February 2016.	312
14.2	Housing market regulation policy options.	313
14.3	Population changes in the county.	315
14.4	Details of a housing estate under construction in 2010.	318
15.1	Descriptive statistics.	339
15.2	Estimation results of benchmark regression.	340
15.3	Estimation results of mechanism test (1).	343
15.4	Estimation results of mechanism test (2).	344
15.5	Estimation results of mechanism test (3).	345
15.6	Estimation results of mechanism test (1), with one-year lag.	346
15.7	2015 data used in the research.	348

Contributors

Chin-Oh Chang is a Distinguished Professor in the Department of Land Economics and the Director of the Taiwan Real Estate Research Centre, National Chengchi University, Taipei, Taiwan. He received his Master's in Architecture from MIT, USA in 1980, and a City and Regional Planning PhD at the University of Pennsylvania, USA in 1986. He was President of the Asian Real Estate Society (AsRES) from 1997 to 1998, and President of the Global Chinese Real Estate Congress (GCREC) from 2009 to 2010. His research has been published in *Urban Studies*, *Housing Studies*, *Journal of Property Research*, *Habitat International*, *International Real Estate Review*, and several other journals. He has concentrated on areas related to housing and land policy, real estate investment, and financial analysis.

Rebecca L. H. Chiu is a Professor and Head of the Department of Urban Planning and Design, and Director of the Centre of Urban Studies and Urban Planning and One Belt One Road Urban Observatory at the University of Hong Kong. She is an elected Fellow of the Academy of Social Sciences in the UK. Her current research interests include housing and urban sustainability in high-density Asian cities, housing in Hong Kong and China, comparative housing and planning studies, liveability in high-density cities and in ageing communities, and urban governance and urban management in China and the Belt and Road Region. She is the Founder Chairman of the Asia Pacific Network for Housing Research. She has been appointed to government boards and committees on housing, urban planning, land, urban renewal, and natural and heritage conservation in Hong Kong and elsewhere. She is co-author of *Politics, Planning and Housing Supply in Australia, England and Hong Kong* and chief editor of *Housing Policy and Social Development in Asia*, both published by Routledge.

Man Cho is a Professor in Real Estate Finance and Regional Development at KDI School of Public Policy and Management, Sejong City, Korea, and is jointly appointed as a Research Fellow at KDI (Korea Development Institute). His teaching and research areas include real estate finance, credit risk management, financial innovations (FinTech), and urban and regional economics. Before joining the KDI School in May 2007, he worked for Fannie Mae in the US, in which he was involved with various R&D projects, such as mortgage and MBS pricing and collateral (property) assessment, and also served in several managerial positions. Prior to Fannie Mae, he worked for the World Bank as a long-term consultant, and taught at Johns Hopkins University as an adjunct professor. He holds a PhD in Applied Economics and Managerial Science from the Wharton School of the University of Pennsylvania.

Huina Gao is an Assistant Professor in the Department of Tourism Management, Beijing International Studies University, China. She also serves as an academic expert to the Beijing Tourism Institute. She was a research assistant at the Peking University-Lincoln Institute Center for Urban Development and Land Policy from 2010 to 2015, a visiting scholar at the Lincoln Institute of Land Policy in 2016, and a postdoctoral researcher in the School of Economics, Peking University from 2015 to 2017. She holds a PhD in human geography from Peking University. Her main research areas are housing and land policy, real estate development, and tourism investment.

Edward Glaeser is the Fred and Eleanor Glimp Professor of Economics in the Faculty of Arts and Sciences at Harvard University, where he has taught since 1992. He regularly teaches microeconomics theory, and occasionally urban and public economics. He has served as Director of the Taubman Center for State and Local Government, and Director of the Rappaport Institute for Greater Boston. He has published dozens of papers on cities' economic growth, law, and economics. In particular, his work has focused on the determinants of city growth and the role of cities as centers of idea transmission. He received his PhD from the University of Chicago in 1992.

Seong-Kyu Ha is an Emeritus Professor of Urban Planning and Real Estate at the Chung-Ang University, South Korea. He also serves as the President of the Korea Research Institute of Housing Management. His current research interests have centered on low-income housing policies, urban residential regeneration, and housing management. He has authored numerous publications on housing policy and urban and community regeneration, including *Housing Policy and Practice in Asia* (Croom Helm, 1987). He has been appointed to government boards and committees related to urban planning and housing policy. Recently, he was elected Chairman of the Korea Housing Service Society, a housing expert consulting and research group in Korea.

Bor-Ming Hsieh is an Associate Professor and Head of the Department of Land Management and Development, and Head of the Bachelor's Program in Real Estate Finance at Chang Jung Christian University, Tainan, Taiwan. He received his PhD in Urban Studies at the University of Glasgow, UK, in 2002. His current research interests center on the housing market, spatial analysis of housing prices, housing policy, and urban regeneration. His research is published in *Journal of Real Estate Research*, *Housing Studies*, *City and Planning*, *Journal of Statistics and Management Systems*, and several other journals. He is on the Board of Directors of the Asian Real Estate Society and the Global

Chinese Real Estate Congress, and a Steering Committee Member of the Asia Pacific Network for Housing Research.

Kyung-Hwan Kim is a professor of economics at Sogang University. Dr. Kim is a fellow of the Weimer Graduate School of Advanced Studies in Real Estate and Urban Land Economics, a fellow of the Asian Real Estate Society (AsRES), and a member of the editorial board of *Journal of Housing Economics* and of the international advisory board of *Housing Studies*. He was a Vice Minister of the Ministry of Land, Infrastructure and Transport of the Republic of Korea, President of the Korea Research Institute for Human Settlements, and President of AsRES. Dr. Kim has taught at Syracuse University, the University of Wisconsin-Madison, and the Singapore Management University. He has published in major journals in urban economics and real estate. He received his PhD in economics from Princeton University in 1987.

Yan Liu is a postdoctoral researcher in the College of Urban and Environmental Science, Peking University, China. She also serves as a research fellow to Peking University-Lincoln Institute Center for Urban Development and Land Policy. She holds a PhD in theoretical economics from Nanjing University. Her main research areas are housing policy, real estate development, and institutional economics.

Zhi Liu is a Senior Fellow and Director of the China Program at the Lincoln Institute of Land Policy, and Director of Peking University-Lincoln Institute Center for Urban Development and Land Policy in Beijing. Previously, he was a lead infrastructure specialist at the World Bank, where he gained many years of operational experience in the infrastructure and urban sectors. His research interests are infrastructure finance, municipal finance, land policy, and housing policy. He serves on several expert committees or advisory groups to the central and local governments in China on national social and economic planning, affordable housing policy, and fiscal policy reform.

Yue Long is an Analyst in the Division of Health and Environment at Abt Associates Inc. She received her Master's in Quantitative Methods in the Social Sciences (QMSS) from Columbia University, USA in 2016, and her Bachelor's in Mathematics from Kenyon College, USA in 2014. She specializes in data analysis and data visualization in fields such as healthcare and environment policy, and has worked with clients such as the National Institute of Health (USA), the Center for Medicare and Medicaid Services (USA), and the Environmental Protection Agency (USA), over the years. She has also co-authored publications in *Journal of Immigrant and Minority Health* and *Journal of Offender Rehabilitation*.

Eiji Oizumi is Emeritus Professor at the Wakayama University, and Professor of Regional Economics at the Otemon Gakuin University, Japan. He also serves as one of the vice-presidents of the Japan Housing Council, an interdisciplinary research organization for housing in Japan. His research interests have centered on land and housing markets and policies. His publications in English include works on *Property Finance in Japan*, *Housing Provision and Marketization in 1980s and 1990s Japan*, *Transformations in Housing Construction and Finance in Japan*, and *Government Mortgage Institutions in Asia-Pacific*.

Soojin Park is a Finance Professor of the Energy Policy and Engineering department at KEPCO International Nuclear Graduate School. His research topic covers infrastructure investment policy and finance. Before joining the School, he worked for the Korea Development Institute, where he was responsible for the pre-feasibility studies on national infrastructure projects, including PPP. He earned a PhD in Development Policy from the KDI School of Public Policy and Management. He obtained a Master's degree from the Baker Program of Cornell University. He is qualified with professional certifications: CPA, CTA, CIA, and CFE. He was a member of the APEC PPP Expert Panel and G-20 PPP working group. He was a member of Body of Knowledge for the PPP program in the World Bank Group.

Sock-Yong Phang is Vice Provost (Faculty Matters) and the Celia Moh Chair Professor of Economics at the Singapore Management University, Singapore. She received her PhD from Harvard University in 1989. She is the author of *Policy Innovations for Affordable Housing in Singapore* (2018) and *Housing Finance Systems: Market Failures and Government Failures* (2013), and has published numerous papers on housing and urban transportation. Professor Phang has previously served as a board member of Singapore's Urban Redevelopment Authority, Land Transport Authority, Public Transport Council, and Competition Commission. She is currently a board member of Singapore's Energy Market Authority, a trustee member of the ISEAS-Yusof Ishak Institute, and a Senior Research Fellow at the Centre for Research on the Economics of Aging.

Chien-Wen Peng is a Professor and Head of the Department of Real Estate and Built Environment at the National Taipei University, New Taipei, Taiwan. He is President of the Chinese Society of Housing Studies, and President of the Chinese Institute of Land Economics. His research is published in *Urban Studies*, *Habitat International*, and several other journals. He has concentrated on areas related to housing and land policy, real estate cycles forecasting, and urban renewal.

Bertrand Renaud is an international consultant on urban development and financial markets development. Formerly, he was Adviser in the Financial Development Department of the World Bank, where he has held various positions. He was the first Head of the Urban Affairs Division of the OECD in Paris. Previously, he was Professor of Economics at the University of Hawaii, where he specialized in Asian urban development. He has taught and done research in US and Asian universities, including MIT, Seoul National University, the University of Hong Kong, and the KDI School of Public Policy in Seoul, Korea. He has published extensively. His latest book, *The Dynamics* *of Housing in East Asia*, was co-authored with Kyung-Hwan Kim and Man Cho, and published by Wiley-Blackwell of Oxford in 2016. He holds MS and PhD degrees from the University of California at Berkeley, and an engineering degree from the Paris Institute of Technology for Life Sciences and the Environment, France. He is a Fellow of the Weimer School of the Homer Hoyt Advanced Studies Institute.

Guy M. Robinson is a Human Geography Professor based at the University of Adelaide in South Australia. He is also a Departmental Associate in the Department of Land Economy, University of Cambridge. He has been the editor of *Land Use Policy* since 2007, and works on environmental management and rural development issues. Recently he has been a Guest Professor at the Chinese Academy of Sciences in Beijing, and GIAN Visiting Fellow at Aligarh Muslim University, India. His major publications include *Handbook on the Globalisation of Agriculture*, *Sustainable Rural Systems*, *Methods and Techniques in Human Geography*, and *Conflict and Change in the Countryside*.

Hannu Ruonavaara is Professor of Sociology and Head of the Department of Social Research at the University of Turku, Finland. His current theoretical and empirical research interests center on historical change of housing policies, regulation of rental markets, and the nature of neighbor relations and disputes. His research has been published in international research journals such as *Housing Studies*, *Housing, Theory and Society*, *International Journal of Housing Policy*, *Journal of Housing and the Built Environment*, *Philosophy of the Social Sciences*, *Sociological Research Online*, and *Sociological Theory*. He is the current editor of the journal *Housing, Theory and Society*.

Coen N. Teulings is Distinguished Professor of Economics, Institutions and Society at Utrecht University. Previously, he was President of the Netherlands Bureau for Economic Policy Analysis CPB, and Professor of Labour Economics at the University of Cambridge. His research interests are labor markets, urban economics, and macro-economic policy.

Ya Ping Wang is Professor and Chair in Global City Futures at the School of Social and Political Sciences, University of Glasgow. He is Director of the UKRI Global Challenge Research Fund Centre for Sustainable, Healthy and Learning Cities and Neighbourhoods (SHLC). He leads several UK ESRC and China collaborative research projects on urban transformation. His research focuses on housing, urban transformation, and the living conditions of rural migrants. He is author/co-author of *Housing Policy and Practice in China* (Macmillan), *Urban Poverty, Housing and Social Change in China* (Routledge), and *Planning and Housing in the Rapidly Urbanizing World* (Routledge). Ya Ping Wang is a Fellow of the Social Science Academy (FAcSS).

Seung Dong You is an Assistant Professor and Head of the Department of Economics and Finance at Sangmyung University, Seoul, Korea. He received his PhD from the Sauder School of Business at the University of British

Columbia, and also received his Master's degree from the Baker's Program of Cornell University. With his strong interests in real estate finance, he has been extensively involved in many aspects of housing finance since 2002. He has authored several publications on housing finance, including *Housing Finance Mechanisms in the Republic of Korea* (UN Habitat, 2009). His current research interests cover strategic financial management, and real estate economics and finance.

Preface

This book grows out of the International Symposium on Land Policy and Housing Market held on May 16–17, 2016 in Hong Kong. It was co-organized by the University of Hong Kong (School of Social Sciences and Center of Urban Studies and Planning), the Lincoln Institute of Land Policy, and Peking University-Lincoln Institute Center for Urban Development and Land Policy. The event was conceived at a time when the soaring housing prices in mainland China and Hong Kong had seriously widened the wealth inequalities of the population, deepened social divides, undermined economic growth, and threatened social stability. The purpose of the symposium was to examine the housing market experiences and housing polices of mainland China and Hong Kong in an international context. The symposium brought together a score of international housing experts from the fields of economics, geography, city planning, public policy, and sociology, across countries and regions.

Having turned the conference papers into rigorous academic discourses, this book presents the latest scholarly understanding of international housing market experiences. Its special hope is to shed light on the rapidly evolving housing market in mainland China, which is going through another important transitional growth period as it moves on as an upper-middle-income economy. Each chapter intends to add, in one way or another to our current knowledge of the different facets of housing markets and systems in different countries, while bearing in mind their relevance to mainland China. However, understandably it does not intend to cover all aspects of the international housing markets and housing policies. The concluding chapter builds on the outcomes of a post-symposium roundtable discussion attended by symposium speakers.

This book would not have been produced without the support of the following people: George C. S. Lin and Simon X. Zhao, both faculty members of the University of Hong Kong (HKU), for their active role in co-organizing the symposium in 2016; John Burns and Chris Webster, Dean of Social Sciences and Architecture, HKU respectively, for their support to the symposium; George W. McCarthy, President of the Lincoln Institute of Land Policy, for his support to the symposium and the book; and Elizabeth Fox and Sandra Mather for their editing and cartographic work respectively. Last but not least, we thank the

Faculty of Social Sciences, HKU and the Center of Urban Studies and Urban Planning of the Faculty of Architecture, HKU for co-sponsoring the symposium, and the Lincoln Institute of Land Policy for co-sponsoring the symposium and providing financial support to the production of the manuscript.

Rebecca L. H. Chiu, Zhi Liu, and Bertrand Renaud

Part 1
Housing challenges, policies, and reforms

1 Global and local housing challenges

Bertrand Renaud

HOMER HOYT INSTITUTE, U.S.

Rebecca L. H. Chiu

THE UNIVERSITY OF HONG KONG

Zhi Liu

LINCOLN INSTITUTE OF LAND POLICY AND PEKING UNIVERSITY-LINCOLN INSTITUTE CENTER FOR URBAN DEVELOPMENT AND LAND POLICY

The grave concerns over rapidly soaring housing prices are not unique to mainland China and Hong Kong. They are widespread around the world and have intensified in the aftermath of the Great Recession incurred by the Global Financial Crises of 2008/2009. What could mainland China learn about the present housing market issues and solutions from different countries and regions of the world, in order to tackle its own? The introductory chapter sets out the international and local contexts for addressing this question, which is the theme of this book. It has two main parts. First, it reviews with some degree of analytical detail the structural changes that have been affecting housing systems with increasing intensity since the Great Recession of 2009. Then it provides a comparative context for the country experiences of the book, presented in three groups: (1) Western experiences (Australia, the Netherlands, the Nordic countries, the UK, and the US); (2) East Asian experiences outside mainland China (Japan, Korea, Hong Kong, Singapore, and Taiwan); and (3) the Chinese experience prior to its present growth transition. The chapter closes with the housing policy lessons of international experience and their value for the housing policy adjustments that China needs to make during the major growth transition that it faces now.

1. Housing challenges since the great recession of 2009

1.1 Major impacts of the great recession of 2009

The Global Financial Crisis (GFC) of 2008 was a seismic shock to the global economy and to the housing systems of most countries. Its immediate impact was the Great Recession of 2009, which was the worst global recession since the Great Depression of the 1930s. These twin crises have reshaped our world.

A decade has passed since the onset of the crisis that Ben Bernanke, former US Federal Reserve Chairman and leading scholar in the subject, has called the "worst financial crisis in global history, including the Great Depression."[1] The severity of the crisis had its origins in the Euro-American banking systems that were the axis of world finance at the time. These banking systems had become extremely large compared to the size of the non-financial economies they were supposed to serve. They were also exceedingly leveraged and deeply intertwined, especially through the rapid growth of financial innovations in unregulated and poorly monitored parts of the global financial system, which gained the moniker of "shadow banking" (Tooze, 2018).

In 2018, the present moment, the global economy has reached a new watershed. Leading central banks have started gradually withdrawing the extraordinary monetary support that they had been providing to the economy for a decade, and both interest rates and inflation rates are rising. The political environment has also profoundly changed. The once-dominant consensus in favor of economic and financial liberalization was shattered by the twin financial and economic crises. The global policy environment that had been shaped for seven decades by high-income economies, and the U.S. in particular, is in a state of considerable flux regarding the future.

A decade after the GFC, there are mixed reviews about whether financial markets have become more stable and more resilient to future shocks (International Monetary Fund, 2018). Remarkably, international comparative indicators show that the stability of China's financial system has reached the highest risk level of all the monitored countries, with a high probability of financial distress. This situation is the direct result of the monetary and fiscal policies aimed at supporting employment and growth that China adopted during the Great Recession of 2009 and has maintained since then.

Part of the growth transition challenges that China now faces is to move away from the credit-driven growth of the past decade that leads to increasing inefficiency and waste. The implications of new financial and fiscal policies are very significant for Chinese housing because they will have a direct impact on local governments and on households. In recent years, Chinese household indebtedness has been rising at a very fast rate, although from a low base; additionally, the deteriorating quality of mortgage credit is often a direct cause of housing busts (International Monetary Fund, 2017).[2]

1.2 Five major structural changes are creating a new housing policy environment

In today's unsettled global environment, predicting how well any given individual housing system is likely to meet its desired social, political, economic, and environmental objectives is hard. Policy challenges exist at the national level because of the transmission of shocks across countries, and also at the local level because housing conditions often differ significantly across large and small cities, and between urban and rural areas.

Why have some national housing systems been more resilient to economic, financial, or demographic shocks than others, such as Canada during the GFC of 2008? Why do Canadian housing conditions now differ across major cities in 2018? Why did East Asian housing suffer so much less from the GFC than from the Asian Financial Crisis (AFC) of 1997? Why do some housing systems serve their populations better than others in terms of access, affordability, physical quality, and socially integrated and inclusive cities, such as Singapore?

Housing is a complex economic and social good to analyze and manage properly for four main reasons. First, its durability is measured in decades. Second, it can be quite heterogeneous in terms of its structural features and unit design. Third, its value is not simply determined by the characteristics of the unit itself, but is also linked to the characteristics of the neighborhood where the unit is located. Fourth, the way housing is financed, built, and taxed is usually extensively regulated by national and local governments.

In addition, national or local structural idiosyncrasies affect housing performance to varying degrees. These result from the path dependence of every housing system. The concept of path dependence refers to the contingent and non-reversible dynamical evolution of a housing system over time (David, 2000). Contingency refers to the historical accidents and the public and private responses to internal as well as external shocks that can change a housing system and lastingly reshape it over time. This path dependence of a housing system has the following five dimensions:

i. Physical, through the interactions between the long life of housing, land-use planning, and other forms of urban planning;
ii. Legal, because new laws have to take into consideration existing laws, especially in the case of constitutional and legal definitions of property rights;
iii. Institutional, through laws, policies, and government regulations that shape the organization and the competitiveness of the local real estate industry, and of the real estate development process;
iv. Economic, as current conditions in the economy support or hinder the accumulation of housing assets attractive to different members of society; and
v. Socio-political, as definitions of affordable housing and social housing policies are shaped by prevalent socio-political views that can vary very significantly across different eras.

Because China has the largest housing system in the world in scale and in value, it is worth keeping in mind the structural legacies from its central planning era that still affect the performance of Chinese housing and the design of today's housing policies. These legacies include the privatization of state-owned enterprises and the housing reforms of 1998 that played a major role in the emergence of high-net-worth households and had a major impact on worsening the household income distribution, and therefore on the composition of the demand for new housing now driven by high-income households. The discriminatory residential registration system (known as *hukou*) limits the access of the low-income rural-urban migrants to housing and other urban services, especially in the largest industrial cities. The major differences in property rights between rural land and urban land have had a major impact on the spatial structure of cities and the efficiency of their land markets. The lasting impact of the intergovernmental fiscal reforms of 1994 created strong incentives for local governments to borrow and invest heavily in land and real estate in order to improve the balance between their obligations and their fiscal revenues.

In addition to traditional issues of housing access and affordability, policy responses have also had to adapt to five new structural changes that are affecting high-income housing systems to varying degrees. These structural changes are shaping national housing development paths. They affect the diagnosis of housing sector performance and the principles of good housing policy design, which are our leading interest in this book. These five structural changes interact with each other in ways that have not been quantified and may differ from country to country.

a. Demographic change and population aging

Population aging is a gradual structural change that is difficult to reverse and has major long-term implications for the internal organization and performance of a housing system. There are three main dimensions of aging in a given society: the decline in total fertility rates (TFR), increases in life expectancy, and variations in birth and death rates across age cohorts that can cause baby booms and echo booms, and also the reverse over time. Such demographic mini-booms and busts have had a direct impact on the level and composition of housing demand in different time periods.

In an increasing number of countries, TFRs have been falling below the replacement rate of 2.1 that would stabilize the population. Two important drivers of falling TFRs are the relatively low socio-economic status of women irrespective of their education level, and the structure of the economy when a low share of aggregate consumption in GDP signals low household living standards and a high cost of living for a given per capita GDP of the economy. In mainland China, the new economic policy of rebalancing the economy in favor of domestic demand and household demand could affect the TFR.

Across East Asian societies (mainland China, Japan, South Korea, Taiwan, Hong Kong, and Singapore), the speeds in years and the magnitudes of change in demographic indicators are the strongest in the world. According to the U.S. Population Bureau, the lowest TFRs in the world are found in East Asia. In the small city economies of Hong Kong and Singapore, these demographic changes can be masked by in-migration that selectively affects different parts of the household income distribution. However, demographic aging is not unique to East Asia. Some large European societies like Germany and Italy are also aging quickly. These societies have fewer and fewer children and live longer and longer. The net effects are a rapidly rising share of people older than 65, a falling ratio of the active labor force available to support the inactive rest of the population, and the shrinking size of households that in turn changes the composition of housing demand toward one-person households. Improving the demographic structure through international migration has become politically contentious in most countries.

Analytically, demographers define an "aging" society as one having at least 7 percent of its population older than 65. An "aged" society has more than

14 percent of its population above 65. Japan has been the fastest aging society in the world so far. It took Japan only 24 years to move from aging to aged, while it took 40 years for Germany to make the same transition. However, South Korea is now on track to age even faster than Japan. In China, the size of the active labor force began declining in 2012. Recent Chinese government policies to reverse the mandatory one-child policy long criticized by demographers are proving unsuccessful because of the very high cost of raising and educating a child in cities, especially large ones. Japan currently has the world's highest life expectancy, thanks to the high quality of its health system and its lifestyle. In parallel, life expectancy is now rising beyond 90 years in South Korea for similar health and societal reasons. Life expectancy is higher for women than men around the world.

Until now, population aging has not been integrated into macro-economic analyses and macro-economic management because of the short-term dimension of political and economic decision-making compared to the slow, gradual, long-term nature of demographic change. It takes a small, very cohesive, long-term-oriented society like Estonia to implement policies that will raise the status of women and improve family living conditions well and coherently enough to raise the TFR. Otherwise, it took the highly stressful demographic changes of Japan to begin to integrate demographic change into national policies. Spectacularly, the Abe administration has created in 2018 a national "Council for Designing the 100-year Life Society."[3] However, concrete, pragmatic actions to improve housing and other living conditions for their residents have remained the initiative of competing autonomous Japanese local governments.

b. Worsening income, wealth inequality, and job insecurity

Economic inequality has become a major policy issue in the aftermath of the Global Financial Crisis because public policies have relied on household savings and taxation to quickly rescue financial institutions that had enjoyed high incomes before the crisis. Meanwhile, programs to assist insolvent households facing mortgage defaults were often slow and incomplete. Reflecting widespread and growing public dissatisfaction, Thomas Piketty's *Capital in the 21st Century* (2014) even became an unexpected bestseller. Piketty's thorough economic study documents the rising shares of income and wealth to the benefit of high-income groups that are the predominant holders of capital in advanced economies.

The main channel of increased *income* inequality has been the structure of labor markets. One duality is the presence of very high productivity and capital-intensive international corporations co-existing with a large number of low-productive small and medium enterprises (SMEs), where the largest share of the labor force works. The other problem is a rapid increase of short-term contracts, part-time workers, and temporary employees affecting young workers since the 2009 GFC, which weakens their access to housing for rental and ownership, as rising housing prices are driven by high-income households.

The distribution of *wealth* is even more skewed than that of income. A startling result of a follow-up evaluation of Piketty's findings by Matthew Rognlie

(2015) was that the net increase of the share of capital in developed economies that has been rising since 1948 was *entirely* due to housing capital gains. Policies concerned about wealth inequality must therefore focus closely on the housing channel. In countries like China, where income and wealth inequality has worsened, empirical evidence shows that in private housing markets, the composition of new housing supply in Tier-1 cities has been dominated by the demand from high-net-worth groups and the top household income deciles during the Great Housing Boom of 1998–2013. Similar outcomes prevail in other countries.

In East Asia, the diversification of household portfolios away from housing investment into financial assets has long been constrained by public finance policies and the relatively late or limited development of public and private pension funds serving most household income deciles. Usually, only the top two income deciles have developed relatively diversified asset portfolios.

Another wealth effect affecting selected international cities is the globalization of real estate markets since 2009, in which high-net-worth households seek housing investments to protect their wealth and raise their expected returns in "superstar cities" that combine high-quality amenities, a stable currency, high returns, and secure property rights. Such investment must, for effect, sharply raise local housing prices from the top down and considerably worsen the average housing price-to-income ratios for residents. Among other superstar cities of the world, this problem differentiates selected Canadian cities in 2018 from those in 2008. In Hong Kong, corporate loopholes debilitate the new tax policies designed to discourage such housing investments.

The price levels of housing markets have recovered fully since the 2008 financial crisis in most economies. However, since the crisis, the housing sector has been a major channel of wealth redistribution across age cohorts. In the U.S., more than 40 percent of housing wealth is now concentrated in the hands of those aged 60 or more, according to a recent study of the Federal Reserve Bank of New York, in comparison with 24 percent in 2006, just prior to the crisis. People under the age of 45 now own only 14 percent of housing wealth, down from 24 percent in 2006. As a 2018 New York Federal Reserve Bank press briefing puts it, "Household formation and transitions from renting into home-ownership are much lower than their pre-crisis levels" (Federal Reserve Bank of New York, 2018).

A very important consequence of the redistribution of housing wealth in the hands of high-net-worth individuals since the 2008 financial crisis is that housing markets are functioning less well. The surplus of empty houses has doubled in many "booming" housing markets around the world. The fact that the number of units may be larger, and may even be rising faster, than the number of households can be totally misleading about the existence of housing shortages. The number of housing units is not "supply" because it says nothing about the composition of the housing stock (i.e., its location, quality, size, age, tenure, transport links, and neighborhood amenities). Similarly, the aggregate number of households says nothing about the composition and characteristics of the demand for "housing services." As people face high and rising housing costs together with greater employment uncertainty, the rate of household formation among young age cohorts has fallen. An increased rate of doubling up in a housing unit means that the consumption of housing space and housing services is falling for younger age cohorts.

Beyond housing issues, income and wealth inequality have negative impacts on the performance of a society. Recent analyses by the International Monetary Fund based on the best comparative data indicate that *high* levels of net inequality (after adjusting for the effects of redistributive policies) impair economic growth by making it more volatile and less sustainable. Ostry et al. (2014) also find that the long-prevailing academic view that there is a "big trade-off" between equality and efficiency or equity and that growth is not supported by the available evidence. In countries like the U.S., such an academic view has been repeatedly and successfully used by wealthy owners of capital to defend their privileged status in political debates.

It is not clear from the available evidence that inequality undermines public governance or that rising inequality corrodes the quality of public institutions and policies. On the other hand, Beesley and Persson (2011) find evidence that inequality limits the demands for social action, which over time reduces the capacity of the state to act and weakens its legal capacity, fiscal capacity, and its capacity to deliver *public goods* (such as environmental policies). They also find evidence that from time to time the values of citizens can shift markedly.

c. Rising overall level of household debt

Before the financial crisis of 2008, there were five main drivers of the running-up of household debt that occurred in high-income economies: financial liberalization and the deregulation of bank lending; financial innovation and the loosening of credit standards; the decline of borrowing costs; the global housing price boom; and the rise in income inequality, with high-income households investing heavily in housing, while middle- and low-income households accumulated debt to gain access to housing as housing prices kept rising (Hunt, 2014).

What surprised financial authorities of many countries is that the level of household debt kept rising after 2008 during the decade-long recovery. Housing loans have usually remained the main component of housing debt and typically represent more than 80 percent of total household debt, often a higher share. However, conditions vary from country to country, and the composition of total household debt can be quite different. Some household debt may be collateralized debt for small business activities. In addition, in aging countries where systems of social protection are not adequately developed, older households may resort to debt to (temporarily) maintain their living standards.

The main policy concern is that rising levels of indebtedness make households increasingly sensitive to changes in their income or rising interest rates, since most loans carry variable and adjustable interest rates. Then the consumption of these households and the stability of their livelihood can be negatively affected by financial shocks. In particular, the high probability of rising interest rates by central banks from 2018 on is likely to induce household problems, unless explicit mitigating policies are adopted.

What matters for financial sector stability is the total size of household debt and the aggregate debt-to-income ratio of the household sector. What matters for social stability and social cohesion is the distribution of household debt across household income deciles, especially when income inequality is rising. Income inequality and aging interact as well. Therefore, a major social indicator to monitor and address pro-actively is the percentage of poor old people. This percentage is high in countries of low social protection like Japan and the U.S., where the ratio of the poor among the old is on the order of 22 percent. The situation is significantly better in Germany or France, where the ratio is less than 10 percent, and in Canada, where it is closer to 5 percent. On the other hand, the situation is particularly serious in South Korea, where the percentage of people older than 65 falling below the poverty line was as high as 45 percent in the mid-2000s (OECD, 2013).

d. Slower rates of GDP growth as the "new normal"

Will the global economy again experience sustained high growth rates of the same magnitudes as those experienced prior to the Great Recession of 2009? Discussions of slower global economic growth as the "new normal" have dominated the media during the decade that the recovery from the twin financial and economic crises has taken. The three main engines of global economic growth have been the U.S. economy, the European Union, and China. There are structural questions about the capacity of each one to grow at sustained high rates again.

Martin Wolf (2015) argues that the global "new normal" economic growth is likely to be characterized by six chronic conditions: a deficient aggregate demand, stagnant productivity, fragile financial conditions, unstable national politics, tense geopolitics, and an overload of challenging issues. The prevailing evidence supports his analysis. Secular decline of nominal and real economic growth rates in the largest Western economies and in Japan have already been occurring under the pressures of demographic aging. The challenge of high economic growth is now that, to compensate for shrinking work forces, the labor productivity rate would have to rise sharply to generate growth rates comparable to historical ones.

In the case of the U.S. economy, some economists have strongly argued that an acceleration in the rate of innovation is unlikely to generate the gains in productivity needed to bring the economy back to the rates it experienced decades earlier (Gordon, 2016). Disruptive, divisive, and myopic policies on many fronts have hardly been conducive to the long-term structural reforms needed internally; further, they do little, if anything, to enhance a global cooperation that is supportive of global growth.

China's contribution to global GDP has risen sharply since the early 1990s. By now the Chinese economy has grown larger than that of the U.S., but the very high growth rates that it has experienced during its industrial take-off, and its massive investment in cities, belong to the past. China is an increasingly mature economy. Its urbanization rate is slowing down considerably now that it is beyond its 50 percent urban inflection point. The demographic age dividend that favored high growth rates is also gone; the Chinese labor force started shrinking in 2012. Chinese policies are now rebalancing the economy toward internal demand and spatially across regions and cities. Rebalancing policies must also control the rising risks of financial instability caused by the very high level of internal debt following years of credit-driven growth that favored low-productivity state-owned enterprises at the expense of highly productive private firms, since the financial stimulus of 2009 of about 12.5 percent of GDP, which was the largest peacetime stimulus on record.

For its part, the European Union (EU) has gone through several episodes of slow growth associated with its inability to fully reform its banking system after the 2009 global crisis. EU decision-making processes have also become slower, with an expanding and heterogeneous country membership from Eastern Europe. In addition, the Brexit crisis has increased growth uncertainties not only for the U.K. but also for the EU, as the U.K. has been a major EU member.

The prospects for high rates of growth among large emerging economies like Brazil, Turkey, and Mexico, and in countries of the Middle East, have also become more uncertain.

e. Climate change and environmental constraints on green growth challenging cities

Scientific modeling of climate change caused by human behavior suggests that change may happening more rapidly than previously expected, and that tipping points of irreversible change may be approaching faster (Voosen, 2018). Most economic growth now originates in cities in high-income economies. Thus issues of green economic growth and green cities greatly overlap.

The core categories of environmental issues facing cities are essentially the same everywhere: forms of energy use, transport and land use, energy-efficient buildings and better forms of construction technologies, coordination in the management of solid wastes, and water pollution and air pollution. However, actions to lower the negative environmental externalities of economic growth differ across cities depending on their specific physical characteristics, spatial organization, the quality of the inventory of existing buildings, and the environmental demands of the economic growth base.

Cities can achieve much on their own. Alternative concepts of smart, green cities relying on intensive information collection and new information and communication technologies (ICT) are developing fast. However, cities also need the active support of higher levels of government to lower the long-term costs and raise the benefits to the entire economy of national environmental policies. Issues of bureaucratic incentives, risk co-management, and information-sharing vertically across levels of government and horizontally across governments and cities always arise.

2. Cumulative lessons of international housing experience

Providing a comparative context for each individual country chapter of this book shows more clearly its specific contributions to our understanding of housing policy across countries. These individual housing experiences should eventually be contrasted and compared with the synthesis of the lessons of international research and housing policy experience that motivates the entire book and closes it.

2.1 Western experiences

Over five decades, the quality of our analytical tools has improved significantly. We have learned how to better evaluate housing performance and the quality of housing policies nationally and locally. However, a major constraint has been, and remains, the availability of housing data of quality in terms of granularity of coverage, comparability, timeliness, and consistency over time across cities of the same country. Constraints have been even more serious across countries. However, the situation is changing in countries where information technology is advancing rapidly. In these countries, "smart city" information technologies are moving from the lab to commercial uses.

Chapter 2, "Taming real estate bubbles with technology" by Edward Glaeser, is a forward-looking discussion of the potential use of new information technology in lowering housing price volatility in U.S. cities and avoiding pricing bubbles. It discusses the new technological opportunities to improve automated housing price appraisals, and the stabilizing potential of centralized real estate appraisals, with benefits for households, local government finance, and credit risk management by lenders.

The diversity of institutions, demand-side characteristics, and supply-side structures is considerable across Western housing systems. The three other chapters on housing in the Nordic countries, the Netherlands, and the U.K. and Australia cannot possibly account for the full diversity of Western housing systems, which a single indicator such as the aggregate tenure choice between renters and owner-occupants would already reveal. These three chapters are just reminders of that vast diversity. Moreover, these three chapters focus on different dimensions of a housing system.

The Scandinavian (or Nordic) welfare model that emerged after World War II became famous for its ability to compensate for the social shortcomings of competitive and efficient capitalist systems of production with universal coverage of social benefits and services. This universal welfare system was financed with high progressive taxation that kept income inequality rather low. However, the five Nordic countries gradually moved away from the ideal Nordic model, and this trend sped up after the 2009 GFC.

In spite of similar welfare arrangements, the housing policies of Denmark, Finland, Iceland, Norway, and Sweden have never been similar. In Chapter 3, "Thinking of the development of housing policy," Hannu Ruonavaara reports on a Nordic comparative project, which found that in spite of important national differences, the five Nordic or Scandinavian housing systems went through four comparable historical stages of development: an introduction stage, a construction stage, a management stage, and a public retrenchment or privatization stage. This paper presents these four stages in the case of Finland. Public retrenchment is surely not the end of the history of housing policy, so what comes after?

The Netherlands constitutes a case of special interest to China. It is the only Western country with a gross population density comparable to countries in East Asia (about 500 people per km^2). Dutch cities function very much like a large megacity, with many people commuting between cities (instead of within them). The spatial dynamics of land rents and housing prices across the Netherlands could be compared with those in one of the 19 large Chinese megacities. In Chapter 4, "Housing prices, land rents, and agglomeration benefits in the Netherlands," Coen N. Teulings focuses on the dynamics and level of land rents for residential real estate across the entire Netherlands, and over time.

The Dutch database of one million housing price transactions covers the entire country over three decades (1985–2013). It shows that in a relatively small and densely populated country like the Netherlands, land rents differ widely from 3,500 euro per square meter in Amsterdam's Canal Zone, to only 17 euro in the northeastern part of the country along the border with Germany. Regression analysis can explain 77 percent of these price differentials. Agglomeration benefits contribute 3 percent to GDP. There is strong persistence in regional population growth rates. Hence, empty land in growing regions carries an option value. Chapter 4 discusses the rule for the optimal construction of new residential real estate, from both a private and a public point of view.

Land use and urban planning have been influenced in every country by the public spillover effects of private activities, which economists call market failures. But there can also be government failures when urban planning fails either in terms of its objectives or in relation to some wide goals, such as efficiency, growth, choice, or environmental sustainability (Bramley, Bartlett, and Lambert, 1995). Government failures as the unintended consequences of housing policy are, or should be, an ever-present concern. Seeing and understanding these failures more clearly and anticipating them is a leading rationale behind comparative housing policy work. Peri-urban fringes are located at the regulated or physical boundary between urban and rural land. In many parts of the world these peri-urban fringes are undergoing major changes as natural and semi-natural landscapes on the edges of cities are converted into built-up environments as part of the process of rapid urban expansion. Increasingly, these peri-urban areas have become the focus for various environmental and landscape protection measures and planning regulations to control urban sprawl.

Differences in the availability of land for development and demand for land for housing, industry, and service provision have contributed to major differences between countries worldwide. In Chapter 5, "Green belts, edgelands, and urban sprawl," Guy M. Robinson focuses on the land and housing pressure growing on the peri-urban fringe in the U.K. and Australia. Focusing initially on the Green

Belt policy in the U.K., Robinson examines recent developments in peri-urban fringes. He highlights the emergence of distinct "edgelands," where one finds a confused and generally unattractive pattern of land use under growing pressures to build more houses on land previously not zoned for development.

Robinson discusses recent attempts in the U.K. to relax the 70-year-old Green Belt system that has been part of a land-use planning system seeking to distinguish between definitive sets of "urban" and "rural" land uses, to sharpen the interface between urban and rural, thereby restricting the extent of the peri-urban fringe. Even in those countries that can be regarded as land-rich, such as Australia, concerns are being voiced about loss of valuable farmland to housing, with growing lobbies seeking to prioritize environmental conservation and presentation of arguments that economic growth is best supported through urban densification schemes as opposed to more centrifugal development.

Starting from this ongoing international land-use experience, Robinson then discusses some of the impacts of the emphasis in China on the creation of substantial "new towns" and the implementation of a Chinese version of Green Belts. Particular aspects of the Chinese context are considered, including farmland preservation issues, the role of local government in encouraging urban development, and the lack of strict enforcement on land-use zoning in the peri-urban fringe.

2.2 East Asian housing experiences[4]

The six East Asian economies of mainland China, Japan, South Korea, Taiwan, Hong Kong, and Singapore share major socio-cultural characteristics that differentiate them from Western economies: a deep Sinic heritage that has had a lasting impact on public and private institutions and norms of public and private behavior in many ways (including high levels of human capital); very high rural and national population densities; and the critical role of the "developmental state" in developing and implementing policies during their growth take-off decades (Wade, 2004). Today, including mainland China, these six East Asian economies produce about one-third or more of the world's housing output in value terms, out of about 190 countries in the world.

The operations of East Asian housing systems are characterized by a much higher degree of government intervention than Western housing systems in four key areas: land-use regulations, housing construction, real estate development processes, and housing finance. Their housing policies have long had a strong, almost exclusive focus on homeownership. Alone in East Asia, Taiwan followed non-interventionist housing policies relying on market mechanisms. However, as we shall see, these six housing systems are far from being identical. Given the path dependence of housing systems, the institutional organization of these housing systems that emerged during the very high-growth urbanization decades of their economic take-off still retains significant explanatory power.

Given the large size of their housing systems, the two-way interactions between housing and the wider economy are crucial to the stability of East Asian economies. The evidence is that both the duration and depth of East Asian housing cycles have differed significantly from Western housing cycles (Renaud, Kim, and Cho, 2016). The duration of East Asian housing booms has been much shorter due to an apparently better degree of market pricing efficiency, and reversion to long-term price trends has also been faster.

During the 1997–98 AFC, East Asian housing systems were negatively and strongly affected. Drawing quickly from the lessons of these bad experiences, East Asian economies were pioneers in the adoption of macro-prudential policies, and they built up their current account surplus. These macro-prudential policies successfully dampened the pro-cyclical responses of East Asian housing to the twin global crises of 2008–09. Compared to the devastating impacts of these twin crises on housing in the U.S., Spain, and Ireland, their impact on East Asian housing was short-lived and shallow. The correlation of East Asian housing price movements appears to have increased during the GFC compared to the AFC a decade earlier. This increased correlation probably reflected expanded channels of contagion through the globalization of financial markets, as well as greater trade and supply chain integration across Asia.

The changing dynamics of Japanese housing markets are of great interest because Japan has been among the very first countries to enter the new housing world created by the interacting forces of aging, changing spatial preferences of households, slow economic growth, rising environmental constraints, and new socio-economic institutional arrangements at the city level. Adding to these challenges, Japan has experienced one of the most severe financial crises on record, which combined with a large real estate bubble with staggering prices that burst in 1991 and left the Japanese economy in a long-lasting deflationary cycle over two decades.

Chapter 6, "Land and housing market dynamics and housing policy in Japanese metropolitan cities" by Eiji Oizumi, analyzes the profound changes that have occurred in the dynamics of land and housing markets and social housing in the metropolitan cities of post-bubble Japan. Oizumi first gives an overview of the major developments in the Japanese economy from the 1990s to the mid-2010s that have had direct impacts on cities. Then he provides theoretical and empirical explanations of the distinctive socio-economic dynamics and spatial relationships across Japanese urban housing. He offers well-structured and up-to-date evidence on Japanese housing problems and explains why the housing market created a bubble, then burst, and eventually stagnated for such a long time. Since the 1991 burst of the real estate bubble, there has been a considerable deregulation and decentralization of land policy and urban planning at the local level.

Oizumi's concern is that the growing housing divides in Japanese cities, as well as the increasing spatial differentiation caused by continued growth in a few segments of the markets and stagnation in the remaining others, are creating important disparities across the housing stock, especially the rental housing stock that is of low quality. These disparities are demanding more intensive social housing policies that will also facilitate a socially effective recycling of the good-quality housing stock. The question is whether the central government has the fiscal resources to revive the subsidized housing policy with its provision of social housing.

In Chapter 7, "The role of government in housing market dynamics in Korea," Seong-Kyu Ha gives a historical and urban planning perspective on the evolution and transformation of the South Korean housing system over five decades as South Korea industrialized and urbanized at the fastest speed on record. He discusses the evolution of housing equity issues over time and across cities, and focuses in particular on the central government's regulation and control of housing prices.

Expanding on Ha's historical overview, Man Cho, Kyung-Hwan Kim, Soojin Park, and Seung Dong You give an informative Chapter 8, "The land and housing delivery system in Korea," about the forces behind the evolution of the system since the early 1960s, and how it works. Instead of providing a very detailed, linear historical account of the evolution of the housing sector, they focus on the key institutional building blocks and the salient features of the Korean housing system affecting land supply, infrastructure development and construction, financing mechanisms, and (on the demand side) the allocation of units. They show how the system was established and went through four major phases over time.

A first phase of Korean institution building occurred between the early 1960s and the mid-1980s. A second short phase was marked by a massive expansion of housing supply from the mid-1980s until the 1997–98 AFC. It was during that second phase that an aggregate balance between the number of households and the number of housing units was finally reached. It was only late during that second period, in 1993, that a guarantee program began to protect households against contract performance failures in the pre-sale contracts used by construction companies to finance their construction projects. During the 1997–98 AFC, Korea was one of the four most severely affected Asian countries.[5] Housing and the real estate sector were not directly impacted by the crisis in Korea, as they were in Thailand and Indonesia, but housing supply and prices later contracted sharply. Korea then went through a rapid V-shaped recovery.

The decade between the 1997–98 AFC and the 2008–09 GFC represents a third phase of major financial liberalization and of real estate deregulation. Cho et al. show how the existing housing institutions had to adjust to a demand-driven, market-based housing system where the nature of risks and the stability of the components of the system had changed. Finally, the fourth and current phase of the Korean housing system started with the 2008–09 GFC, whose quantitative impact on housing was small and short-lived.

An important event of relevance to other countries is that the five post-2009 structural changes appear to have changed the *expectations* of Korean households about housing as investment, which in turn is inducing major changes in the composition of housing demand, and also causing the declining use of the uniquely Korean and socially dominant rental contract known as *chonsei*. In addition to the institutional evolution of the Korean housing system, Cho et al. provide performance indicators to follow the evolution of the housing system over five decades, and, to a lesser degree, the equity and efficiency of the present mature system.

Next, the Taiwan case is a powerful demonstration of the unintended consequences and eventual high social and economic costs of a housing policy that has focused exclusively on homeownership over decades, as opposed to the lessons of international experience to adopt housing policies that prefer to focus on the consumption of housing services and that are also tenure-neutral (Chapter 16), which also implies that housing policies be designed and implemented for the benefit of citizen households, and not that of the corporate real estate industry.

In Chapter 9, Chien-Wen Peng, Bor-Ming Hsieh, and Chin-Oh Chang give an update entitled "Housing price, housing mobility, and housing policy in Taiwan." The Taiwanese housing system is the most market-oriented housing system of East Asia. Taiwan is the only East Asian economy to have experienced four genuine demand-driven housing cycles since the 1970s. During the growth take-off years, housing was a low priority in the development agenda (Chang and Chen, 2012). Different from other East Asian housing systems, the Taiwanese housing markets have operated under very limited government interventions within a legal and regulatory framework derived from Henri George land taxation principles. However, these land tax rules exempted residential real estate, and the unintended long-term consequences of a seemingly harmless and costless government decision made in the 1960s are now in full view.

Peng et al. argue that previous Taiwanese policies and indirect subsidy programs that favored home buying did not distinguish between owner-occupants and investors, which has led to unreasonable phenomena such as high house price-to-income ratios, very high homeownership rates (above 80 percent), high housing vacancy rates (above 20 percent) in Taiwan's housing market, and a very small, low-quality residual rental market. The rapid rise in Taiwanese housing prices since the 2009 GFC has caused severe problems of housing affordability. It has made the quality of life worse, and it has reduced housing mobility.

As occurs everywhere else, the younger generation is paying the main price, and housing divides are deepening. Many households are forced to lower their demand for housing in order to increase their chances of becoming homeowners. These phenomena are the driving political forces behind the very strong citizen demand for housing market reforms. Over the past two decades, the central government's efforts with housing policy reform have been limited and late. The causes are that the government sees the real estate industry as a leading industry in economic growth, and that many real estate developers are major donors to the ruling party. Those developers thus have powers to influence the government's housing policies and strategies and gain very large profits from them. Peng et al. recommend that the government speed up the implementation of urban regeneration; increase the number of social housing units; secure the rental housing market; and make reforms on real estate taxation, on real estate industry management, and on real estate finance to achieve the Taiwanese policy goal of "helping every household to live in a decent home."

The dynamics of Hong Kong's dualistic housing system hold lessons for the large Tier-1 and Tier-2 cities of mainland China, in strategic areas such as the supply of urban land and the *de facto* preference for non-residential land use; the behavior of the oligopolistic real estate industry; and the ability of housing policies to meet the needs of the bottom four income deciles of the population, especially young people. Hong Kong is also one of the statistically best-documented housing systems in the world to investigate specific issues. The 2016 census shows a dualistic housing system, where the public sector houses 46 percent of the population (31 percent in public rental housing and 15 percent in subsidized owner-occupied housing) and the private sector houses 54 percent (21 percent in rental housing and 33 percent in owner-occupied housing).

In Chapter 10, "Housing challenges in Hong Kong's dualistic housing system: Implications for Chinese cities," Rebecca L. H. Chiu shows why and how Hong Kong housing is a system under rising stress. In Hong Kong, the growing housing divides encountered in so many cities of the world since the 2009 GFC have been widening quickly and are particularly severe. Chiu discusses the "overt" and "covert" challenges in managing Hong Kong's dualistic housing system, whose market housing prices have become the highest in the world since the 2008–09 GFC, and where housing is the least affordable. Affordability is declining, and space consumption standards for younger people are eroding. The "overt challenges" refer to the increasing public policy difficulties in achieving minimal housing mobility across six main housing tiers across the dualistic system.

The "covert" challenges refer to non-housing structural factors that are contributing to high housing prices and an inadequate supply response. Probably the most significant and least frequently discussed is the collusive incentive structure between the Hong Kong Special Administration Region (SAR) government and the oligopolistic real estate industry to maintain high real estate and housing prices. Another major systemic distortion is the inability to manage interest rates according to local economic cyclical conditions because of the currency peg between the Hong Kong dollar and the U.S. dollar, in place since 1983. Local Hong Kong interest rates are thereby dictated by U.S. Federal Reserve policies. The zero-interest policies of the U.S. Federal Reserve since the 2009 GFC have been totally inconsistent with the booming economic conditions in Hong Kong and have further fueled rising housing prices.

For comparison with a mainland Chinese city, Chiu chooses neighboring Shenzhen municipality, and she examines the policy tools that the municipality uses to dampen housing price increases in comparison with those of Hong Kong. Indeed a full comparison of the similarities and differences of the housing systems between Hong Kong and Tier-1 Chinese cities would include quite a number of structural factors: the top-down diffusion of housing price increases from the highest market segments; the structure and dynamics of land supply, where preference is given to non-residential land use; the structure of the oligopolistic local real estate industry, which dominates the Hong Kong economy and its politics, as it also does in many Chinese cities where local governments are leading investors; Hong Kong's asymmetry in the allocation of financial risks between banks and households in the name of "free markets"; the high frequency of "hot and cold" interventions in volatile housing markets; the contrast in social housing policies between Hong Kong and Chinese cities; the conflicting roles of the government's fiscal reliance on "land public finance" in both cases; and the negative impacts of the lack of alternative investments to housing assets. Most of these factors are currently widening the housing divides on both sides.

Singapore has become an important point of reference in international comparative discussions of housing policy because of the remarkable success of its housing policies over five decades. Yet none of this success was inevitable, even less foreseeable, as Singapore was thrown into total uncertainty in August 1965 when forced to leave the Malaysian Federation, of which it had been a member state. Singapore abruptly became an independent city state deprived of its natural hinterland. There was an intense sense of vulnerability because the post-colonial economy was in shambles and serious ethnic tensions between Malays, Chinese, Indians, and other nationalities were leading to confrontations. The youthful state government, in power since 1959, reached the strategic conclusion that to have any chance of surviving and moving forward, Singapore should become a nation of homeowning citizens with a personal stake in the success of their society. Singapore became the only East Asian state to fully and actively integrate housing in its overall development policies from the start. Today, as Sock-Yong Phang says, Singapore is "a global multiracial city-state with a land area of 720 square km and a population of 5.6 million. Singapore has managed to develop a framework for affordable housing that has resulted in a high homeownership rate of 91%" (Phang, 2018).

Some may wonder why we classify Singapore as an East Asian economy given its location at the strategic center of Southeast Asia. The evidence is that institutions (and the culture that shapes them) are much more important than geography and physical characteristics in explaining economic development (Phang, 2018). It has been legitimately argued that Singapore managed to combine the best of the Chinese and the British traditions. On one hand, there is the strong emphasis on education leading to high levels of human capital, meritocracy, the primacy of the general interest, the prestige of public service, private thrift, and public fiscal balance, and the primary social focus on the family. On the other, there is the respect for the rule of law and due process, transparency, democratic processes, and good governance. A significant difference in the design and implementation of housing policies for ordinary citizens between Singapore and Hong Kong is that Singapore adopted from the first day of its independence the model of British democratic institutions and governance, while Hong Kong residents still labor under a system only partially modified from its colonial era since its return to China in 1997.

Chapter 11, "Building an equitable and inclusive city through housing policies" by Sock-Yong Phang, first reviews the economic debates and the evidence on the rise of inequality in income and wealth since the end of World War II. Phang provides spatially detailed and up-to-date quantitative comparative evidence on the equitable and socially inclusive city that Singapore has built. Then she details the inter-related housing policies that have shaped the growth and performance of Singapore's housing system. She shows that Singapore has the best and most equal housing wealth distribution in the world, with the share of housing wealth of the lowest 40 percent of its residents being about five times the share of housing wealth in other OECD (Organisation for Economic Co-operation and Development) countries. However, no socio-economic system is ever static, and Singapore faces most of the same pressures that other countries have faced since the 2009 GFC. In closing, Phang reports the ongoing current debates on social protection in Singapore.

2.3 China's growth transition and its implications for housing

Economic development and urban development are functionally and deeply interwoven. Four decades after the historical reforms launched by Deng Xiaoping in 1978, China is a transformed society with a middle-income per capita and an economy soon to be equal that of the U.S. This growth transition is a critical and risky transition on the way to becoming a high-income economy. Successfully crossing this stage of development has proven so difficult for most emerging countries that it has become known among development economists as "the middle-income trap."

The reason for repeated growth transition failures around the world is that this transition requires major political and economic adjustments to what worked well during the take-off decades, when the country moved successfully from a low-income to a middle-income economy. The political ability to adopt—and sustain—the institutional and economic reforms necessary to achieve a high-income, advanced level is the most important. Without such a political ability, the interest groups that played a key positive role during the take-off decades may oppose the required changes that could diminish their own economic position. The political travails of middle-income Turkey, Iran, or Saudi Arabia in the Middle East; Argentina, Brazil, and Mexico in Latin America; or Malaysia and Thailand in Southeast Asia are familiar. Remarkably, since World War II, the five East Asian economies of Japan South Korea, Hong Kong, Singapore, and Taiwan are the *only* five societies to have avoided the middle-income trap and become high-income economies (International Growth Commission, 2008).

There are multiple reasons for expecting slower economic growth rates during China's growth transition. One is the ongoing rebalancing of the economy away from investments and exports toward consumer domestic demand. Another is that China's urbanization went through its urban inflection point when it passed its 50 percent urbanization level in 2011, and the progressively slower rates of growth of cities with tapered infrastructure investments. A third factor is the urgent need to stabilize the financial system and lower risks of financial instability after a decade of credit-driven growth that led to a very risky debt-to-GDP ratio. The five structural changes affecting housing discussed in Section 1.2 are also in play.

Given the expected slower economic growth and the heightened environmental pressures during the growth transition, Chinese housing policies must somehow adjust from the quantitative results of the past decade to a higher quality of housing output in terms of economic efficiency, spatial accessibility, social inclusiveness, and environmental sustainability. An additional challenge is to prevent or at least modify the role of housing as a channel of wealth redistribution upwards to the richest households. Can international experience help in these policy adjustments?

The four chapters of Part 3 aim to provide a better understanding of the current performance of the Chinese housing system, and where Chinese housing policies could go from today considering the lessons of international experience presented in the book's closing chapter.

Chapter 12, "The long-term dynamics of housing in China" by Bertrand Renaud, gives a short macro-history of China's housing from the pre-industrial era of Chinese cities to the present growth transition. The urban system of China has a very long history measured in millennia. Very large cities of over one million people existed centuries before the start of the Industrial Revolution. This pre-industrial system of cities has been the matrix within which rapid industrialization and technological change has taken place. Between 1978 and 2015, China's socio-economic transformation has likely been the most profound in China's entire history.

Chapter 12 has four parts. First, it shows how three important economic models of very different vintages give complementary insights into the long-term dynamics of Chinese economic development. It also shows that the Chinese ratio of annual housing investment to GDP follows the Burns-Grebler hypothesis and is currently peaking at around 7–8 percent of GDP. Beyond the present growth transition stage, this aggregate housing investment ratio is expected to decline to 2–3 percent. The second part presents China's development from an East Asian perspective and compares China's implicit housing strategy with those of the five other economies.

The third part focuses on the emergence of China's modern housing system during the growth take-off decades and highlights four important features of its growth: China's overinvestment in housing and real estate after 1994, the historical housing privatizations between 1988 and 2013 that altered China's long-term economic development path, the Great Housing Boom of 1998–2013, and the spatial dimensions of the housing boom across Chinese cities. The final part notes four clusters of factors that interact and currently distort urbanization. Their successful correction would significantly raise the quality of forthcoming Chinese urbanization and housing performance. These factors are: the integration of land institutions and land operations across urban and rural areas that could yield important benefits for the spatial efficiency and the preservation of very scarce land; the *hukou* registration system that creates two classes of urban residents with different access to housing and other urban services; reforms of urban finance and intergovernmental fiscal relations that would change local government incentives to overinvest in land and real estate and reduce risks of financial instability; and local governance reforms with improved decentralization and public accountability to improve the performance of housing estates and the livelihood of urban residents.

Chapter 13, "Urban housing challenges in mainland China" by Zhi Liu and Yan Liu, argues that while China's growth-oriented housing policy has contributed to the significant improvement of housing conditions for the majority of the urban population, it has also resulted in a number of unintended economic, financial, and social consequences. Housing the migrant households and the new families will remain a major challenge for the near future. As the conditions that made possible the "golden period," which others call "the Great Housing Boom," fade away, a new housing policy framework must be developed to put housing sector development back on a healthy, sustainable, demand-responsive track.

Chapter 13 also adds more specificity to the macro-history of housing presented in Chapter 12. After a series of housing reform actions, China decided to complete the transition of urban housing provision to the market in 1998. This was followed by a government economic policy in 2003 that promoted the real estate sector as one of the pillar industries of the national economy. The policy triggered a rapid growth of housing markets and sharply rising housing prices that lasted from 2004 to 2013. The best methodology using the national Tsinghua University Housing Price Index shows a powerful real housing price increase of 17 percent per year between 2006-Q1 and 2010-Q4 across top cities (Wu, Deng, and Liu, 2011). Peak rates of urbanization and peak rates of economic growth went together in China during its take-off. Indeed, international experience shows that "the relationship between GDP and urbanization is tight"; what remains ambiguous is "what is cause and what is consequence" (Duranton, 2014).

While housing demand in the most vibrant major cities remains strong, and prices continue to rise, demand in most other Chinese cities has cooled down significantly. The rapid increases of housing prices in the most vibrant cities have widened the wealth inequalities between homeowners and those who do not own a housing unit. This chapter analyzes the factors that contributed to the "golden period" of housing market growth, including economic policy, land policy, municipal finance, and affordable housing programs.

Significant regional variations have emerged recently. Housing prices in Beijing, Shanghai, Shenzhen, Guangzhou, and some second-tier cities have experienced severe inflation, some of which has been confirmed as bubbles, with prices moving sharply away from market fundamentals. Meanwhile, market stagnation with slow sales of the oversupplied housing stock has occurred in other cities and smaller towns. Why is the Chinese urban housing market so volatile? Why do the main causes of urban housing price inflation in some cities differ from the main causes in others? Why does this property boom in some Chinese cities not bust, as many experts have expected? Answers to these questions are very important for designing the right policies to manage the urban housing market in the future. Early analyses of the new Chinese housing markets have focused on the largest Tier-1 and Tier-2 cities, where the data were more easily available (and changes were most rapid).

Chapter 14, "Migration and urban housing market dynamics in China" by Ya Ping Wang, adds a new dimension to our understanding of Chinese housing and shows the importance of county-level housing markets for a full understanding of Chinese housing. Chapter 14 focuses on the dynamics of local housing markets and the impact of migration at the bottom of China's urban housing system, the county town, where market relations are simpler and migration patterns are easily identifiable. Based on the understanding of the local market in small towns, the general features of housing market dynamics at higher levels are discussed. Wang emphasizes the importance of linkages in housing markets at different levels in the country and the relationship between the housing market and migration. The first section briefly reviews urban housing market development in China and highlights the short cycles of new housing production and price inflation under government control efforts. This section is followed by a case study of the dynamics of commercial housing development and the market in a county town named Qishan. This case study is based on several field trips; continuous monitoring of the progress of housing development in the county; and interviews with residents, officials, and developers. Using the findings, Chapter 14 discusses the patterns of migration in the country and their impacts on urban housing markets.

Housing market fluctuations and price inflation are problems faced by all countries in the world, but the urban housing market in China shows some very distinct features. As already mentioned, the housing market was established recently through a series of housing reform policies that commercialized the socialist welfare housing system in the late 1980s and the 1990s. Ever since its introduction, the Chinese urban housing market has been volatile, with short periods of output increases soon followed by an output slowdown. The drivers of these short, shallow cycles are Chinese housing policy makers who have been alternating periods of "adjustment and control" and "market rescuing" in the last 15 years. When market support policies were introduced in one year, property markets could overheat, and price inflation could accelerate in the following months. This would then lead to the adjustment and control policies that often resulted in a price slowdown and a decline in property development activities and investment. In South Korea, analysts have called comparable government interventions in the housing market "hot bath, cold bath" housing policies.

Regardless of these alternating control and support policies, Chinese markets have seen continuous housing price increases and the fast accumulation of housing (and other) assets by the richest urban residents. Indeed, commercial housing prices in major Chinese cities have increased so dramatically over the last 10 years that they have made some of the first-tier Chinese cities such as Beijing, Shanghai, and Shenzhen among the most expensive places to live (relative to income) on Earth. At the same time, the urban housing market in China did not follow the expectations of many housing experts, and housing price inflation in many cities has become very serious over time. However, the Chinese market has not experienced housing busts similar to those that occurred (as triggers or as impacted victims) during the AFC in Thailand, Indonesia, Malaysia, and Korea.

Housing prices have increased rapidly over the past 15 years in many large cities in China, but how much do we know about what drives these price hikes? Are housing prices driven primarily by household demand-side factors or by the structure of the housing supply, and in a particular the supply of land? This is a hot topic of public and academic debate. Some observers attribute the price hikes to the robust housing demand, while others consider the government's monopoly on supply of urban land to be a key factor. Chapter 15, "Residential land supply and housing prices in China" by Huina Gao, Zhi Liu, and Yue Long, focuses on these questions. Using panel data on 31 large cities for the boom years 2000 to 2015, they analyze the effects of residential land supply on housing prices and the mechanism through which land supply influences housing prices. They find that housing price increases in their sample cities resulted not only from strong and rising demand, but also from decreasing supply of residential land. They also find that reduced residential land supply contributed to housing price hikes mainly through increases in residential land prices and public *expectations* of future housing price increases, not through the decreasing amount of newly supplied housing.

2.4 Value and use of international experience in improving Chinese housing policies

Chinese housing policies must adjust to China's new and risky "growth transition" environment. Chapter 16, "A roadmap for housing policy: Lessons of international experience" by Edward Glaeser, Rebecca Chiu, Zhi Liu, and Bertrand Renaud, synthesizes the international housing policy lessons that have accumulated over the previous four decades and then discusses their value for improving China's current housing policies.[6] This chapter is the single most important part of the book for housing policy makers and others who are interested in the design and implementation of new housing policies to improve the socio-economic performance of the sector.

The chapter targets policy makers. It states four core principles for diagnosing a national or local housing system and making housing policy choices: to prioritize housing consumption over asset ownership, to promote widespread affordability, always to consider jointly the housing structure and its location, and to promote tenure neutrality and robust rental markets. Then policies must address three efficiency issues: efficiency in resource use, the efficiency of subsidies (demand side vs. supply side), and the efficiency of land-use regulations. Four institutional qualities underlie a well-performing housing system: uniform property rights, transparent rules, inclusive but balanced planning, and free entry and competition. The lessons synthesized here reflect international collective experience, but we stress that not everyone may agree with every idea of this "roadmap" in every housing context.

The second part of Chapter 16 focuses on policy actions that China needs to consider during the present growth transition. It divides them into two groups based on the likely difficulty of their implementation. The less difficult institutional reforms include establishing a universal property register and cadaster, adopting internet-accessible zoning and permitting processes, drawing on international best practice for joint ownership structures, and improving information systems for demand-side subsidies. More difficult institutional reforms include integrating rural and urban land-ownership systems, reforming public property development institutions, reducing local regulatory barriers and lowering the volatility of local housing supply, and changing the financing structure of local government away from land-based finance. In the case of urban planning reforms, difficult reforms include planning for long-term urban change, considering alternative infrastructure provision mechanisms, and managing rural-urban conversions. Among the needed housing finance reforms are adopting dynamic real estate price evaluation for bank (lender) oversight and providing better alternatives to real estate for households' retirement savings.

Notes

1 Mr. Bernanke was quoted making the statement in a document filed on August 22, 2014 with the U.S. Court of Federal Claims as part of a lawsuit linked to the 2008 government bailout of insurance giant American International Group, Inc.
2 This major report is the result of two years of analyses and evaluations carried out jointly by international experts and Chinese financial and regulatory authorities.
3 See the editorial of the *Financial Times*, "Adapting to the world of the 100-year lifespan," August 11–12, 2018, which highlights the fact that "a child born in the west today has a better than 50–50 chance of living beyond 105." The *Times* also notes the influential impact of the 2016 book *The 100-Year Life* by L. Gratton and A. Scott of the London Business School.
4 Comparative elements of this section draw from Renaud, Kim, and Cho (2016).
5 For an excellent discussion of how Korea and the other Asian countries went through the AFC, see Sheng (2009).
6 Chapter 16 reflects the policy workshop that immediately followed the International Urban Symposium held in Hong Kong in May 2016.

References

Beesley, T. and Persson, T. (2011). *Pillars of Prosperity: The Political Economy of Development Clusters*. Princeton, NJ: Princeton University Press.
Bramley, G., Bartlett, W., and Lambert, C. (1995). *Planning the Market and Private Housebuilding*. London, U.K.: University College Press.
Chang, C. and Chen, M. (2012). Taiwan: Housing Bubbles and Affordability. In A. Bardhan, R. H. M. Edelstein, and C. A. Kroll (Eds.). *Global Housing Markets: Crises, Policies, and Institutions*, Kolb Series in Finance: Essential Perspectives. Hoboken, New Jersey: John Wiley & Sons, Inc.
David, P. (2000). Path Dependence, its Critics and the Quest for 'Historical Economics'. Keynote address to the European Association for Evolutionary Political Economy, Athens, 1997, rev. In P. Garrouste and S. Ioannides (Eds.). *Evolution and Path Dependence in Economic Ideas: Past and Present*. Cheltenham, U.K.: Edward Elgar.
Duranton, G. (2014). The Urbanization and Development Puzzle. In Shahid Yusuf (Ed.). *The Buzz in Cities: New Economic Thinking*. Washington D.C.: The Growth Dialogue. www.growthdialogue.org
Federal Reserve Bank of New York. (2018). Press Briefing on Homeownership and Housing Wealth, May 17, 2018, www.newyorkfed.org/press/pressbriefings/homeownership-and-housing-wealth.

Duranton, G. (2014). The Urbanization and Development Puzzle. In Shahid Yusuf (Ed.). *The Buzz in Cities: New Economic Thinking*. Washington D.C.: The Growth Dialogue. www.growthdialogue.org

Gordon, R. J. (2012). *Is U.S. Economic Growth Over? Faltering Innovation Confronts the Six Headwinds*. Working Paper 18315. Cambridge, MA: National Bureau of Economic Research.

Gordon, R. J. (2016). *The Rise and Fall of American Growth: The U.S. Standard of Living Since the Civil War*. Princeton, NJ: Princeton University Press.

Hunt, C. (2014). Household Debt: A Cross-Country Perspective. *Reserve Bank of New Zealand Bulleting*, 77(4). pp. 1–13.

International Growth Commission. (2008). *The Growth Report: Strategies for Sustained Growth and Inclusive Development* (also known as the Spence Report). Washington D.C.: World Bank.

International Monetary Fund. (2017). People's Republic of China. Financial System Stability Assessment. IMF Country Report No. 17/358.

International Monetary Fund, Global Financial Stability Report April 2018: A Bumpy Road Ahead, www.imf.org/en/Publications/GFSR/Issues/2018/04/02/Global-Financial-Stability-Report-April-2018.

Jones, R. S. and Urasawa, S. (2014). Reducing Income Inequality and Promoting Social Stability in Korea. OECD Economics Department Working Paper 1153.

OECD. (2013). Getting Older, Getting Poorer. *OECD Factbook*. Paris, France: OECD, pp. 264–265.

Ostry, J., Berg, A., and Tsarangides, C. G. (2014). Redistribution, Inequality and Growth. IMF Staff Discussion Note. Washington, D.C.: International Monetary Fund.

Phang, S.-Y. (2018). *Policy Innovations for Affordable Housing in Singapore: From Colony to Global City*. New York, NY: Palgrave-Macmillan.

Piketty, T. (2014). *Capital in the 21st Century*. Cambridge, MA: Belknap Press.

Renaud, B., Kim, K., and Cho, M. (2016). *The Dynamics of Housing in East Asia*. Oxford, U.K.: Wiley-Blackwell.

Rodrik, D., Subramanian, A., and Trebbi, F. (2002). Institutions Rules: The Primacy of Institutions over Geography and Integration in Economic Development. IMF Working Paper WP/02/189, www.imf.org/external/pubs/ft/wp/2002/wp02189.pdf.

Rognlie, M. (2015). Deciphering the Fall and Rise in the Net Capital Share: Accumulation or Scarcity? Brookings Papers on Economic Activity, 46(1), 1–69.

Sheng, A. (2009). *From Asian to Global Financial Crisis*. Cambridge, U.K.: Cambridge University Press.

Tooze, A. (2018). *Crashed: How a Decade of Financial Crises Changed the World*. London, U.K.: Allen Lane.

Voosen, P. (2018). The Earth Machine, *Science*, 361(6400), 344–347.

Wade, R. (2004). *Governing the Market. Economic Theory and the Role of Government in East Asian Industrialization*, 2nd ed. Princeton, NJ: Princeton University Press.

Wolf, M. (2015). Chronic Economic and Political Ills Defy Easy Cure, *Financial Times*, January 21, 2015.

Wu, J., Deng, Y., and Liu, H. (2011). Housing Price Index Construction in a Nascent Housing Market: The Case of China. IRES Working Paper IRES-2011–047, National University of Singapore.

Part 2
Housing challenges of developed economies

2 Taming real estate bubbles with technology

*Edward Glaeser**

HARVARD UNIVERSITY

1. Introduction

Between 1998 and 2011, the United States experienced a great housing convulsion that made and destroyed fortunes and left financial wreckage in its wake. According to Federal Housing Finance Agency Data, U.S. markets experienced 48 percent real price growth between 1998 and 2006, and 28 percent real price decline between 2007 and 2012. Aggregate statistics mask the far more extreme fluctuations that can occur within individual metropolitan areas. Las Vegas experienced 78 percent real price growth between 2002 and 2006, and 67 percent real price decline between 2006 and 2012.

There is more debate over Chinese statistics than U.S. statistics. Fang et al. (2015) document that annual real price growth in China's first-tier cities averaged 13.1 percent from 2003 to 2013, a truly spectacular performance. Even third-tier cities experienced an average of 7.9 percent real price growth annually between 2003 and 2013. Official statistics often show more modest gains, and the Organisation for Economic Cooperation and Development (OECD) data imply that real Chinese housing prices have been substantially flat since 2010. Yet even the most conservative data show that China experienced impressive housing-price growth between 2000 and today, and many analysts allege that China is a bubble waiting to burst.

This chapter will make no claims about whether China is likely to experience a major housing-price correction in the near future. Typically, busts do follow booms, but Chinese housing growth has also been accompanied by spectacular growth in income. Moreover, unique institutional features of the housing market make it possible that a price correction will come in the form of a soft landing rather than a hard crash. Nonetheless, prudent public management still requires reforms that will reduce the risk of future property convulsions and the damage that future convulsions may create.

In response to the U.S. crash, many studies emerged debating the causes of the price swings. Some authors emphasized the rise of subprime lending (Mian and Sufi, 2009), while others focused on extrapolative beliefs (Glaeser and Nathanson, 2015). It seems unlikely that a consensus will be reached any time soon. Yet, as many authors have emphasized, we did not need to understand the physics of iceberg locations to improve the safety of ocean liners by requiring more lifeboats. Similarly, housing policies can increase economic safety even if we do not fully understand the causes of housing bubbles.

This chapter focuses on technology, better real estate price evaluation, and its relation to three specific parts of the housing market. The ability to appraise land and property properly is intimately related to three distinct aspects of housing bubbles: property taxation, the use of eminent domain, and bank oversight. New technologies are coming online that can significantly improve our ability to automate the business of property evaluation.

In Section 3, I briefly discuss the innovative new approaches to property evaluation using computer-vision techniques developed by computer scientists. In much of the world, but not in China, we increasingly have external images of buildings provided by Google Street View and other providers. Internal images are also occasionally available on real estate database websites such as Zillow, but these exist only when sellers have voluntarily shared them. In both cases, these images provide the means of significantly improving our ability to assess property values without relying on human appraisers.

Typically, economists have assessed property values using standard housing price hedonic estimates, which rely on regressions in which sales prices are connected with listed attributes of the home, such as the number of bedrooms and bathrooms. The images of the home are essentially just a new explanatory variable, but they provide a considerably richer set of information that can promise a significantly higher amount of explanatory power. In this chapter, I recommend an approach in which the predictive power of visually enhanced regression estimates is compared with the predictive power of human appraisers and listed attributes of homes.

I then discuss the uses of this technology for eminent domain, property tax evaluation, and bank oversight. The use of eminent domain is frequently criticized in both the U.S. and China. Typically, the critics allege that eminent domain undervalues property and that this leads to an excess of takings. If property valuations are performed nationally, this can potentially reduce local abuses. I also discuss other potential eminent domain-related reforms, including compensation that exceeds market value and requiring supermajority approval among current property owners.

The issue of property valuation is also critical in the successful implementation of local property taxes. Such taxes are a natural means of funding local government, which can fund local service delivery and reduce the tendency to use land expropriation as a tool for local revenue generation. In many cases, the optimal property tax is actually a land tax, as suggested by Henry George over a century ago. In that case, standard regression techniques will function well. If more standard real estate taxes are to be used, then visual valuation techniques can be helpful as well.

Finally, the issue of property valuation looms large in the area of bank oversight. Accurate evaluation of bank balance sheets can both help reduce the likelihood that a bubble will emerge, and also reduce the damage that occurs after a bubble bursts. Better property valuation can help improve the quality of bank regulation. In some cases, property valuation may also correct for the documented tendency of housing prices to mean revert over time (Glaeser et al., 2014).

Real estate bubbles have a long history (Glaeser, 2013), and they will always occur. Yet we can reduce their frequency and costs by improving the process of eminent domain, instituting effective land and property taxes, and strengthening bank oversight. Visual valuation techniques can help in all three areas.

2. The property booms in the U.S. and China

America's great housing convulsion has been well studied. While important questions remain, many of the facts of the boom and the bust have been well established. China's great housing boom is far more opaque. Diligent scholars, such as Hanming Fang, Wei Xiong, and Joseph Gyourko, have established excellent documentation of the changes in Chinese housing (Fang et al., 2016) and land (Deng et al., 2015) prices. Yet we still do not fully understand Chinese vacancy rates, the exposure of the banking sector to real estate, or the ultimate fate of the boom.

The general arc of the U.S. boom is well known. In the early 1990s, the United States was still recovering from the bust of the late 1980s, and for many years real estate markets were almost static. Gradually, prices began to pick up, especially in coastal markets. In metropolitan areas, such as New York, Los Angeles, San Francisco, and Boston, demand was high because of the success of knowledge-intensive industries, such as finance and technology, and a web of regulations restricted housing supply. Price growth gradually spread into the American interior, following a process that almost seems like a geographic contagion of a mania (Ferreira and Gyourko, 2001).

The growth of housing prices within the interior seems to have generally ignored the far more elastic housing supply that exists in places like Phoenix and Las Vegas.[1] In such places, the boom was also associated with a vast increase in housing supply leading to a large overhang of excess housing in many areas. That excess inventory has helped ensure that U.S. construction activity is still far below the levels reached in the years before 2006.

The boom was concentrated in warmer housing markets, but there is no relationship between local economic growth and local home-price growth over the 2000–2006 period. The role of easy credit seems to have been part of the boom, but there is little consensus about its overall importance or the channel through which easy credit operated. At the height of the boom, Himmelberg, Mayer, and Sinai (2005) argued that low interest rates and modest expectations of rising rents could justify the high prices then being paid. Glaeser, Gottlieb, and Gyourko (2014) argue that the power of interest rates to drive prices in a rational pricing model will be muted by the mean reversion of rates and, consequently, low interest rates on their own cannot explain the boom in prices. Other authors, most notably Mian and Sufi (2009), provide evidence linking the growth in subprime lending with the rise in prices. While the empirical link between subprime lending and price growth seems clear, it is unclear whether this link is compatible with standard models of rational behavior.

The primary alternative explanation to the credit market-based view of the boom is that the boom reflects egregiously optimistic expectations, possibly fueled by extrapolative expectations. Case, Shiller, and Thompson (2012) document the extraordinarily high growth rates that homebuyers claimed to expect during the boom. Glaeser and Nathanson (2015) provide a model where extrapolative beliefs come naturally from using a natural approximation in interpreting past prices. This model broadly fits the key moments of recent U.S. housing markets.

By contrast, the Chinese housing price boom has not yet ended, and there is considerable debate over the time path of prices. Official statistics typically show impressive but moderate gains, such as 100 percent price growth between 2007 and 2014 (Chivakul et al., 2015). By contrast, Fang et al. (2015) look at sales data for new homes using development project fixed effects to control for changing quality. Their work shows real price growth of 15.9 percent per year in Tier-1 cities between 2003 and 2013, despite a slowdown in price growth during 2008 and 2009. While this period also experienced substantial income growth, price growth was still higher. Even in Tier 3, real price growth averaged 7.9 percent per year between 2003 and 2013. Deng, Gyourko, and Wu (2015) report that real land prices in 35 Chinese cities have increased fivefold between 2005 and 2015.[2]

A construction boom of truly epic proportions has accompanied the rise in housing prices. For example, the National Bureau of Statistics tells us that, in 2013, 120 billion square feet of floor space was under construction, and 42 billion square feet of floor space was completed. Over the five-year period from 2008 to 2013, Chinese builders produced 170 billion square feet of floor space, or 125 square feet per person.[3] On average, the Chinese consume about three times that much floor space (Chivakul et al., 2015), so builders created about one-third of China's total space needs over a five-year period.

China's boom has also been accompanied by a major shift in the nature of the construction industry. As recently as 2000, 73 percent of Chinese floor space was produced by state-owned enterprises and government organizations (for their employees) and housing management bureaus. In 2013, only 13 percent of floor space was produced by such public entities. Over those 13 years, state-produced space has actually risen, but collective-produced space has fallen. Both categories are now dwarfed by the growth in commercial sector-produced floor space.

Fears of excess inventory have accompanied this building boom. Chivakul et al. (2015) estimate that builders had over 10 billion square feet of inventory in 2013. Moreover, the true extent of China's excess inventory is virtually unmeasurable because much of the sold housing stock is in the hands of absentee investors. Like many individual American investors during the U.S. boom, Chinese investors have also bought individual apartments, expecting future housing-price appreciation. Unlike the American investors, Chinese investors often choose to leave these apartments vacant, presumably because rents are not large enough to offset the depreciation costs created by having a renter living in the unit. This fact suggests that the new housing stock has been built at a quality level that is far above the level appropriate to current market conditions.

The nature of individual housing investors is also quite different in China because Chinese investors are not typically over-borrowing to finance their purchases. Whereas the stereotypical American homebuyer buys on credit and thereby puts the banking system at risk, the stereotypical Chinese homebuyer pays a large cash down payment. Consequently, a drop in housing prices in China is unlikely to lead to an American-style ocean of defaults by individual households.

The Chinese banking system does face risks from the default of real estate developers, who are typically debt-financed in China, as in the United States. Chivakul and Lam (2015) report that more than 60 percent of Chinese real estate and construction firms have leverage ratios greater than three. In that sector, the ratio of bonds and loans outstanding to annual revenues has increased from about 70 percent in 2007 to almost 150 percent in 2015. Business debt, not household debt, provides the channel through which a housing-price correction might have systemic implications.

Yet there are other critical differences between the Chinese boom and the American bubble, which may mean that the Chinese experience ends up looking quite different. Most importantly, China is still a quickly growing economy, which may be much richer in 2030 than it is today. Consequently, the demand for vast amounts of urban space may materialize. Nineteenth-century America had its real estate bubbles, but typically demand eventually caught up with supply. Chinese growth may eventually catch up with supply as well, as long as the supply of new real estate starts to slow down.

China's public sector also has vastly more power than the U.S. public sector. China's government can essentially eliminate a housing glut by moving units off the market and using them for social housing. The Chinese government can temporarily make it much more difficult to build, which would reduce the supply of new housing and prop up prices. The broad array of tools that are available to the Chinese government makes it less likely that China's housing boom will end in catastrophe. Better property evaluation is one tool that can help reduce future real estate risks.

3. The theory and technology of property evaluation

The process of property evaluation is as important as it is unexciting. Eminent domain, property taxation, and appropriate banking oversight all rely on solid assessment of the values of property. In this section, I briefly discuss the path forward using machine-learning algorithms and computer-vision methods.

The most basic means of assessing property values is through comparable sales prices. Using either a computer or an assessor's judgment, comparable sales methods essentially use other recent sales to form an estimate of the market value of a home that has not recently been sold. If the assessor literally has 20 recent sales of homes that are essentially identical to the home in question, then this procedure is both easy and accurate.

More typically, the comparable sales homes are not precisely identical to the home in question, and consequently adjustments must be made based on structure and location. Hedonic regressions enable these adjustments to occur in a reliable statistical fashion. Recent techniques, such as Artificial Neural Networks, make it easier to estimate non-linearities in the relationships, but these are still limited by the number of variables included in standard housing data sets. Typically, however, hedonic estimates only capture a modest fraction (30 percent) of the variation in housing prices, which means that there is continued use of human assessors who can presumably evaluate many more of the idiosyncratic elements of a particular home.

The difference between a human assessor and a hedonic estimate is that the human assessor has eyes and can use his or her vision to make adjustments to the estimated value. The disadvantage of the human assessor is that he or she has limited cognitive capacity and will make mistakes that a regression model will not. The great advantage of computer-vision methods is that, in principle, they can take in all of the images perceived by the human assessor, but do so without the biases and limitations involved in human assessment.

The extreme computer-vision method would occur when the computer has access to every image perceived by a human assessor. In that case, the human's information set and the computer's information set would be completely equivalent, and the computer's ability to search for patterns should make it unambiguously superior to its human counterpart. The computer will have the same inputs, but will be able to use machine-learning algorithms to process that data, while the human will just be using his or her best judgment.

Nikhil Naik, Michael Kinkaid, and I are using the City of Boston's register of sold homes; we have connected those homes with Google Street View images of the housing. We are in the process of evaluating the added explanatory power brought by these external images.

We have used a similar procedure (Glaeser, Kominers, Luca, and Naik, 2015) for predicting income across New York City block groups. In this sample, the images were able to explain over 80 percent of the variation in income within the testing sample. The algorithm produced with New York City images and income was also able to explain over 70 percent of the variation in income across Boston-area block groups. Consequently, we are optimistic about our ability to explain prices in Boston and elsewhere.

Location, of course, also matters profoundly for real estate evaluation. Any image-based assessment will also want to take into account location, as well as other known characteristics of the house. Moreover, it may well be that important interactions occur between location and the visual characteristics of the house. To properly use location, machine-learning algorithms must include these variables. Unfortunately, if the interactions are important, it will be harder to export a valuation algorithm produced with data from one metropolitan area to a second.

The rise of computer-vision techniques has the capacity to fundamentally revolutionize non-human real estate evaluation. We may no longer face a trade-off between in-person evaluation, with rich visual cues but imperfect human cognition, and hedonic estimates, with strong statistics and limited explanatory variables. Instead, we may be able to combine all the visual knowledge of a human assessor with a great deal of computing power. The combination makes it possible to do mass, accurate valuation cheaply and quickly. That prospect makes it possible to imagine a brighter future for eminent domain, property taxation, and banking oversight.

4. Towards a better system of eminent domain

The most familiar complaint with the use of eminent domain is that it does not really "fairly" compensate those who are evicted from their homes. The word "fair" is itself quite debatable. One natural meaning is that the owner receives fair market value for their property, meaning the price the property could get on the open market. The common criticism of this justification is that the resident could have presumably sold the property at any point in time and preferred not to do that. Consequently, we must infer that the resident preferred the house to its fair market value.

An alternative approach is to define "fair" as the amount of compensation that is needed to ensure that the owner is at least as well off after the taking than before it. The natural means for assuring that result is that the owner voluntarily relinquishes their property. As is well known, however, the process of bargaining can lead to a holdout problem, where one or more of the residents at the end of the bargaining chain demands vast sums to vacate, and that can destroy the entire process.

A further set of issues around "fairness" is raised in settings where the resident does not own the property. Even if a landlord is made whole by the taking, his tenants may not be. Squatters who are evicted from their homes will also presumably be made worse off by the taking. Philosophical debates rage about whether it is just or wise to compensate such residents for such takings.

The taking is economically efficient if its value to the taker is greater than the opportunity cost of the land—its value to alternative users. For this calculation, it is worthwhile considering the well-being of non-owners, although an argument can be made against compensating non-owners because such compensation may encourage future squatting. Following standard economic logic, I will assume that the goal is to establish a reasonable method of estimating the value of the property to its current residents, and to ensure that this reasonable method is used.

Within the U.S., the standard debate about eminent domain swirls around its uses. Can public takings be used for private purposes or only those with clearly public uses? In a pragmatic fashion, the Supreme Court in Hawaii Housing Authority v. Midkiff (1984) established that the public could take private property just to redistribute it to private holds. This view was essentially upheld in the more famous Kelo v. City of New London Case in 2005. The Supreme Court's view seems to have been that it is not capable of really establishing what is and is not a public purpose, and it is the job of locally elected government to make that choice. This seems like a sensible recognition of the limits of juridical competence, but it does increase the need to effectively evaluate the alternative value of the property to its current users.

Broadly speaking, there are two approaches to this valuation task. The first can be called "enhanced value assessment." The second can be called "weakened property rights." The idea of enhanced value assessment is to improve the quality of the value assessment and enhance it in some value to reflect the extra surplus. In some cases, the two approaches can be merged.

Enhanced value assessment means keeping the basic structure of eminent domain takings unchanged, but improving the functioning of the system with better evaluations; more professional, independent implementation; and premia to reflect mobility costs. The methods discussed in Section 3 provide some ideas about improvements in valuation. There will surely be others. The key is to begin with the state-of-the-art evaluation techniques that are professionally administered.

If a nation plans to allow a significant amount of eminent domain takings, then it makes sense to have a professionally staffed, centralized office of property evaluation, rather than relying on local, ad hoc property valuations. Small, individual jurisdictions typically lack the expertise to use statistically sophisticated models in the U.S., and I am sure that this is true in China as well. A national evaluation squad should perform the evaluations. Fees paid by entities that want to engage in eminent domain can finance the costs of these services.

A single national evaluation team can be matched with an independent national adjudication body to handle disputes over these valuations. Presumably, it is possible for either party in a taking to dispute the valuations put forward. A review board that is somewhat independent of the initial evaluation team might consider whether such complaints have merit. Again, the costs of this board need to be included in the fees charged to the entity that does the taking.

Non-local evaluation and adjudication is also important in maintaining the integrity of the process. It makes little sense to have values assessed by the entity that is doing the taking, or by any entity that is related to the taker. The price of the taking should be established by a completely independent entity to reduce potential conflicts of interest and subversion of justice. The best way to fight subversion is to have strong national oversight over the compensation that needs to be paid for takings.

The third element of enhanced compensation is to provide premia over market value to reflect the fact that owners presumably value their land over the market, or else they would have sold the land. The principle of a premium over market value exists in many countries. India's 2013 Land Law, for example, required a 300 percent premium over market value for rural land and a 100 percent premium for urban land.

Such premia are somewhat arbitrary in the absence of social science evidence on the premium that owners would require to be made whole. One simple means of ascertaining a national average figure is to ask people about their willingness to accept money to move. This is probably a sensible thing to do, but I would worry about people overstating their fondness for their house or just having difficulties dealing with an abstract question. The alternative is to use mobility rates of owners after price increases to get some indication about the distribution of unobservable tastes for location.

The approach that I have called weakened property rights maintains some popular ability to refuse participation if the compensation offered is unsatisfactory. The extreme version of this refusal has no weakening and allows any person to block the entire deal, which can create a holdout problem. Weakened property rights means that the deal can go through despite some opposition, but a supermajority of impacted residents must agree to the deal. For example, India's 2013 Law requires either 70 percent (in the case of a public–private partnership taking) or 80 percent (in the case of a taking for private ownership) approval of the residents for the deal to go through.

Heller and Hills (2008) champion the idea of Land Assembly Districts within the U.S. These districts elect representatives to negotiate a total sales price on their behalf. The members of the district vote to approve the final offer, and the sale goes through only if there is a sufficient supermajority. The final sales price is split up among the members of the district based on pre-existing rules. This concept seems to set up a system whereby holdouts cannot destroy a transaction, but the basic structure of property rights is maintained.

The basic structure of these districts can be mimicked almost anywhere. First, a developer makes a bid on a plot of land. The impacted residents are informed of how much each would get in the event of a sale. The simple way of sharing would be based on the amount of land that each resident holds as a share of the total. A more complex arrangement would be sharing on the basis of independently assessed property values. The deal goes through only if a supermajority agrees to the sale.

The structure still requires details. What supermajority is optimal? What is the basis of sharing receipts? Can the developers raise their bid if it is rejected? Will things function better if the residents have some designated bargaining representative to work out the initial details? These can also be worked through over time.

The enhanced eminent domain structure is likely to be somewhat smoother and less difficult to operate. The weakened property rights regime is somewhat more cumbersome and more likely to break the speed of development. If the great fear is local obstruction of good projects, then perhaps enhanced eminent domain makes more sense. If the great fear is subversion of the evaluation process, then weakened property rights are the more natural tool.

Finally, it is quite possible to imagine an intermediate system that uses both tools. For example, the bidding might start at the enhanced eminent domain value. Developers would then have the opportunity to raise the bid in order to secure a supermajority of residents. This system would be even more cumbersome, but it would offer more protection to residents against abuse of the system. In all cases, it makes sense for there to be non-local oversight of the system's operation.

5. Towards a better system of property taxation

We now switch from eminent domain to property taxation. A long tradition of economists supports the notion that property taxation is a relatively efficient means of financing local government (e.g., Arnott and Stiglitz, 1979). Other means of financing local government can create significant spatial distortions if tax rates differ across space. For example, a location with high local income taxes might repel the wealthy. A location with high local sales taxes might induce retail to relocate. Property is relatively immobile, which means that the spatial distortions created by using property taxes are typically less socially costly than other taxes. Property taxes can also create helpful incentives for local governments because, if government revenues increase with the value of local property, this increase may induce them to undertake actions that raise the demand for living in the locale, and those actions will typically have positive social value.[4]

The one major distortion created by a property tax is that developers will have less incentive to build if their tax bill scales up with the value of the physical structures that they create. The natural fix for this problem is to tax land, rather than structures. While this idea was strongly advanced by Henry George more than a century ago, in practice property taxes remain the more common means of financing local governments worldwide. In 2008, more than two-thirds of the tax revenue received by U.S. local governments came from property taxes.

China has also historically made use of a heavy property tax, but the Chinese tax is a tax on transactions rather than a tax on value. This tax structure is heavily distortionary because it keeps owners in place and reduces mobility. It is tied not to the total value of property in the area, but just to the property that changes hands, which may lead to revenue shortfalls as Chinese cities eventually become less dynamic. This tax also creates an incentive to constantly rebuild the housing stock to squeeze revenues as property changes hands after redevelopment.

Property taxes relate to housing bubbles in two distinct fashions. First, if there is an actual over-building glut, then property taxes (as opposed to land taxes) can help deter over-building. Since over-building is far from a universal problem, one possibility is to have a tax that is a flexible function of land values and property values, and that is oriented more towards property values as the amount of new construction increases. A related tool could be a property tax that is capped by value, a property tax with different rates on land and the property above it. With this structure, areas with high property values, that presumably need more development, would depend more on the land tax, while areas with lower values, that presumably need less development, would depend more on the property tax.

Second, well-designed property taxes could reduce some of the incentive for localities to push for over-building. As many authors have argued, a steady stream of revenues from property taxes might reduce the need for local governments to use eminent domain as a tool for raising revenues. A property tax based on the value of property, rather than a transaction, would reduce the incentive to redevelop in order to gain more transaction-tax revenues.

Better property value assessment is also an advantage to property taxation. The tools discussed in Section 3 are useful for property-tax assessment as well. Moreover, since property taxes require special expertise and also have the potential for abuse, it can make sense to centralize property assessment. In this case, the central office would be responsible for evaluating property and potentially maintaining a court for adjudicating disputes over property values. The local government would actually set the tax rates. The costs of maintaining the central office would come from appropriating a small share of the tax revenues themselves.

The problem of evaluating land values is somewhat distinct from the problem of evaluating property, partially because land values are so tightly tied to the legal rights to build on that land. For example, Glaeser, Gyourko, and Saks (2005) estimated land values in two ways. First, we started with housing values and then subtracted the estimated physical value of the structure based on construction costs. Second, we used a hedonic estimate to value land by essentially comparing similar houses with big lots and small lots. The two procedures generated values that often differed by an order of magnitude.

Why was land that extended the size of a lot worth so much less than the land that sat under a home? The owner did not typically have the right to build a home on that extra land, and it is the right to build a home that actually carries the most value. When Henry George wrote, there were few limits on new construction. Consequently, land that extended a lot was worth the same as land that sat under a new house, for otherwise the owner would build upon the extra land. Today, land-use limitations essentially determine the value of land, which makes it considerably harder to actually say what any plot of land is worth.

Nonetheless, there are simple and easy ways to implement a land tax that can also reduce the disincentive to add new structures. For example, the tax might begin with a full property appraisal and include a deduction for the physical cost of building the structure on the property. The owner might be required to provide an estimate of the cost of the structure, or alternatively and more simply, the cost could simply be estimated based on the square footage of the structure and average construction costs. This second approach would not completely eliminate adverse disincentives. Taxes would still be higher on higher-quality buildings, which would reduce the incentive to upgrade quality levels. Yet this property-value-minus-construction-cost approach would still provide better incentives to build than a standard property tax.

Perhaps the simplest way to implement a land tax is to form an estimate of the value of land in different parts of the city and then apply that tax to each property based on its location. This approach might start with property values and then subtract estimated construction cost. Importantly, since this approach will be based on average value, not the property's own value, there will be no adverse disincentives. The downside, however, is that it may not accurately assess very local differences in the value of property, either because of micro-location or because of local land-use rules.

Property taxes are an attractive and highly feasible means of financing local government. Transaction-based taxes are not particularly fair and strongly disincentivize transactions. A better approach is to tax the value of the property annually, or better still to tax the value of the land. A centralized assessment office that uses cutting-edge estimation techniques can improve the quality of land or property value estimation.

6. Towards better financial oversight

Property value assessment is also an important element in banking oversight. In both the U.S. and China, real estate often represents an important element in financial balance sheets. Real estate serves as collateral for consumer loans in the U.S. and for real estate development loans in both the U.S. and China. Real estate-backed securities played an extremely important role in America's 2007 financial meltdown. As the value of these assets plummeted, systemically important financial institutions, such as Lehman Brothers, put the entire U.S. banking system at risk. Earlier crises, in the U.S. and elsewhere, also had their roots in the real estate exposure of financial entities.

I have too little expertise to opine on Chinese banking oversight, so I will constrain my discussion to the oversight of U.S. banks. I hope that this discussion has some relevance for China as well. Within the U.S., there are two distinct ways in which property value assessment relates to bank oversight. First, mortgage origination typically requires a property appraisal. Second, real estate appraisal must form a part of balance-sheet oversight, when real estate-related assets provide a major form of that balance sheet.

Real estate appraisal for banking oversight has a slightly different purpose than real estate appraisal for eminent domain purposes or for property taxation. The goal of bank oversight is to ensure that lending institutions remain solvent in the future. Hence, it is at least as important to determine the value of property in the future as it is to determine value at the moment of lending. Moreover, sensible bank oversight needs to consider the distribution of future housing prices, not just the mean expected future housing price. The key question is whether there is a considerable probability that the house will be worth much less than it is today.

One natural approach is to begin with an appraisal of the value of the home and then adjust that value based on aggregate patterns of price changes and price risk. There is not yet enough knowledge about neighborhood price dynamics to use neighborhood-level facts about changes and volatility. We do, however, have more than enough knowledge about price changes at the metropolitan-area level to embed these in the appraisal process.

For example, Case and Shiller (1989) document that housing prices display strong momentum at one-year frequencies. Glaeser et al. (2014) document a momentum coefficient that may be as high as 0.7, suggesting that a one-year positive price change of 1,000 dollars in the metropolitan area in one year will be associated with a 700-dollar price change in the next year. Cutler, Poterba, and Summers (1991) also document that there is mean reversion over longer time horizons. Glaeser et al. (2014) find a mean reversion coefficient of -0.3, meaning that if prices have increased by 1,000 dollars over the past five years, then they will decline by 300 dollars over the next five years.

Adjusting for the dynamic path of the mean price might mean forming an expected price change over the next 10 years and then basing the assessment on a weighted average of those 10 years. If the oversight authority designates an appropriate discount factor (denoted β), then the weighted average price at time t could be $\dfrac{\sum_{i=0}^{10} \beta^i \hat{P}_{t+i}}{\sum_{i=0}^{10} \beta^i}$, where $\hat{P}_{t+i}$ represents the expectation of price at time $t+i$ based on a dynamic prediction model. This structure would tend to reduce appraisal values if the growth rate of past prices has been high over the past five years, assuming that the momentum effect does not dominate. This will tend to make real estate less valuable as collateral at the peak of a boom, and that will tend to mitigate against further credit expansions during a real estate bubble.

A more sophisticated approach is to recognize that future upside risk has little value to the bank, since it will typically not receive the benefits of housing-price appreciation, but that downside risk is important. In the event of a default, if the house price is worth more than the outstanding debt, then the natural result is a sale, where the creditor receives the outstanding debt and the borrower receives the excess. Consequently, the expected value of the collateral can be seen as $(1-F(D))D + \int_0^D Pf(P)dP$, where D is the debt owed, P is the price, and $f(.)$ and $F(.)$ are the density and cumulative distribution functions of price at some future period.

Using the current price, the path of debt, and the past distribution of price changes, this quantity can be estimated. One approach is to assume a particular distribution of price changes, such as the log normal distribution. A second approach would be to estimate the distribution non-parametrically. In both cases, the past history of price changes at the metropolitan-area level can be used to estimate these series. In either case, it should be possible to adjust the value of the property downward to reflect the possibility that the price will be worth less than the outstanding debt level in future years.

Just as in the case of the estimation of the current price, estimating the expected future distribution of price changes involves significant expertise. Local bank examiners cannot be expected to perform such operations on their own. A far better approach would be for a centralized office to estimate correction terms that reflect housing dynamics and downside risks, and then these correction terms can be imposed on the static housing-price estimates.

7. Conclusion

Real estate appraisal may not be glamorous, but it is important. The equitable and efficient use of eminent domain requires accurate assessment of the evaluation of taken property. Good property (or land) taxation requires taxing authorities to know the value of properties. Banking oversight likewise depends on proper evaluation of the risks of future housing-price drops.

This paper has argued that improvements in technology make it possible to better evaluate housing and land prices. Computer vision means that we can now, in principle, duplicate the eyes that on-the-ground assessors bring to the valuation problem. Such tools mean that property assessment can be done centrally through computer models that combine data on past sales, housing characteristics, and images. Central evaluation makes it possible to reduce local errors and any possibility that local property value assessments cater to local policy objectives, such as expanding property development.

In all of these settings, better property value assessment is only a first step. In the case of eminent domain takings, it may be desirable to augment valuations with the requirement that a supermajority of current owners vote to accept the proposed deal. With property taxation, it may beneficial to move towards land taxation in order to reduce the disincentives to development created by taxing structures. With bank oversight, it is crucial to take housing dynamics and downside risk into account, since bubble-level prices may be a poor guide to the value of a house after a bust.

Improvements in eminent domain, property taxation, and bank oversight will not eliminate real estate bubbles. Moments of irrational exuberance will always occur. However, reform may reduce the number of bubbles, and the suffering created when bubbles burst. For example, increasing the cost of using eminent domain may reduce over-building. Better use of property taxation may limit the tendency of local governments to use real estate development as a financing tool. Improved bank oversight should reduce the impact of declining real estate prices on the stability of the banking system.

China's housing boom may not end in a crash, but even if China escapes that fate, appropriate reform is still sensible. A national system of effectively monitoring the value of housing prices provides one tool for enhancing economic stability for decades to come.

Notes

* The author thanks the Taubman Center for State and Local Government for financial support. Parts of this paper reflect ongoing research with Nikhil Naik, Yueran Ma, and Andrei Shleifer.
1 Nathanson and Zwick (2015) present an interesting argument that the land-intensive nature of speculation in Las Vegas helped encourage the bubble, since in land acquisition, as opposed to housing purchases, a few optimistic buyers can dominate the market.
2 http://real.wharton.upenn.edu/~gyourko/chineselandpriceindex.html.
3 www.stats.gov.cn/tjsj/ndsj/2014/indexeh.htm.
4 The one prominent exception is that restricting land development can potentially increase the total value of property by limiting access to a location with idiosyncratic amenities. Essentially, localities can act like local monopolists, and that will not be Pareto optimal for society as a whole.

References

Arnott, R. J. and Stiglitz, J. E. (1979). Aggregate Land Rents, Expenditure on Public Goods, and Optimal City Size. *The Quarterly Journal of Economics*, 93(4), 471–500.

Case, K. E. and Shiller, R. J. (1989). The Efficiency of the Market for Single-Family Homes. *American Economic Review*, 79(1). 125–137.

Case, K. E., Shiller, R. J., and Thompson, A. K. (2012). What Have They Been Thinking? Homebuyer Behavior in Hot and Cold Markets. Brookings Papers on Economic Activity (Fall 2012).

Chivakul, M. and Lam, W. R. (2015). Assessing China's Corporate Sector Vulnerabilities. International Monetary Fund Working Paper WP/15/72 (March 2015).

Chivakul, M., Lam, W. R., Liu, X., Maliszewski, W., and Schipke, A. (2015). Understanding Residential Real Estate in China. International Monetary Fund Working Paper WP/15/84 (April 2015).

Cutler, D. M., Poterba, J. M., and Summers, L. H. (1991). Speculative Dynamics. *The Review of Economic Studies*, 58(3) 529–546.

Deng, Y., Gyourko, J., and Wu, J. (2015). The Wharton/NUS/Tsinghua Chinese Residential Land Price Indexes (CRLPI) White Paper, Updated to the 4th Quarter of 2015. http://real.wharton.upenn.edu/~gyourko/chineselandpriceindex.html

Fang, H., Gu, Q., Xiong, W., and Zhou, L. (2016). Demystifying the Chinese Housing Boom. *NBER Macroeconomics Annual 2015*, 30(1), 105–166.

Ferreira, F. and Gyourko, J. (2001). Anatomy of the Beginning of the Housing Boom: U.S. Neighborhoods and Metropolitan Areas, 1993–2009. NBER Working Paper No. 17374 (August 2011).

Glaeser, E. L., Gyourko, J., and Saks, R. (2005) Why is Manhattan So Expensive? Regulation and the Rise in Housing Prices. *Journal of Law and Economics*, 48(2), 331–369.

Glaeser, E. (2013). A Nation of Gamblers: Real Estate Speculation and American History. *American Economic Review*, 103(3), 1–42.

Glaeser, E. L., Gottlieb, J. D., and Gyourko, J. (2013). Can Cheap Credit Explain the Housing Boom? In E. L. Glaeser and T. Sinai (Eds.). *Housing and the Financial Crisis*. 2013 NBER Conference Report. Chicago: University of Chicago Press, pp. 301–359.

Glaeser, E., Gyourko, J., Morales, E., and Nathanson, C. G. (2014). Housing Dynamics: An Urban Approach. *Journal of Urban Economics*, 81, 45–56.

Glaeser, E. and Nathanson, C. (2015). An Extrapolative Model of House Price Dynamics. NBER Working Paper No. 21037 (March 2015). DOI:10.3386/w21037.

Glaeser, E. L., Kominers, S. D., Luca, M., and Naik, N. (2015). Big Data and Big Cities: The Promises and Limitations of Improved Measures of Urban Life. NBER Working Paper No. 21778 (December 2015). DOI: 10.3386/w21778.

Heller, M. and Hills, R. (2008). Land Assembly Districts. *Harvard Law Review*, 121(6), 1467–1527.

Himmelberg, C., Mayer, C., and Sinai, T. (2005). Assessing High House Prices: Bubbles, Fundamentals, and Misperceptions. *Journal of Economic Perspectives*, 19(4), 67–92.

Mian, A. and Sufi, A. (2009). The Consequences of Mortgage Credit Expansion: Evidence from the U.S. Mortgage Default Crisis. *The Quarterly Journal of Economics*, 124(4), 1449–1496. DOI:10.1162/qjec.2009.124.4.1449.

Nathanson, C. G. and Zwick, E. Arrested Development: Theory and Evidence of Supply-Side Speculation in the Housing Market. NBER Working Paper No. 23030 (January 2017). DOI: 10.3386/w23030.

3 Thinking of the development of housing policy

Hannu Ruonavaara

UNIVERSITY OF TURKU

1. Introduction

During the decades since the 1980s, there has been a decisive change in European housing policies. In almost all countries there has been a shift from a housing policy regime where the state plays an extensive role in the production and consumption of housing, to a system where market mechanisms play a more central role. The uniformity of the development across many countries raises a question of whether there is some more general regularity in this development, perhaps even a *law of development* of housing policies applicable to all societies, including the East Asian ones. It is, however, quite questionable whether general laws valid in all socio-historical contexts can be detected in social and political development (Elster, 1993). If, indeed, no general laws of development exist in societies, we should generalize in more modest ways, as generalize we must if we are to be social researchers. One form of such generalizations is the creation of models that attempt to capture how a given process typically happens. Such models are often developed inductively, by looking empirically at how processes have happened in various cases.

In the classic Chicago School urban sociology from the early twentieth century, these kinds of models were called "natural histories." Writing about the theoretical contribution of the Chicago sociologists, Abbott characterizes natural history as follows: "A natural history was a temporal pattern that followed a relatively predictable course. It could be diverted or shaped by environing facts, but its general sequence could be understood as a whole beyond the contingent details" (Abbott, 1999).

In this paper I first present and discuss a model that describes the natural history of housing policies in the Nordic countries. By Nordic countries I am referring to the three Scandinavian countries of Denmark, Norway, and Sweden, and two of their neighbors, Finland and Iceland. These are all rather small countries in the northwest semi-periphery of Europe. In the past these countries have been part of the same state formations, though never all at the same time. They share the Protestant religious faith, though all are quite secular now (despite the high church membership). Due to the long-standing historical links between these countries, many cultural similarities (e.g., customs and ways of seeing the world) exist between them. In the three Scandinavian countries the main language spoken is quite similar, and residents can understand the language in their neighboring countries. All of the Nordic countries are relatively prosperous, though for some of them (Finland and Iceland), prosperity really came only after the Second World War. Most importantly, all of them conform more or less to what is internationally known as the Scandinavian (or Nordic) welfare model.

The Scandinavian welfare model is known for its universal coverage of social benefits and services, generous benefit levels, and well-developed services – all financed with a relatively high progressive taxation that keeps income inequality relatively small. Of course, the broad similarity is associated with a number of specific differences, and in recent decades all five of the Nordic countries have moved away from the ideal typical Nordic model. In spite of the similarity of the welfare arrangements in the Nordic countries, their housing policies have never been similar except on a very general level: "All the Nordic countries seek to provide housing of a decent standard for the whole of their population" (Lujanen, 2004). What all Nordic countries share is also a similar *policy theory* in regard to housing, a view about the proper division of responsibility between the state and the market in housing provision (Bengtsson, 1995).

The dominant policy theory in these countries is what Bengtsson calls the "state as a corrective to the market." This means that in all of these countries a constitutive assumption of housing policy is that housing is provided mainly in the market, and the role of the state is to correct market imperfections and failures. This is, however, not very distinctive of Nordic countries, as in practically all European societies with market economies the policy theory is the same. Though Nordic countries share the same general policy theory, the ways these countries have tackled housing problems politically have been quite different (Bengtsson, 2013b). Sweden and Denmark have had *comprehensive* housing policies targeted to all households. Finland and Iceland have had *selective* policies targeted to those who need assistance most, and to special groups. Norway used to have a general orientation in housing policy but has moved towards a selective one.

The forms of housing tenure in the Nordic countries also differ: various national forms of rental and owner-occupied housing exist in the countries (Karlberg and Victorin, 2004). Types of housing tenure are distributed differently across the population and housing stock of each country. Iceland, Norway, and Finland are nations of homeowners, where the percentage of households living in owner-occupied housing is high (from 70 to 90 percent of households at the highest), whereas in Denmark and Sweden the percentage of owner-occupier households has been lower (at the highest around 60 percent). In these two last countries, publicly supported rental housing has been a major influence on the housing system, though not numerically dominant. In contrast, in the homeowning societies of Iceland and Norway, such housing has been relatively insignificant. A special feature of both Sweden and Norway is that co-operative tenure is more common than elsewhere in Europe or North America (see, e.g., Sørvoll, 2014). For example, in 2015, 21 percent of Swedish and 14 percent of Norwegian households were living in co-operative housing as "tenant-owners."[1]

Table 3.1 Types of housing tenure in 2014 in the Nordic countries (unit: percentages).

	Rental	Ownership	Other
Denmark	39	58	3
Sweden	38	62	0
Finland	33	67	0
Iceland	25	74	1
Norway	23	77	0

Source: Grunfelder, Rispling, and Norlén (2016, 130); Statistics Iceland home page.

Table 3.1 shows the distribution of tenures in the Nordic countries on the level of *types* of tenure. In different countries there are many different forms of tenure. For example, renting in the private market and public or social renting can be classified into two major types, that of owner-occupation and that of renting (Ruonavaara, 1993b). Looking at the matter from this perspective, co-operative tenure also falls under owner-occupation. After all, in co-operative housing the residents' rights over the dwelling are based on an ownership relation, not that of a rental contract. The division of housing stock between two types of tenure, renting and owner-occupation, is shown in Table 3.1 for the Nordic countries:

That housing policies are so different in countries that are otherwise similar, and have a broadly similar orientation to how the state is to provide welfare for its citizens, requires explanation. This was the key question in a collaborative research project involving one researcher from each of the five Nordic countries. The project was called, in English, "Why so Different? Nordic Housing Policy in a Comparative Historical Light." Each of the researchers was in charge of a national case study and contributed to the overall approach, the comparisons between the national cases, and drawing conclusions.

The project searched for the explanations behind the very different histories of housing policy across the five Nordic countries, applying a historical institutionalist perspective (Jensen, 2013). A central idea of the project was the *path dependence* of housing policies. This is a now-fashionable term emphasizing a point many social scientists would generally accept, but perhaps not take seriously in their research practice: *history matters*. There is quite a lot of methodological discussion about how to understand path dependence, what types of path dependence occur, and so on. Here it is sufficient to get a general idea of the kind of approach we are talking about. The sociologist-historian William H. Sewell defines path dependence as a process whereby "what happened earlier at an earlier point in time will affect the outcomes of a sequence of events occurring at a later point in time" (Sewell, 2005). This is an often-quoted broad definition of path dependence – too broad for some researchers. Therefore it is important to be able to account for *how* and through what kinds of mechanisms history matters. Usually it is thought that path dependence keeps things as they are, rather than promoting change. The literature on path dependence lists several mechanisms supporting stability, the most well known of these being "increasing returns" (Pierson, 2004). Does path dependence then make changes in housing policy impossible? No, the theory assumes that in the course of history, every now and then *critical junctures* occur whereby policy makers face a situation where an alternative for the present policy opens and could be chosen.

The idea of path dependence was, however, not the only organizing theoretical insight of the research. First, the focus of the research was on housing institutions, and it was argued that housing tenures are the central institutions in housing policies. Therefore, most critical junctures analyzed concerned housing tenures in one way or another. Second, an important theoretical and methodological tool was an ideal typical periodization of housing policies. According to this periodization, four stages of development of housing policy can be distinguished: (1) the introduction stage, (2) the construction stage, (3) the management stage, and (4) the public retrenchment or privatization stage (Jensen, 2013b). In the next section I explain the periodization in more detail, as well as discuss the central term of the paper, "housing policy."

2. Four stages of housing policy

By "housing policy" I refer to lines of action taken by public organizations with the intention of influencing the availability of housing, the variety of housing options, the material and environmental quality of housing, and the cost of housing for a population settled or settling in some defined area (neighborhood, locality, nation).[2] The actions that fall under "housing policy" are varied indeed, including creating institutions (e.g., housing tenures), establishing actors to the housing system (e.g., state housing banks), regulating the housing market (e.g., rent control), creating norms for housing quality (e.g., building regulations), subsidizing the material production of housing (e.g., renovation grants), evening out the housing costs (e.g., rent subsidies), improving the residential environments (e.g., neighborhood renewal), and preventing residential segregation (e.g., social mix policies). This definition excludes the actions of private-sector actors, like employers (important in many times and places, e.g., in Finland up to the 1970s), as well as those of third-sector organizations, like churches. Thus, I focus here on public housing policies.

Housing policies follow from political decisions taken with consideration of a population's housing conditions. Some of these are explicitly and directly targeted at housing conditions. However, there are also policies that are not primarily policies in the housing field, but have an important housing policy aspect and are often motivated by housing policy considerations, like the tax subsidies that homeowners enjoy in many countries. For example, homeowners may have a right to deduct some of their mortgage interest payments from their taxable income. For the most part, then, housing policy is an intervention in housing conditions, but we must not forget the other policies that have important housing policy aspects.

Here the basic assumption is that housing policy is a historical phenomenon. It enters the historical stage after the *housing question* has been politicized – that is, some actors have brought housing conditions into the political agenda as something that the local or national state should do something about. What this implies is that *laissez-faire* (government does nothing) is a housing policy only after the housing question has been politicized. It is a housing policy if it results from a decision made after the issue of state intervention in the housing situation is raised to the political agenda, or there has already been public intervention in the workings of the housing market. Therefore, not all countries have always had a housing policy. There is housing policy if public intervention in housing is debated regardless of the extent of public intervention in housing conditions. The end of housing policy happens only after the housing question has completely disappeared from the political stage. As housing is such a central part of people's lives and well-being, the end is not in sight.

In Europe the term "housing question" usually refers to the late-nineteenth- and early-twentieth-century concern about the housing situation of the urban and rural working classes (one of the most famous texts about this is Friedrich Engels' *Zur Wohnungsfrage*, written in 1872–3). Bourgeois reformists, as well as activists in the emerging labor movement, were appalled by the unsanitary, unhealthy, and crowded housing conditions of the poor. Poor housing conditions were considered physically detrimental, but also morally harmful (Ruonavaara, 1996). Congested housing drove men to the street and pubs, and the presence of subtenants in the same rooms with the family was seen as contributing to illicit sexual relations. Apart from humanistic and moralistic concerns, bourgeois reformers were also concerned about the possibility that inhuman housing conditions might radicalize the urban working class.

Throughout history, poor people's housing situation has probably been a privately experienced concern for them. It becomes *the* housing question only after some actors publicly raise it to be a social problem that the "society" should do something about. This *politicization* of the housing question happens in the introduction stage of housing policy. That stage happened in the Nordic countries in the late nineteenth and early twentieth centuries, when industrialization and urbanization really began to accelerate. In this stage the housing question transforms from being a series of recurrent crises raising concern every now and then, to something that is constantly kept on the political agenda (Bengtsson, 2013a). However, in the introduction stage, both public subsidies to housing production and consumption and interventions in the workings of the housing market remain limited. The concern for the housing question is, in this stage, more talk than action.

In the *construction stage* the political decision-makers start putting into practice the ideas raised to political debate in the introduction stage. The main concern of housing policy in this stage is to eliminate housing poverty by producing more housing. By "housing poverty" I mean the shortage of housing of a standard considered adequate in the society, at a price affordable to low-income people. In the construction stage local and national governments start supporting housing production by different means, for example by starting to provide subsidized financing for housing construction and by giving tax concessions for housing production. But the construction stage is not only about the physical construction of housing. In this stage, the system of housing policy that is to take care of the succession of housing problems raised on the political agenda is built up (Bengtsson, 2013a). In the Nordic countries the construction stage happened from the early post-Second World War period to the early 1970s.

Gradually, housing poverty starts to be eliminated as a problem concerning large numbers of low- and middle-income earners, and it becomes a problem that concerns only special groups and people in specific difficult life situations. Now the concern shifts from producing as much housing as possible to the management and improvement of existing housing and neighborhoods. In the construction period the focus on elimination of housing poverty by increasing new production overrides concerns about the quality of housing and the residential environment. In the management stage these come to the fore: the political agenda shifts from production to questions of management, housing renewal, tenant participation, and the physical and social environment (Bengtsson, 2013a). The housing mass-produced during the construction stage needs to be renewed and updated to meet new needs and standards of space, quality, and amenities. In this third stage the physical and social environments of the neighborhoods that were built quickly at the height of the construction period are increasingly criticized. Tenant participation in the management of housing is raised to the political agenda. Concerns about the environmental sustainability of housing, as well as social and ethnic segregation, enter housing policy debate. In the Nordic countries the beginning of the management stage can be dated to the late 1970s.

The last of the four stages is the public retrenchment or privatization stage. In this stage the policies and institutions produced in the previous stages are seriously questioned, re-evaluated, remodeled, and partly discarded. The old policies and institutions are argued to be overgrown, bureaucratic, and expensive. This re-evaluation results in changes in the system of housing policy, gradually shifting the responsibility of housing from the public powers to the market and the consumer. The cost effectiveness of the public investment in housing, the removal of regulations experienced as too bureaucratic, and the avoidance of harmful public intervention in the workings of the market become major issues in the housing policy. The public retrenchment stage started in the Nordic countries, as well as in other parts of the Western world, in the 1990s (Bengtsson, 2013a).

The stage model is an *ideal type*, partly abstracted from the histories of Nordic housing policies and partly the result of theoretical reflection upon the "logic" of housing policy development. Like any model, it is not intended to be a one-to-one description of how the Nordic housing policies have developed. Its function is to provide a framework for looking at the development of housing policies in Nordic countries. Both correspondence to the model and difference from it are interesting. It is a verbal model, and there is no sense in trying to test whether it is true or not, but as with statistical models, we can discuss to what extent it fits the reality. The fruitfulness of the stage model lies in its capacity to make sense of the history we are analyzing. Let us now look at how the stage model works in one case: that of Finland.

3. The stages of Finnish housing policy[3]

3.1 Introduction stage: Late nineteenth century to Second World War

As in many other countries in Europe, in Finland the housing policy started with early industrialization during the late nineteenth and early twentieth centuries. Before the Second World War, the housing situation was poor in Finland (Ruonavaara, 1993a). In Finnish cities the majority of households lived in one-room dwellings, housing on average 2–4 persons per dwelling (depending on the census year and location). The level of housing amenities was low. In the capital city, Helsinki, running water and inside toilets were already common in 1920; in dwellings in smaller urban centers, they were found much less often and became common much later.

Finland industrialized later than many other countries, and also the development of its housing policy lagged behind more advanced European societies. Along with the "social question" in general, the "housing question" was eagerly discussed among middle-class reformers, as well as in the new workers' associations. However, not much happened in public housing policy until after the Second World War. The public powers' involvement in housing provision remained marginal, and they limited their activities to acute crisis situations, like the housing shortage after the First World War, when a temporary rent control was enacted. This was no different from the general welfare policy orientation during the 1920s and 1930s: the existing social policy was generally of a residual kind and control-spirited. There were, however, some enabling measures (in contrast to restrictive ones, like rent control) in interwar housing policies. There was, for example, a modest state housing loan system for building non-profit rental housing, as well as self-built owner-occupied housing. In the early twentieth century, the self-builders' movement was close to many housing reformers' hearts (Ruonavaara, 1999).

The central critical junctures of the introduction stage had to do with legal reforms concerning housing tenures, rather than subsidies or financial policies concerning housing. In the mid-1920s two Finnish forms of housing tenure were institutionalized by new legislation. A special feature of the Finnish system of housing provision was the emergence of a specific form of owner-occupation in multi-family housing. Since the late nineteenth century, various kinds of "housing companies" had existed; these provided residents flats in multi-family housing. Sometimes the terms of possession were quite similar to renting; sometimes they resembled owner-occupation. Many of the early housing companies were a result of workers' self-help. A group of workers came together to establish a joint-stock company that would act as a developer of a new residential building. The house was built, sometimes self-built, and the shareholders in the company moved in as residents who were collectively responsible for the management and maintenance of the house. To legally regulate this kind of activity, and to root out speculative activity operating under the name of a housing company, a law on housing companies was passed in 1925. This law consolidated an urban form of owner-occupation that was to become very important in the period after the Second World War. During that later period, the housing company lost its character of enabling housing consumers' self-help and became a form of organization that commercial developers also made use of (see Lujanen, 2010 for an up-to-date description of this form of housing tenure).

The law on housing companies was passed with little discussion and with great agreement between the parties in the Finnish parliament. After all, it was a law that promoted the popular self-help of people trying to become homeowners. This goal appealed to both Left and Right. Such unanimity was not reached in the making of the first law on rental housing in Finland, also passed in 1925. The disagreement between the Left and the Right concerned the position of the tenant: were tenants the weaker party in the rental relation who had to be guaranteed special protection by the law? The landlords who were at first positive about the law started to see it as a continuation of rent control and a violation of their constitutional rights. In spite of strong resistance from the landlords' camp, the 1925 law guaranteeing some security of tenure for tenants was passed. At the time, both of these laws predominantly concerned a minority of the total Finnish population, who lived in urban and semi-urban municipalities. Housing companies existed only in urban localities, and households in the countryside were already mostly homeowners in the interwar period.

3.2 The construction stage: From the Second World War to the 1970s

Finland properly urbanized, and also industrialized, only after the Second World War. The housing situation after the war was extremely difficult. Finland had lost the war against the Soviet Union, large areas of Eastern Finland were annexed by the Soviet Union, and most of the Finnish population in those areas was evacuated to Finland's reduced territory. So Finland had an evacuee housing problem at the same time as the problem of housing the veterans coming home from the front. Building materials were hard to get, capital was scarce, and skilled labor was in great demand in public works and other branches of the industry. The post-war period was also a landmark era in the development of the Finnish welfare system. Many histories of the welfare state in Finland consider the reconstruction period after the war as a decisive break with the previous residual and control-spirited welfare state policy. Housing policies also experienced a change in the reconstruction period. The role of the state became more active: rents were controlled and the building of new housing was subsidized by taxation. A particularly interesting critical juncture was the founding of the state housing loan system, ARAVA, in 1949.

The aim of the ARAVA system was to provide subsidized loans to promoters of housing construction in urban municipalities. Originally, ARAVA was to have been an organization that would not only finance housing production, but would also have wide-ranging planning powers concerning housing production. In the political process that led to the establishment of ARAVA, the organization's assigned tasks were reduced to those concerning financing. The bourgeois political forces suspicious of large-scale state intervention in housing provision emphasized ARAVA's nature as a temporary organization for relief of the housing crisis that followed the war. However, ARAVA was not abolished, even though private market production recovered. By the 1960s it had become fully consolidated as part of a repertoire of policy instruments in Finnish housing policy, as well as a means to exercise state control over a part of housing provision.

The early post-war period marks the transition between the introduction stage and the construction stage. New institutions were created and consolidated, while residential construction got an extra boost with the new state housing finance. However, as private housing construction recovered from the wartime crisis by the end of the 1950s, the role of state housing finance started to diminish. Rent controls were also gradually removed in all localities. This, however, did not mean a return to the *laissez-faire* situation, as some important policies were started in the early 1960s – for example, the housing allowance system for families with children.

The construction period began for real in the mid-1960s. The housing standards in Finland were low by European standards in the early 1960s, and the country was experiencing rapid changes in economic and occupational structure, with accompanying large-scale migration from the countryside to the urban municipalities (as well as emigration to Sweden). There was a pressing need to build more, and this need was largely satisfied by building concrete suburbs at the outskirts of towns and cities. These were inspired not only by the new techniques of building with pre-fabricated elements, but also by new ideas concerning social planning. The new techniques made it easier to plan and construct larger residential areas. With the possibilities of large-scale planned housing development, housing policy was also to be centrally planned and equally large-scale. These prospects created co-operation between big construction companies and state housing policy makers. The boost in housing production that peaked in 1974 succeeded in relieving the housing shortage in Finland. During the period 1971–5, about 55 percent of the dwelling units produced had received public housing finance; during the period 1976–80, that share rose to 60 percent.

This development of housing policy followed the guidelines for a new housing policy that two parliamentary committees had designed at the end of the 1960s. In their reports the state was considered to bear the responsibility for the workings of the housing provision system and its distributive outcomes. A minimum level of housing standard that every citizen should be able to enjoy was defined, and the state was understood as being responsible for guaranteeing that minimum to every citizen. The conceptual triad of steering by the state, social justice, and effectiveness characterized the new housing policy. The scope of the public financing of housing production was to be expanded radically. When benefiting from subsidized housing finance, the developers also had to comply with various state regulations concerning the standards of housing produced, house prices and rents, and choice of residents. Social justice was to be guaranteed by favoring the least well-off. Had the ideas in these two committee reports been fully put into practice, the resulting housing system would have been strongly regulated by the government and the public housing authorities. However, such a system was not developed.

3.3 Overlapping management and public retrenchment stages

The planning ideology of the 1960s started to wane at the end of the 1970s. The 1980s were a time of rapid increase in the standard of living and the income level in Finland. Most of the new housing production was privately financed, and the share of households in homeownership was increasing. The shift in the housing policies from supporting housing production to supporting housing consumption signals the gradual passing of the construction period. Selective and direct forms of subsidy increased in importance in the state subsidy budget. However, at the same time the tax subsidy enjoyed by indebted homeowners became more important than any of the subsidy forms in the official housing budget. This was the consequence of the heavy growth of homeownership during the period. In terms of the four-stage model, the period from the early 1980s had characteristics of both the management and public retrenchment stages. The orientation towards a more market-based housing policy, typical of the retrenchment stage, had emerged already in the 1980s, whereas some of the typical features of the management stage, like the concern for residential segregation and the renovation of the large housing stock resulting from the construction stage, emerged more recently.

The central "critical juncture" of the public retrenchment stage was the deregulation of the rental housing market in the mid-1990s (Kettunen and Ruonavaara, 2015). In the early post-war period, rents had been strictly controlled. This control was first removed completely, and later replaced by a more lenient form of rent regulation. This new regulation system was one where the government decided the yearly maximum rent increase based on new housing development costs. The idea was both to keep rents affordable for the tenant and to guarantee the landlord a reasonable return on their investment. This system was seen by many to contribute to reducing the private rental housing stock.

The first steps of the deregulation were taken during the severe economic recession that Finland experienced in the early 1990s. An external shock, the recession opened a window of opportunity that made deregulation of rental housing relatively politically painless. The Finnish parliament debated the gradually expanded reform for years, occasionally rather heatedly. The non-socialist parties unanimously supported deregulation of rental housing, whereas the Left was divided on the question. The Left socialists were so strongly against the deregulation that they condemned it as privatization of rental housing, whereas social democrats had more mixed feelings about the reform. The rent deregulation happened in three stages, and it was supported by the majority in parliament.

Another sign of the public retrenchment stage is the gradual withdrawal of the state from financing housing construction. The ARAVA system that was once the flagship of Finnish housing policy was terminated by a government decision not to grant new ARAVA loans after 2007. The state support for housing finance was now channeled only through interest subsidies on bank loans. At the same time, state financing of the production of owner-occupied housing that once was quite substantial was practically ended. State financial support for housing production is now targeted mainly at social rental housing. Land policies were deregulated by giving market actors more freedom to plan land use for their own business purposes. The government has gradually cut to extinction the substantial "universal subsidy" for homeownership resulting from the right to deduct a part of the mortgage interest payments in taxation.

4. Scrutinizing the stage model of housing policy development

We found the stage model useful in our comparative research on Nordic housing. It does seem to help organize the analysis of the development of housing policies in the Nordic countries. In all of the countries the different stages could be distinguished; in the special case of Finland, two of them were overlapping. In the beginning of this chapter I referred to the model as a "natural history" depicting a predictable succession of stages from which "environing facts" (Abbott, 1999) can cause deviations. By environing facts, Abbott means features of the larger social and political environment that relate to the development of the process in question. The context that made the two last stages overlap in Finland should perhaps be studied in more detail.

The stage model has attracted interest among researchers, and others have applied it (see articles in Holt-Jensen and Pollock, 2009) or found it useful in developing their own approach (Sørvoll, 2014). Bengtsson and his co-authors suggest that the stage model might also be more widely applicable to analyzing housing policy "in other similar countries" (Bengtsson, 2013a). The notion of similarity is not explicitly discussed by Bengtsson, but "similar countries" refers to European capitalist industrial societies with a history of state intervention in the workings of the housing market. A question is whether the model could be even more general. Is it indeed an ideal type that can be fruitfully used to analyze the development of housing policy in any society where the housing question becomes politicized – including East Asian societies?

Three of the four stages follow from "the internal structural dynamics of housing provision" (Bengtsson, 2013a). The idea of dynamics is rather easy to understand. First, there is a need to raise housing poverty on the political agenda. The shortage of housing is an obvious beginning to housing policy. When the discussion reaches a political consensus sufficient for putting words into action, housing policy moves to the second stage, where the building of housing policy institutions and the elimination of housing poverty through publicly subsidized mass production of housing begin. After a period of intensive construction, the most urgent need for housing is satisfied. When the quantitative shortage of housing is no longer the first priority, the attention shifts from production to management and to the quality of the housing and the residential environment.

It seems that the stages follow each other according to an internal structural logic: the concern for housing shortage must transform into a concern for production; need for production will be saturated sooner or later, and then the concern shifts to the quality of what is produced. However, the fourth and last stage does not seem to fit into this internal dynamic: it appears to be more ideologically influenced and linked to "the general public retrenchment of the welfare state" (Bengtsson, 2013a). But is this disparity so? Could it be that public retrenchment is also structural? Or, conversely, are the previous stages also ideological? If we take into account the old sociological insight about bureaucratic systems' tendency to preserve themselves and grow, the interests of the personnel employed as professionals in the systems, the political stakes invested in maintaining the old systems, and the possible rigidity of the machinery created in tackling the emerging housing problems, we can fit public retrenchment into the structural logic framework. The housing policy systems become too large, expensive, and rigid, and therefore create an incentive to restructure and deconstruct them.

For housing policy to change, there must be some sort of grounds for dissatisfaction with the old housing regime, and not only an ideological interest in transforming it. If we look in this direction, the historically changing and relative nature of housing problems to which policies are responding would have to be taken into account more systematically. The ideal model does not ignore this concern: it starts from housing poverty. There is room for an approach conceptualizing housing policy periods (like stages) as sequences of problem-solving (Haydu, 1998). The evolution of housing conditions creates issues of quantity, then issues of quality, and finally issues of economic sustainability.

But the analytical lens can be turned the other way around. Is it not possible to think that the stages before public retrenchment are nearly or just as ideologically influenced? The initial stage, the politicization of the housing question, requires a shift in thinking about housing poverty such that it is not people's personal problem, but one in which the society, represented by the local and national state, has a moral duty to intervene. Drawing from the Finnish case, we could subdivide the construction period into the construction of institutions and the construction of housing units. The construction of housing policy institutions could also be seen as part of the introduction stage. Institutions can be established, but it is not evident that they are used for a state-promoted mass production of housing on a large scale. In the Finnish case, the policy tools created in the immediate post-war period were put to extensive use from the mid-1960s onwards. It can be argued that in Finland, as well as in other Nordic countries, the extra effort of "building out of the housing shortage" (a slogan used in Sweden) was influenced by a modernist welfare ideology best embodied in Scandinavian social democracy, but shared also by other political forces at the time.

The shift of concerns from production to management feels quite structural, but a counterfactual question can be posed: why had public retrenchment not already happened by then? After all, the state had done the job it was asked to do: eliminate extreme housing poverty. Somehow quality issues were now moved up on the political agenda.[4] One contributing factor is probably the rise in people's housing standards. When the population's level of living rises, issues about the *quality of life* become more pressing. In the housing sphere, this means that people's ideas about what housing standard is acceptable change. Having a roof over one's head is not enough anymore, and issues of the social and physical characteristics of the residential environment rise on the housing policy agenda. If all stages of housing policy are more or less ideologically grounded, public retrenchment is not an outlier in the stage model.

What the preceding discussion suggests is that it might be possible to elaborate the stage model by linking it more to the changing ideological environment and cultural expectations about public responsibilities concerning housing provision. That linkage would mean that the stage model is a more historically grounded ideal type than the idea of "structural logic" would suggest. Historical grounding would also undermine a suggestion to develop the stage model into a cyclical one, where the public retrenchment stage would mark the end of one cycle leading to another introduction stage, beginning another cycle of housing policy.[5] Though attractive, this idea is questionable if the cycle is understood as a return to the original situation. Housing problems never recur in exactly the same way; the housing situation is an evolving one, and this evolution creates new problems to be solved. The more historically grounded the stage model is, the more difficult it becomes to develop an abstract cyclical model of housing policy change.

One aspect that is worth thinking about is the relation of the stage model to the institutionalist approach to researching housing policies. In our research project, the theoretical approach is actor-based, not structural. Paraphrasing Marx, people are seen to be making the history of housing policy, albeit not in the conditions they have themselves chosen. Past policies frame current conditions, shaping the environment for policy choice and the ways societies see their housing problems and policies. Ultimately, housing policies are the result of actors' actions and interactions. So the structural dynamics of the stage model should at least in principle be decomposed into actions and interactions by different categories of actors: government ministers, politicians, bureaucrats, developers, building companies, residents' interest organizations, and so on. However, the stage model is not so presented. It is presented in a passive voice, as shifts occur from one stage of housing policy to another. An actor-centered elaboration of the stage model could be possible, and it might offer theoretical insights missed while doing research relying on a structural model. If the model is developed towards a succession of problem-solving situations, an actor-based "translation" of the stage model would have to be made.

It is possible to read our analyses of the national histories of the five Nordic countries as what Andrew Abbott calls an "ancestors' plot" (Abbott, 2016): accounts mapping out what we think are the important causal factors leading to one outcome, public retrenchment. In fact, the case study of Denmark was explicitly written so that it starts from the description of the failed attempts to privatize the Danish social housing sector, and then goes on to discuss the historical development of the factors that contribute to the sector's inertia against policy change (Jensen, 2013a). However, the intention of the research was not to map the road to public retrenchment, but rather to account for the fact that housing policy regimes in the Nordic countries are different. The idea was that, during the post-war period in each of the Nordic countries, a housing policy regime developed gradually that reached its high point in the 1980s and thereafter was gradually restructured and decomposed. The research outcome is actually that of a relatively stable constellation of national policies and institutions that evolved through the post-Second World War period and now faces pressures for change.

A reader of the findings of comparative research on Nordic housing policies might get an impression that public retrenchment is the end of the history of housing policy. Surely this is not so, even if aspects of the approach chosen could lead the reader to think so. The focus on path dependence might excessively direct our attention towards stability. Change happens at "critical junctures" in which we move from one stage to another, and there is no going back. When the problem of housing shortage is solved, we no longer have to worry about it – unless external events like wars or natural disasters create it anew. But stability is not characteristic of housing conditions. Housing is a social process where change is endemic (cf. Abbott, 2016). To consider the demand side, population grows (or diminishes), its age structure changes, it changes place in the country, and its level of living changes, as do its housing-related values. On the supply side, production technology changes, economic conjunctures boost or restrain production, financing instruments develop, markets emerge, and others wither away. Change is also path dependent, and path dependence does not preclude change. The housing question keeps moving and evolving, acquiring new forms in different times and places.

Problems encountered in earlier stages of housing policy can re-emerge but in new forms. An example is the quantitative shortage of housing. In present-day Finland, the most debated housing policy problem is the shortage of affordable rental housing in the capital city, Helsinki, and its metropolitan area.[6] Job opportunities attract people to move to the Helsinki region. However, rents and housing prices there are so high that ordinary people in service occupations, like nurses or police officers, cannot easily afford to rent or buy housing there. The situation is even worse for low-paid manual workers, not to speak of new immigrants. What is the policy response to the problem? Housing policy experts and politicians are desperately seeking ways to increase housing production in the Helsinki region. We are back to the construction-stage problem of a numerical shortage of housing, but this time it is not a national but a local one. There is hardly any discussion of housing problems outside the Helsinki region and other growth centers. In this situation, where the ideology of public retrenchment is dominant, some policy instruments used during the original construction stage are ruled out. There is no serious discussion of reintroducing any form of regulation of rents. The government is not considering taking the lead in housing provision and expanding its role as a financer of housing production, as the 1960s committee reports envisioned. The solution is now sought through partnerships between the government, the municipalities, and the private sector, as well as through dismantling regulations that owners of social rental housing are operating under, to make investment in new production more attractive.

5. Concluding thoughts

A central issue discussed above is whether the development of housing policies basically follows an inbuilt internal structural "logic" or is driven mainly by outside forces, defined by Abbott as "environing factors." There are all kinds of environing factors, ranging from the general economic situation to the movement of the population between different parts of the country, but one important factor is the ideological discourse that suggests and legitimizes particular policy developments.

The example of the present Finnish situation suggests that there are both internal dynamic elements of the housing question and external ideological frameworks within which solutions to the problems can be sought. However, there is a tension between explaining the development of housing policy by ideological influences and explaining it by inbuilt dynamics. It is quite reasonable to say that both have a part to play in the development of housing policies, but the real challenge is to try to specify what part each of them plays. Is one or the other more important in explaining the observed development of housing policies? I have argued that the politicization of the housing question is an ideological choice, something that is self-evident only to those who live in countries with a history of housing policy. The idea that the government should intervene specifically in housing provision was once a new way of thinking about housing, and it marked an ideological rupture with how housing policy had been conceived in the past. There are also other frames.

The Nordic countries all share a similar policy theory about the role of public housing policy: the role of the state is to correct market imperfections and failures. This is probably so obvious to people in these five countries that it is scarcely recognized as a particular ideology concerning housing provision. The stage model used in the research on Nordic housing policies rests on this policy theory, but it is clear that when this ideological choice has been made, there is also structural logic (described above) that constitutes the "natural history" of housing policy that the stage model basically is. The stage model discussed here is one step in developing a more nuanced approach to analyzing housing policy development.

Notes

1 The percentages are calculated from statistics available at the Statistics Sweden (www.scb.se/en/) and Statistics Norway (www.ssb.no/en/) home pages. "Tenant-owner" is a term invented by Stefan Svensson to characterize the housing tenure situation of residents in co-operative housing (Svensson, 1998). Co-operative housing is often seen as an intermediate form of tenure between owning and renting. On the level of *types* of housing tenure it should be classified as owner-occupation.
2 This is a reworked version of the definition given in my doctoral dissertation (Ruonavaara, 1993a). In that work I was concerned *not* to define housing policy in a way that would imply that it always improves housing conditions. I was thinking

of privatization policies in the United Kingdom and elsewhere, for which the positive net welfare consequences were doubtful: for some people they increased housing welfare; for others they decreased it. Therefore, I tried to develop a definition that did not assume that housing policy always improves housing conditions, something I still think is right. However, it is probably true that housing policies in most cases are and need to be *justified* with an argument that this particular policy indeed increases the population's housing welfare immediately or in the long run.

3 The account is taken largely from my country chapter on Finland in the book on the puzzle of Nordic housing policies (Ruonavaara, 2013). The reader interested in the sources is advised to consult that chapter.

4 Visiting Sweden in the late 1980s or early 1990s when I was beginning as a housing policy researcher, I was baffled by some housing researchers' view that in Sweden no research on housing policy is needed anymore, only research on housing management. Nowadays, researchers in Sweden would not agree with this.

5 This point was suggested to me by the Finnish philosopher and sociologist Petri Ylikoski in a seminar.

6 This is also a problem in Sweden's capital city and other large urban areas there. Possibly other Nordic capitals also suffer from similar problems.

References

Abbott, A. (1999). *Department and Discipline. Chicago Sociology at One Hundred*. Chicago: The University of Chicago Press.

Abbott, A. (2016). *Processual Sociology*. Chicago: The University of Chicago Press.

Bengtsson, B. (1995). Politics and Housing Markets – Four Normative Arguments. *Scandinavian Housing and Planning Research*, 12(3), 123–140.

Bengtsson, B. (2013a). Varför så olika? Om en nordisk gåta och hur den kan lösas [Why So Different? On a Nordic Puzzle and How It Can Be Solved]. In B. Bengtsson et al. *Varför så olika? Nordisk bostadspolitik i jämförande historiskt ljus* [*Why So Different? Nordic Housing Policy in Comparative Historical Light*]. Malmö: Égalité. pp. 13–47.

Bengtsson, B. (2013b). Fem länder, fem regimer, fem historier – en nordisk jämförelse [Five Countries, Five Regimes, Five Histories – A Nordic Comparison]. In B. Bengtsson et al. *Varför så olika? Nordisk bostadspolitik i jämförande historiskt ljus* [*Why So Different? Nordic Housing Policy in Comparative Historical Light*]. Malmö: Égalité., pp. 401–436.

Bengtsson, B. and Ruonavaara, H. (2011). Comparative Process Tracing in Housing Studies. *International Journal of Housing Policy*, 1(4), 395–414.

Bengtsson, B., Annaniassen, E., Jensen, L., Ruonavaara, H., and Sveinsson, J. R. (Eds.) (2013). *Varför så olika? Nordisk bostadspolitik i jämförande historiskt ljus* [*Why So Different? Nordic Housing Policy in Comparative Historical Light*]. Malmö: Égalité.

Elster, J. (1993). *Political Psychology*. Cambridge: Cambridge University Press.

Grunfelder, J., Rispling, L., and Norlén, G. (Eds.). (2016). *State of the Nordic Region 2016*, 2nd ed. Stockholm: Nordregio. www.nordregio.se/Global/Publications/Publications%20 2016/State%20of%20the%20Nordic%20Region%202016/sotnr2016-2nd-editon-DIGI.pdf

Haydu, J. (1998). Making Use of the Past: Time Periods as Cases to Compare as Sequences of Problem Solving. *American Journal of Sociology*, 104(2), 339–371. DOI:10.1086/210041.

Holt-Jensen, A. and Pollock, E. (Eds.). (2009). *Urban Sustainability and Governance. New Changes in Nordic-Baltic Housing Policies*. New York: Nova Science Publishers.

Jensen, L. (2013a). Danmark – lokal boendedemokrati och nationell korporatism [Denmark – Local Residents' Democracy and National Corporatism]. In B. Bengtsson et al. *Varför så olika? Nordisk bostadspolitik i jämförande historiskt ljus* [*Why So Different? Nordic Housing Policy in Comparative Historical Light*]. Malmö: Égalité, pp. 49–117.

Jensen, L. (2013b). Housing Welfare Policies in Scandinavia: A Comparative Perspective on a Transition Era. *LHI Journal of Land, Housing, and Urban Affairs*, 4(2), 133–144.

Karlberg, B. and Victorin, A. (2004). Housing tenures in the Nordic countries. In M. Lujanen (Ed.). *Housing and Housing Policy in the Nordic Countries*. Nord 2004, 7. Copenhagen: Nordic Council of Ministers, pp. 57–78.

Kettunen, H. and Ruonavaara, H. (2015). Discoursing Deregulation: The Case of the Finnish Rental Housing Market. *International Journal of Housing Policy*, 15(2), 187–204.

Lujanen, M. (2004). Main Lines of Nordic Housing Policy. In M. Lujanen (Ed.). *Housing and Housing Policy in the Nordic Countries*. Nord 2004, 7. Copenhagen: Nordic Council of Ministers, pp. 15–22.

Lujanen, M. (Ed.). (2004). *Housing and Housing Policy in the Nordic Countries*. Nord 2004, 7. Copenhagen: Nordic Council of Ministers.

Lujanen, M. (2010). Legal Challenges in Ensuring Regular Maintenance and Repairs of Owner-occupied Apartment Blocks. *International Journal of Law in the Built Environment*, 2(2), 178–197.

Mahoney, J. (2000). Path Dependence in Historical Sociology. *Theory & Society*, 27(3), 507–548.

Pierson, P. (2004). *Politics in Time. History, Institutions, and Social Analysis*. Princeton, N.J.: Princeton University Press.

Ruonavaara, H. (1993a). *Omat kodit ja vuokrahuoneet. Sosiologinen tutkimus asunnonhallinnan muodoista Suomen asutuskeskuksissa 1920–1950* [*Owned Homes and Rented Rooms. A Sociological Study of Forms of Housing Tenure in Finnish Population Centers*]. Turku: University of Turku.

Ruonavaara, H. (1993b). Types and Forms of Housing Tenure: Towards Solving the Comparison/Translation Problem. *Scandinavian Housing and Planning Research*, 10(1), 3–20.

Ruonavaara, H. (1996). Home Ideology and Housing Discourse in Finland 1900–1950. *Housing Studies*, 11(1), 89–104.

Ruonavaara, H. (1999). The State and Self-Help Housing in Urban Finland, 1920 to 1950. *Housing Studies*, 14(3), 329–339.

Ruonavaara, H. (2013). Den dualistiska bostadsregimen och jakten på det sociala [The Dualist Housing Regime and the Quest for the Social]. In B. Bengtsson et al. *Varför så olika? Nordisk bostadspolitik i jämförande historiskt ljus* [*Why So Different? Nordic Housing Policy in Comparative Historical Light*]. Malmö: Égalité, pp. 277–346.

Sewell, W. H., Jr. (2005). *Logics of History. Social Theory and Social Transformation*. Chicago and London: Chicago University Press.

Sørvoll, J. (2014). *The Politics of Cooperative Housing in Sweden and Norway – The Swedish Deregulation of 1968 and the Norwegian Liberalization of the 1980s*. Oslo: University of Oslo.

Statistics Iceland. *Tenure Status [of] Households and Individuals by Sex 2004–2014*. www.statice.is/statistics/society/quality-of-life/housing/.

Statistics Norway. *Households, by Tenure Status, Contents and Year*. www.ssb.no/en/statbank/table/11081/tableViewLayout1/?rxid=4bbaead1-a8b0-4938-ab54-8bd25b24902d.

Statistics Sweden. *Housing Expenditures and Number of Households by Tenure and Type of Household, 2015*. www.scb.se/en/finding-statistics/statistics-by-subject-area/household-finances/household-expenditures/housing-costs/pong/tables-and-graphs/housing-expenditures-and-number-of-households-by-tenure-and-type-of-household/.

Svensson, S. K. A. (1998). *Transformation of a Tenure: The Evolution of Tenant-Ownership in Sweden*. Gävle: Institute of Housing Research.

4 House prices, land rents, and agglomeration benefits in the Netherlands

Coen N. Teulings

UTRECHT UNIVERSITY & UNIVERSITY OF CAMBRIDGE

1. Introduction

Land has traditionally been viewed as the fixed factor in the economy that puts a constraint on the level of a country's GDP. That view has become increasingly untenable. The growth of the capitalist economy has led to massive urbanization. In highly developed countries more than 70% of the population lives in cities. Countries that are currently able to make the transition from a traditional agricultural economy towards a modern industrial economy achieve this through a massive increase in urbanization. Due to this urbanization and the high population density in urban areas, the traditional view of land as the scarce factor has become irrelevant. To the contrary, many countries suffer from large regions where the population density has fallen to such low levels that many services can no longer be provided at an efficient scale.

One would expect that this depopulation in large parts of the country would have led to a fall in the share of land rents in GDP. This has not happened. The rent on agricultural land has indeed fallen, but instead the land rent in cities has increased dramatically over the past decades. Cities have something to offer that makes them attractive places to live. Everybody wants to live and work close to the centers of economic activity, which makes land at these locations highly desirable. Agricultural land might be abundantly available; land in city hot spots becomes increasingly scarce and hence expensive. We present a simple framework for analyzing the mechanics of city development, and we derive some simple economic rules of thumb for city planners and other policy makers to use when developing a city.

This paper presents a summary of the outcomes of a number of research projects that document this process for the Netherlands, and which were undertaken by a team of four researchers: Henri De Groot, Gerard Marlet, Wouter Vermeulen, and the author of this paper. Our main focus is a detailed analysis of land rents for residential real estate across a fine network of some 2,500 ZIP codes covering the whole country. Even in a relatively small and densely populated country like the Netherlands, land rents differ from 3,500 euro per square meter in Amsterdam's Canal Zone, to 17 euro in the northeastern part of the country. Land rents are important for policy makers as they allow the valuation of local public goods, in particular in cities. A regression analysis shows that both the production and consumption factors play important roles. On the consumption side, the number of monuments, the amount of specialist luxury shopping, and the proximity of cultural performances have a large impact on land rents. As networks of these activities are difficult to create from scratch by city planners, these networks should be preserved as much as possible.

We extend this analysis by looking at the dynamics of agglomeration benefits. Land rents not only embody current agglomeration benefits but also include the expected growth of these benefits. We show that inter-regional differentials in population growth rates are strongly persistent. Amsterdam is the most successful region in the country and also the region with one of the highest growth rates. We show that this high growth rate can be expected to persist for the next 20 years. A competitive market will price these expectations. We analyze how this mechanism affects house prices. We derive a simple rule for the optimal moment of construction of new homes, and we apply it to our data on the Netherlands. Finally, we address the issue of the role public policy must play to take maximum advantage of agglomeration benefits.

The setup of the paper is as follows. Section 2 discusses a simple framework for policy evaluation as a background for our analysis. Section 3 discusses the pattern of land rents for 2,500 ZIP codes in the Netherlands. Section 4 discusses the factors that explain the wide variety in land rents that we report in Section 3. Section 5 looks at the persistence of inter-regional differentials in population growth rates and uses this insight to analyze the effect of various expectations of land rents, and derives simple rules for the optimal investment in new real estate.

2. A simple framework for policy evaluations

We provide a simple framework for the analysis of cities previously presented in De Groot, Marlet, Teulings, and Vermeulen (2015) that is related to models of the spatial structure of cities, like that of Lucas and Rossi-Hansberg (2002) and Rossi-Hansberg (2004). Consider a city with some public goods in the city center, either on the production side of the economy, like a central business district (CBD), or on the consumption side, like an opera or a shopping mall. In practice, both mechanisms tend to be active at the same location. However, the benefits of a CBD can be measured from excess wages, while those of consumption amenities cannot. Hence, we focus the presentation on a CBD for the production side. For sake of the ease of exposition, we perform this analysis in a one- instead of a two-dimensional land surface. The analysis can be easily extended to two dimensions. Moreover, we discuss the city as if the land is used for residential purposes only, while the CBD itself does not require any land as an input. Hence, the CBD is a single point in the city center, while a city can be represented by a line. Every worker uses one unit of land to live on. We assume that the commuting costs from the home location to the CBD increase linearly with the distance from the CBD.

64 *Coen N. Teulings*

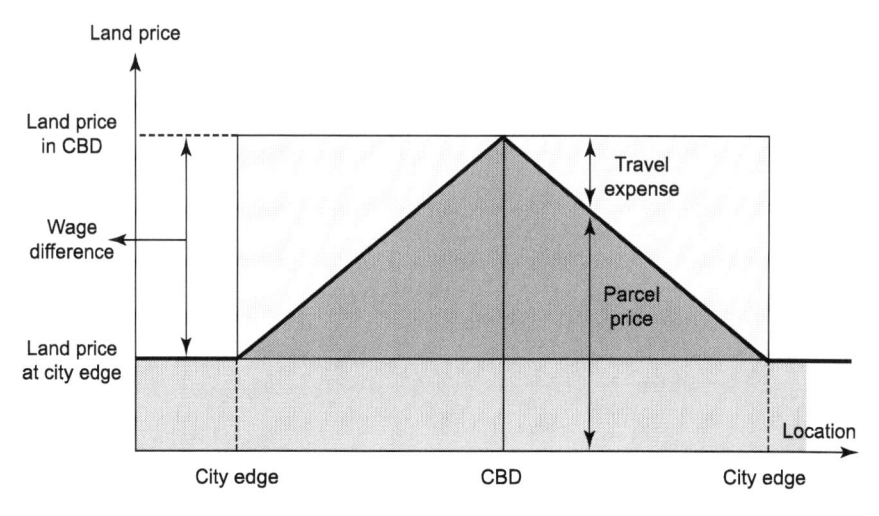

Figure 4.1 The CBD model of a city.
Source: De Groot et al. (2015).

Figure 4.1 offers a graphical representation of this situation. The horizontal line above the light grey area is the value of agricultural land. The rectangle on top of that represents the wage surplus in the CBD above the agricultural wage times the size of the city. The slope of the diagonal lines from the top of the CBD measures commuting cost. People living next to the CBD have no commuting cost. The farther away they live, the higher the cost of commuting. At the edges of the city, this cost is equal to the wage surplus in the CBD. Hence, people are indifferent about working in agriculture and working in the CBD because the wage surplus is exactly offset by the cost of commuting. At locations farther away from the CBD, commuting costs exceed the wage surplus. Hence, people prefer to remain employed at locations in the agricultural industry.

Land rents within the boundaries of the city adjust such that people are indifferent about various home locations. The surplus land rent above the value of agricultural land is therefore equal to the differential value of locations within the city, represented by the dark grey triangle. Next to the CBD, the surplus land rent is equal to the wage surplus of the CBD. At the edges of the city, the surplus land rent is zero. The total land rent surplus is therefore equal to the net value of the CBD—that is, the total wage surplus of all people living in the city and working in the CBD (the surface of the rectangle above the agricultural land rent) minus the commuting cost that people have to spend to reach the CBD (the surface of both very light grey triangles with travel expenses). Hence, the social value of the CBD is equal to the total land rent surplus. This fact leads us to the following general conclusion for local public goods, like CBDs, shopping malls, or transport infrastructure:

Conclusion 1: The total land rent surplus measures the social value of local public goods.

This result is an important aid for policy makers, as it provides them with a hands-on tool for the cost/benefit analysis of public goods. As long as the costs of the project are less than the additional land value that is generated by the project, the project is a worthwhile investment.

Next, we ask whether the outcome of a free market process of city formation is socially desirable (in economist language: whether it is efficient), or whether public intervention is needed. If the wage rate in the CBD is fixed, the boundaries of the city are evidently set efficiently. It does not make sense to work in the CBD if the commuting cost exceeds the wage surplus. Similarly, land use in residential areas is efficient. In the model, as discussed, this issue is irrelevant since land use per worker is fixed and all land is used for the same purpose. However, if workers could choose how much land to consume, then their choice would be efficient. By consuming double the amount of land that somebody else does, a worker would deny somebody else from benefiting from the wage surplus in the CBD. However, the higher land rent the worker has to pay for using that land would exactly compensate for this wage surplus. Private and public cost/benefit assessments coincide and hence the outcome is efficient.

Conclusion 2: When wages are fixed, land use and city size are efficient.

Let us consider the case where people can choose how much land they prefer to use to live on in somewhat more detail. If all workers have the same utility function, with land and other consumption as the only inputs, and if all workers are equally wealthy, then workers who live next to the CBD will consume less land than workers at the edge of the city. The price of other consumption is equal across the city, while land rents are higher in the center. This relative price differential will lead to lower consumption of land close to the CBD. Since real estate is a combination of land and construction, a higher land rent implies that the density of construction is higher close to the CBD.

Conclusion 3: Land consumption per person will be lower close to the CBD, while the population and construction density will be higher.

Again, this conclusion, jointly with Conclusion 2, has important policy implications: land in the proximity of local public goods will be more valuable. Hence, land use per person will be lower. Equivalently, the population density will be higher, which allows more people to benefit from the proximity of the public good. Since Conclusion 2 states that land use is efficient, this process of higher concentration in the proximity of public goods is also efficient. Hence, when creating new local public goods, the density of construction in the neighborhood should be increased. It is therefore very important to define the boundaries of an investment project sufficiently widely, so that potential externalities to a neighborhood can be internalized and the existing real estate can be optimized.

In practice, wages are not fixed in the CBD. Workers in the CBD benefit from agglomeration. Ahlfeldt, Redding, Sturm, and Wolf (2015) use the particular history of Berlin as a natural experiment to estimate this effect. In 1936, Berlin was a large, undivided city; in 1986, it was divided by the Berlin Wall; in 2006, it was reunited. Using this exogenous variation, they show that the elasticity of productivity with respect to local density is about 7%: 10% more workers at a location leads to 0.07% higher productivity at that location. The decay rate of these advantages within a CBD is high: a two-minute walk halves the spillover. For consumption spillovers the half-life is even shorter: one minute.

When wages depend on the number of other workers in the CBD, land use and city size are no longer efficient. The intuition for this is simple. Consider the same example as before, where one worker consumes twice as much land as others, thereby denying another worker the benefit of the higher wage in the CBD. The higher land rent exactly compensates for that. However, the higher land rent does not compensate for the knowledge spillovers that this additional worker in the CBD would have generated for all other workers in the CBD. Hence, workers consume too much land. Similarly, the city is too small. At the edge of the city, a social planner would like to add workers, since each additional worker generates positive externalities to the other co-workers in the CBD. Hence, one would like to subsidize living in the city and to tax land use.

Conclusion 4: *When wages are positively affected by the agglomeration benefits of having more workers working in the proximity, urban density is too low and the city is too small.*

This conclusion runs counter to Conclusion 2. Where Conclusion 2 states that there is no role for city planners to intervene in real estate construction, Conclusion 4 shows that this restriction is not true if there are agglomeration benefits. The density in cities tends to be lower than is efficient. Hence, there is a role for public policy: first, to provide the public goods that make the city an attractive place to work, and second, to make sure that the city is extended to a sufficient scale and that urban density is kept sufficiently high. Both activities are costly. The question is therefore how the city should raise the revenues to pay for that. Land is the fixed factor. Hence, a tax on land is the most efficient way of financing the public good of a city. This is the so-called Henry George taxation, named after the economist who invented this tax more than a century ago.

Conclusion 5: *Taxation of land is the best way to pay for the public goods a city provides.*

This conclusion extends Conclusion 1 to address the question of who should pay for the public goods: let landowners in the proximity pay for them by extracting the land rent surplus, since they are the main beneficiaries of the project. However, not all agglomeration benefits in city centers are the result of deliberate planning and public intervention, as is stressed by Jacobs (1961). Networks of activities that reinforce each other emerge gradually by a process of experimentation and organic growth. These networks cannot be installed from scratch by spatial planning policies because they require combinations of specific entrepreneurial skills (e.g., specialist shops) and spatial configurations (e.g., an historic city center) that can emerge only by a process of trial and error. Hence, existing networks of agglomeration benefits should be treated with care. This principle applies in particular for neighborhoods that are highly attractive for living, like Greenwich Village in New York or the French Concession in Shanghai. They can be destroyed easily and are unlikely to re-emerge afterwards, since various parts can survive only in combination. For this reason, city planners should operate with great care when intervening in these networks.

Conclusion 6: Existing networks of agglomeration benefits in cities should be treated with care.

This analysis has concentrated on the cities with one CBD in the center. The analysis becomes more complicated for cities with more dispersed structures. Glaeser, Kahn, and Rappaport (2000) and Lucas and Rossi-Hansberg (2002) provide arguments for why such structures might emerge. However, the logic of the above arguments about the role of land rents prevails.

3. Land rents and the value of construction

The theoretical framework discussed in the previous section shows why it is interesting to focus on land rents. In this section we discuss the empirical framework that allows us to calculate land rents for about 2,500 ZIP codes in the Netherlands. The Netherlands is a small country with 16 million inhabitants and a land surface of 35,000 square kilometers. ZIP codes cover a smaller area in cities than in the countryside. A typical ZIP code in a city center covers about one square kilometer. Our inference is based on a dataset of about a million house transactions over the period between 1985 and 2013, kindly provided by the Organization of Real Estate Brokers (NVM). The data contain for each transaction the transaction date, the location (by ZIP code), the transaction price, the land surface, the surface of the building, and some characteristics of the building like the number of rooms, whether it is on a corner or a stand-alone construction, the year it was built, and so on.

The valuation of public goods requires an estimate of the land rent. The challenge is to decompose the transaction price of houses into the value of the land on the one hand, and the value of construction on top of that on the other. In the Appendix we explain how we accomplish this task.

Figure 4.2 shows a map of the Netherlands with land rents per ZIP code. The dark lines are the boundaries of the city regions (one can think of these regions

68 *Coen N. Teulings*

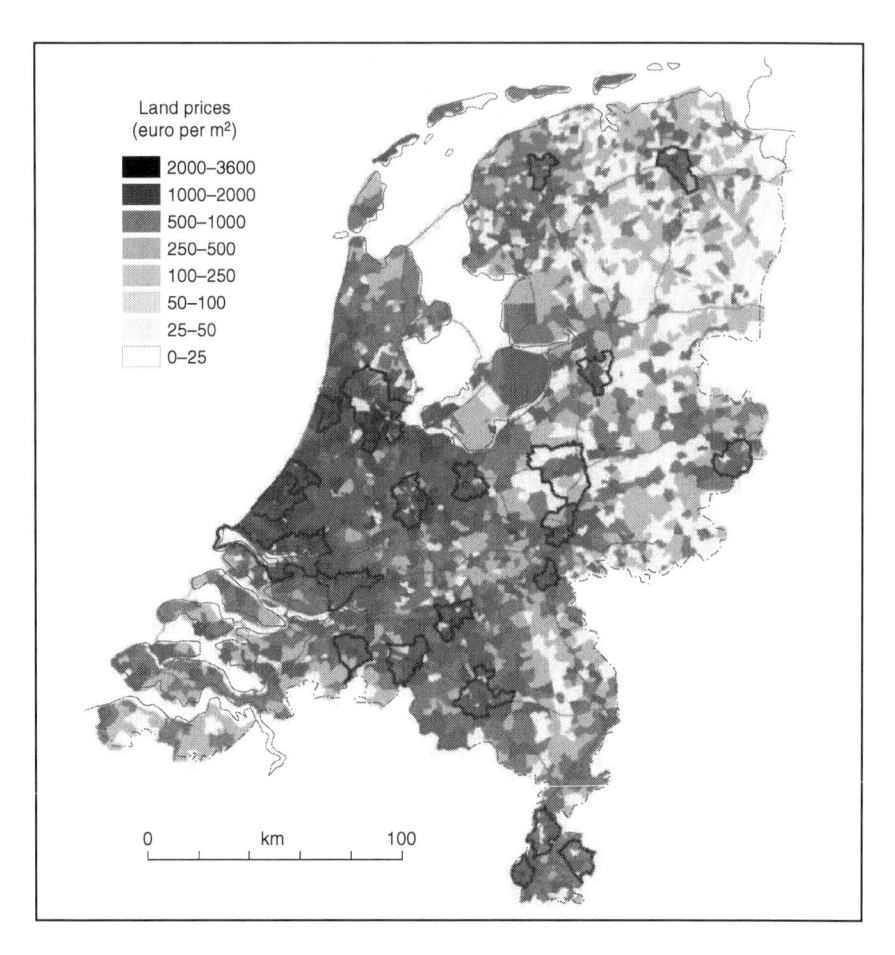

Figure 4.2 Land rents per square meter for each ZIP code.
Source: De Groot et al. (2015).

as the equivalent of a commuting zone). The highest land rent is in Amsterdam's Canal Zone, some 3,500 euro per square meter. The lowest land rent is along the northeastern border with Germany, just 17 euro per square meter. Hence, the ratio between the highest and the lowest rent is a factor of 200. Even without knowing the geographical structure of the Netherlands, it is easy to detect the location of cities. Overall, land rents are much higher in the western part of the country. The small islands in the upper north and the seacoast in the west command rather high land rents since these locations are popular tourist destinations. Figure 4.3 zooms in on the Amsterdam metropolitan area. The figure provides a clear confirmation of the monocentric model of the city: land rents are declining in areas away from the city center.

House prices in the Netherlands 69

Mean land prices (1985–2007)
- 2,000 to 3,600
- 1,000 to 2,000
- 500 to 1,000
- 250 to 500
- 100 to 250
- 50 to 100
- 25 to 50
- 0 to 25

Figure 4.3 Land rents in Amsterdam: a confirmation of the monocentric model.
Source: De Groot et al. (2015).

Figure 4.4 documents the validity of Conclusion 3. Land rents are negatively correlated to lot size and positively correlated to population density.

4. Land rent differentials and the social value of agglomeration

Section 3 has documented a wide dispersion of land rents even in a rather homogeneous country like the Netherlands. It turns out that most of the variation

70 Coen N. Teulings

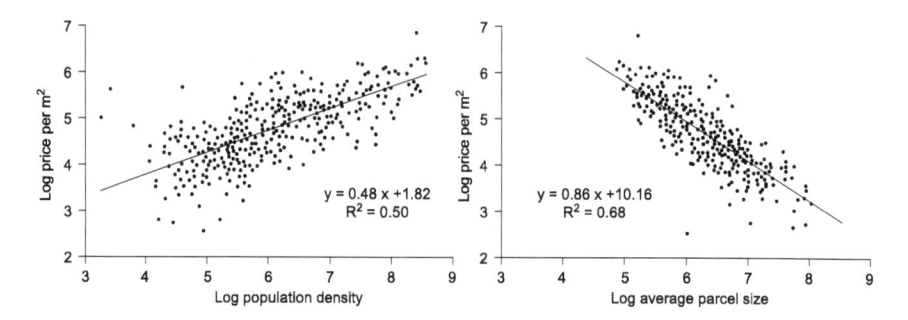

Figure 4.4 Plot size negatively and population density positively related to land rent.
Source: De Groot et al. (2015).

in these land rent differentials can be explained by a rather limited number of variables. These variables are grouped in three clusters: the production side, the consumption side, and disamenities that emerge in declining cities where existing facilities have had to close down and where many shops in shopping centers are vacant. Interestingly, the total contribution of the production and the consumption sides is about equal: 34% for production, 38% for consumption. The results are summarized in Table 4.1.

On the production side, we account separately for the number of jobs that can be reached within commuting times that are actually observed in the data, either by car or by public transport. The contribution of jobs accessible by car is four times larger than that accessible by public transport. However, Table 4.2 presents a decomposition of the land rent surplus in a number of cities. Public transport turns out to play a greater role in the prime cities, Amsterdam, Rotterdam, The Hague, and Utrecht. Land is scarce in these cities, while the car is a very land-intensive means of transport, partly due to land use for roads and partly due to the required parking space, at both the home and the work location. Figure 4.5 shows the geographical distribution of the land rent surplus due to the accessibility of jobs across the country. The value of the accessibility of jobs by private transport is more spread out than the accessibility by public transport. The latter gains are located in the neighborhood of main rail connections and in the main cities, which have an extensive local public transport network. In that sense, public transport is much more important for spatial differentiation of economic activity than its mere contribution to explain the variance of land rents suggests.

On the consumption side, a larger number of variables play a role: a historical neighborhood (measured by the number of monuments), luxury shopping and restaurants, cultural performances, nice scenery, and proximity to the coast. Again, the decomposition by city in Table 4.2 is revealing. For each city, we explain the excess land rent above the average land rent in non-city areas. Cities turn out to have very different profiles. In Amsterdam, by far the country's most successful city, the surplus is 400 euro per square meter. In relative terms,

Table 4.1 Which factors explain the wide variety in land rents per square meter across ZIP codes?

Dependent variable : land prices (per m² at PC-4 level)	Average over the period of 1985–2007	Explained variance (without co-variance)	Explained variance (with co-variance)
Gross hourly wage (in €)	6.43	1%	1%
Access to jobs, by car (adj. for traffic jams)	0.18	13%	25%
Access to jobs, by public transport	0.09	3%	6%
Proximity to railway station	67	1%	2%
Subtotal production side			**34%**
Proximity to natural amenities	0.15	2%	4%
Proximity to city park	213	1%	3%
Location at the seaside	75	1%	3%
Historical inner city	1.70	4%	8%
Location within Amsterdam's Canal Zone	1.491	0%	0%
Proximity to performing arts	0.16	4%	7%
Proximity to quality restaurants and bars	6.63	1%	2%
Proximity to luxury shops	0.71	6%	11%
Subtotal consumption side			**38%**
Bad facilities for daily shopping	−7.73	2%	3%
Public nuisance, degeneration, lack of safety	−1.14	1%	2%
Total		**41%**	**77%**

Source: De Groot et al., 2015; see source for dimensions of the variables.

the city derives most of its attractiveness from its consumption amenities: its vibrant cultural activities (119 euro), its shopping facilities (80 euro), and its monumental status (39 euro, not listed in Table 4.2). These three factors alone explain some 60% of the city's land rent surplus. Note that these numbers are averages across the city. Within the Canal Zone and close to the location of the cultural performances, these effects are even larger. Rotterdam is specializing more in production. In general, the most successful cities in terms of land rents stand out for their strong consumption amenities. But there are more extreme examples: Maastricht, far south in the country, is very successful in consumption, but not at all in production. Dordrecht is relatively strong in production, but not at all in consumption. Finally, Heerlen and Geleen/Sittard in the south are declining cities. These cities were the center of the mining industry, until the mines were closed down in the seventies of the previous century. The effects of the demise of this industry are still felt today: no jobs and the degradation of facilities make these cities unattractive. Similarly, Enschede and Tilburg were the locations of the textile industry, a low-skilled industry that could not compete with imports from low-wage countries. Tilburg recovered rather well, largely due

Table 4.2 Decomposition of the average land rent surplus for a number of cities.

City	Actual land price	Expected land price	Wage	By car	By public transport	Historical center	Performing arts	Restaurants	Luxury shops	City park, nature, sea	Degraded
Amersfoort	89	94	5	21	29	-2	8	2	16	16	-1
Amsterdam	396	379	12	58	38	49	119	19	80	4	0
Apeldoorn	16	26	5	9	13	-6	0	-2	4	12	-9
Arnhem	41	101	3	28	21	-3	21	-6	27	16	-7
Breda	79	60	2	14	14	-1	10	-2	28	1	-7
Dordrecht	55	68	3	29	23	1	5	-6	12	2	-1
Eindhoven	56	66	3	23	13	-7	14	-1	22	3	-3
Enschede	-9	-21	0	-20	1	-7	10	-5	12	-1	-10
Geleen/Sittard	-24	-26	2	-13	-1	-3	0	-5	10	-9	-6
Groningen	81	48	4	-11	1	1	34	-1	24	-6	2
Haarlem	255	193	3	36	27	8	32	8	44	23	11
Heerlen	-34	-24	-1	-18	-4	-4	4	-6	12	6	-13
Leeuwarden	42	-11	0	-17	-3	6	10	-5	15	-11	-4
Leiden	237	135	6	37	28	4	16	0	29	12	3
Maastricht	86	60	3	-16	-3	25	13	13	38	-3	-10
Nijmegen	44	61	7	23	14	-5	19	-2	20	-3	-11
Rotterdam	101	139	6	29	34	-7	30	3	44	6	-4
Hertogenbosch	110	86	4	24	23	0	18	-1	23	-4	-2
The Hague	254	237	11	48	43	-3	45	12	60	16	6
Tilburg	56	71	1	18	18	-6	33	-4	26	2	-16
Utrecht	169	181	6	44	39	1	44	4	42	12	-10
Zwolle	100	37	1	2	11	1	9	1	11	10	-11

Source: De Groot et al., 2015.

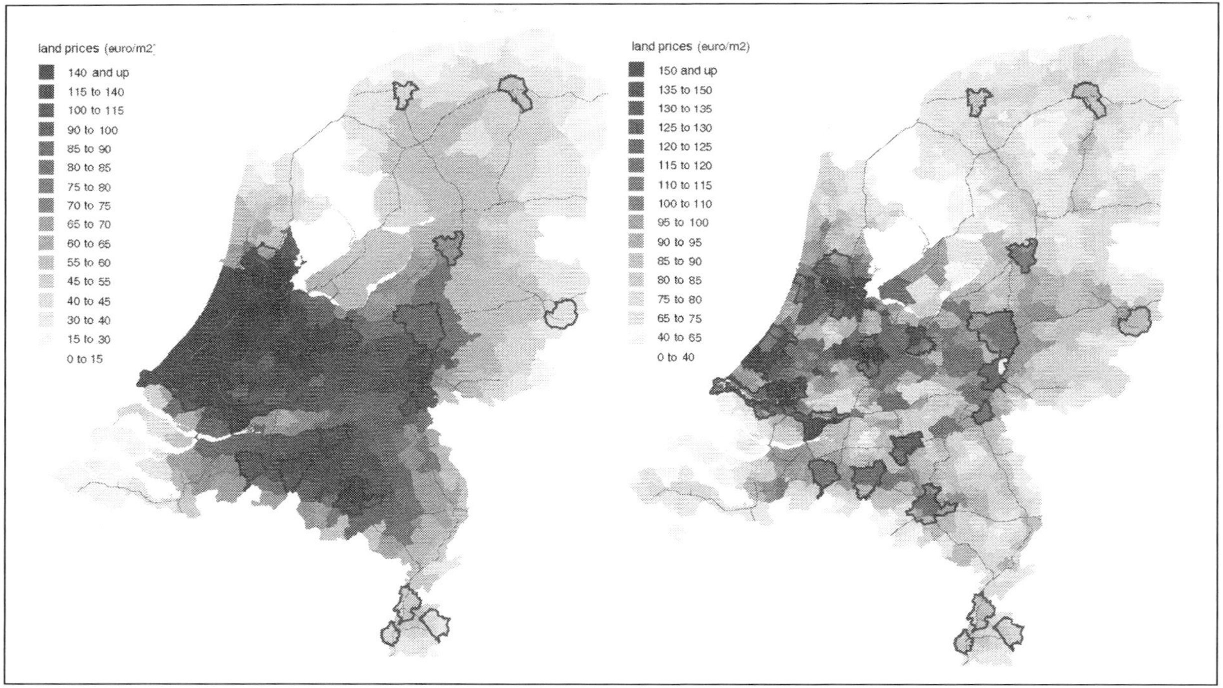

Figure 4.5 Land rent surplus due to production side externalities: private vs. public transport.
Source: De Groot et al. (2015).

to its location close to other successful cities, and due to its university; Enschede, located at the far east of the country, is too far detached from other locations to develop new economic activity.

These results stress the importance of consumption amenities and the value of historical city centers, next to the contribution of more traditional factors like job availability and economic dynamism (the production of the economy). An attractive city center cannot just be constructed by social planners. It emerges by trial and error; see Conclusion 6 in Section 2 and the discussion in Jacobs (1961). In rapidly growing cities—not an uncommon phenomenon in China—it is tempting to scrap low-density, inefficient structures—like *hutongs* in Beijing— and replace them with high-density skyscrapers. However, this change destroys neighborhoods that are potentially quite attractive living areas, and it destroys networks of small shops, which will only be replaced by standard shopping malls with much less interesting variety.

The Netherlands made many mistakes in this regard during the sixties, when the country's population was growing rapidly and the automobile invaded the urban space. Old neighborhoods were demolished and replaced by new high-rises. Usually, that had a negative impact on the city's consumption amenity. A special example in this regard is Rotterdam. The city was bombed by the Germans during World War II and all its historical buildings were destroyed. As Table 4.2 reveals, land rents are much lower in Rotterdam than in Amsterdam even 70 years after the event, though both cities differ very little in size. Historical monuments as evaluated by their effect on land rents play an important role in explaining this difference.

In De Groot, Marlet, Teulings, and Vermeulen (2015), we present evidence on the role of universities and the average level of education. In general, university cities (e.g., Amsterdam, Utrecht, Leiden, Nijmegen, Groningen, Maastricht, Rotterdam, Tilburg, Enschede, and Eindhoven) do very well, where the more specialized technical and economic universities (the last four cities in the list) tend to have less impact on the well-being of the city than the broad universities. Also, the average education level of the work force is a good predictor of the success of a city. There is a clear divide. Cities with more than 42% of their work force having higher education, like Amsterdam, Utrecht, Den Haag, Leiden, Nijmegen, and Eindhoven, do much better than cities with 42% or less having higher education, like Heerlen, Geleen/Sittard, Dordrecht, or Apeldoorn. In general, this trend is consistent with the idea that cities are locations of knowledge spillovers; see for example Glaeser, Saiz, Burtless, and Strange (2004), or Lucas and Rossi Hansberg (2002). Table 4.2 reveals that cities with a non-specialized university and/or a high share of highly educated workers seem to do particularly well on the consumption side. One might conjecture that students provide an externality to the city in the form of a vibrant nightlife and a lot of cultural activities. In an analysis on Danish data, Gautier, Svarer, and Teulings (2010) have shown that the city plays an important role as a marriage market, in particular for the highly educated. The Danish city landscape is dominated by its capital, Copenhagen. People move into Copenhagen when single and move out as soon as they have found a partner. They even move back to the capital after a divorce, apparently in search of a new partner. Again, this underlines the importance of consumption amenities, in particular for highly educated workers. Though it is hard to establish formally the direction of causality, the data for the Netherlands—in particular the importance of an historical city center—suggest that the quality of the consumption amenities of a city contribute a lot to its ability to reap the agglomeration benefits of high-end R&D (research and development) activities.

Since the land rent surplus is a reflection of the value of agglomeration benefits, one can use this surplus to measure the contribution of agglomeration benefits to GDP. We take the value of land used for residential real estate in the 20 largest cities in the Netherlands as a point of reference. This surplus value is 112 billion euro in 2012 above the value of land outside these city areas. Amsterdam alone accounts for 45 billion, which is 37% of the total amount, while its share in the total population is about 10%. However, by focusing on these 20 cities only and by using the administrative definitions of city areas, one underestimates the land value in cities and one overestimates the value of agricultural land. De Groot et al. (2015) correct for these biases, which yields a more realistic estimate of the surplus value of land due to agglomeration benefits of 305 billion euro. Using a rate of return of 4.5%, one obtains a flow value of approximately 15 billion, which is equivalent to 3% of GDP.

5. When to extend a city?

It would be useful if data on land rents and house prices could be used as a source of information on where to build new housing. For that purpose, we have to understand how cities evolve over time. If the growth of cities were highly unpredictable, then developers could build only in response to realized demand, since the current growth of demand does not say anything about its future growth. However, as we discuss below, the growth rate of cities can be predicted to some extent. This has profound consequences for spatial planning, for the role of markets for land, and for the information content of land rents.

As is well known, the size distribution of cities is governed by Zipf's law: the second city in a country is approximately half the size of the largest city, the third city is one-third the size of the largest city, the fourth one-quarter, and so on. Relative to the normal distribution, for example, this distribution is very fat-tailed—that is, there are some extremely large cities, much larger than one would expect on the basis of the normal distribution. This steady-state distribution of size is the result of another law, which governs the evolution of city size and is known as Gibrat's law: the growth rate of a city is independent of its current size. Hence, large and small cities have the same chances to grow or shrink. The current prosperity of a city says nothing about its future prospects of growth.

Growth might be uncorrelated to a city's current size, but this does not preclude the fact that today's growth rate of a city might be correlated to its future growth. In other words: the growth of a city may persist over time. Suppose that

76   Coen N. Teulings

a city is currently growing rapidly. Can that growth be expected to continue for some time? Or is the future growth of a city uncorrelated not only to its current size but also to its current growth? This idea can be captured by the following simple model:

$$g_t = d + r \cdot g_{t-1} + \text{error term}$$

where g_t is the growth rate of the city at time t, and where d and r are parameters. If the parameter r is equal to zero, then there is no persistence in growth: past growth rates have no predictive power for future growth rates. If the parameter r is equal to unity, there is full persistence in growth rates. If a city A grows faster than city B, this growth differential will last forever.

An empirical analysis of the persistence in a city's growth rate requires a proper definition of a city. Administrative definitions tend not to work very well because growing cities quickly use up all the empty land within the boundaries of their jurisdiction. Further growth has therefore to be accommodated outside these boundaries, until a reclustering of jurisdictions merges neighboring municipalities into the central city. Hence, using administrative boundaries tends to understate the growth of the fastest-growing cities.

In order to avoid this problem, we use a definition of cities based on actual outcomes instead of administrative boundaries—e.g., based on commuting flows, on the use public services, or on the catchment area of the main shopping mall used for more specialized shopping. Based on this type of criteria, Marlet and Woerkens (2014) have constructed a classification of the Netherlands in 57 city regions, each consisting of about eight municipalities. These 57 regions cover the full landmass of the Netherlands (except for the islands in the north). We have administrative data available for the evolution of the population of these 57 city regions from 1950 to 2012. Even though these data are highly reliable, there are unexpected jumps in the data, related to administrative changes in the legal boundaries of municipalities, implying that part of the population shifts from one to the other. This measurement error in the population data leads to an underestimation of the persistence in the growth rate because one-time reclassification shocks exhibit no persistence. Since these shocks can be quite large, they lead to an underestimation of the persistence. In the Appendix we discuss how this underestimation can be avoided.

Our statistical analysis shows that $r = 0.95$, implying an "average" persistence of a shock in the structural growth rate of $1/(1-r) = 20$ years. If the annual growth rate of the population of a city is 1% above the nationwide average, one can expect the city eventually to become 20% larger than it would be according to the nationwide average growth rate.

Table 4.3 lists the current growth rates for some agglomerations. The Amsterdam region is currently growing at 1.1% per year, and the population of Heerlen is shrinking at a rate of 0.2% per year. Not surprisingly, cities with a highly educated work force tend to grow faster. Apparently, these cities are more popular among the highly educated. Since the average level of education is

Table 4.3 Population growth rate for various agglomerations in the Netherlands, 1985–2012.

Region	Growth rate
Amsterdam	1.1%
Rotterdam	0.5%
Eindhoven	0.5%
Heerlen	–0.2%

Source: Vermeulen et al., 2016.

still rising because the older cohorts are much less educated than the young, this implies that cities that are popular among the highly educated will tend to grow in the next decades.

Note, however, that even though a persistence of the growth rate over a 20-year period is a major phenomenon, this does not preclude a reversal of fortunes at longer horizons. After all, the city of Amsterdam, which is doing so well currently, lost 25% of its population between 1960 and 1988 (exactly the same occurred for Rotterdam). By 1988, house prices in Amsterdam were at a long-time low, and many houses were occupied by squatters simply because there was no profitable use of the buildings. Many other cities that are striving today went through similar episodes of decline. In the long run, Gibrat's law, which implies that current success is uncorrelated to future growth, holds. Only uncertainty prevails.

This phenomenon of the randomness of city growth in the longer term raises an important policy question: what should a country do when a city is in decline and there is no demand for its existing stock of housing? Should one abandon the city and let it deteriorate until it collapses, or should one preserve the city, hoping that the course of Gibrat's law will revert and the demand for the city's real estate will recover? Currently, this question is less relevant for China as most of its cities are growing rapidly, but one can expect this policy issue to surface shortly.

The persistence of city growth rate differentials has important implications for the interpretation of land rents. Until now, we have applied an essentially static framework. However, when city growth is persistent, land rents reflect not only the current value of land but also the expected future change of that value. If the growth of a city were uncorrelated over time, there would be no way to predict the future evolution of the value of land. However, since growth has been shown to be persistent, the value of land in a particular region will depend on the current growth rate. Consider the simple model depicted in Figure 4.1. The growth of the city can be represented as an increase in wage surplus in the CBD. When the commuting cost per mile is constant, this variable will increase the size of the city proportionally. Hence, the user value of all land within the city will increase. At the edge of the city, some agricultural land will be transformed into new suburbs.

78 Coen N. Teulings

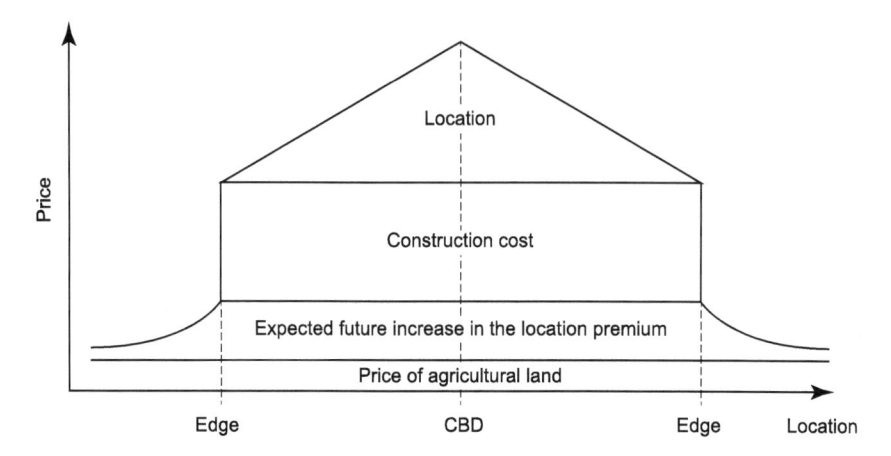

Figure 4.6 Decomposition of the value of land at various distances from the CBD.
Source: Vermeulen et al. (2016).

In a competitive land market, one can expect traders to have reasonable (a theorist would even say, rational) expectations about the future user value of land. Since the current market price of land will be equal to the present value of all future user values, the current price will depend on both the current user value and its expected growth. The consequences of this idea are reflected in Figure 4.6; see Capozza and Helsley (1990). The figure depicts the value of land with or without real estate at various distances from the CBD. We start our explanation from the top, located in the CBD.

The house shape in the middle reflects the built area. The lower floor of the house reflects the cost of construction of the real estate on top of the land. The roof of the house is the premium for locations closer to the CBD, which require less commuting. The hill shape on which the house rests reflects the value of the expected future increase in the user value of the land (for the built area) and the option value of the future transformation into built area (for the agricultural area). This expected increase is the same at every location, since the value differential between locations depends on the commuting cost only. For the agricultural land beyond the edge of the city, the option value decreases gradually. Just beyond the edge, the option value is equal to the value of the expected increase within the city, because the land is expected to be transformed into a suburb soon. At distances farther away from the city edge, the option value is smaller because the owner has to wait longer before this transformation from agricultural land into suburb is worthwhile. In the meantime, the value of the option has to grow at a rate equal to the interest rate, since apart from this wealth gain, the owner of the option receives no other return.

Clearly, the option value must be increasing in the growth rate of the city, since the growth of cities is shown to be persistent. The higher the current growth rate of a city, the more valuable is the option to build in that city, since after the date of construction, the rents will increase over time. Hence, as a first conclusion, having high land prices for vacant land in the proximity of fast-growing cities is not necessarily a sign of sluggishness in the planning procedures.

Conclusion 1: *The land price at the optimal moment of construction is higher for faster-growing cities.*

The counter-intuitive policy implication of this rule is that one should be more reluctant to build new suburbs in fast-growing cities, even though the option value of new construction is very high. The intuition is that a high option value of the future cash flow of rents does not imply that the current rent is already sufficiently high to cover the interest bill. Moreover, by postponing investment one receives more information about the future evolution of the city and hence one can bring the design of the new construction more in line with the most recent information. The higher the growth rate, the more valuable waiting is.

The precise relation between the market price of land and the optimal moment of construction is not known yet.[1] However, the market value of an additional square meter of floor space at the moment of construction is perfectly known: it should be equal to the cost of construction. If the value of the marginal unit is less than the construction cost, then the construction could have been either of lower density or postponed until a later moment. Otherwise, construction is a waste of money. The owner would have been better off by not investing and putting the money in the bank. The reverse holds for the case that the value of the marginal square meter of floor space is higher than the cost of construction. Hence, for any empty plot of land, the value of construction is less than the cost of construction if the owner is free to choose when and where to build. Hence we arrive at the following:

Conclusion 2: *The value added of an additional unit of floor space should be greater or equal to the cost of construction.*

The latter condition might not hold: either regulation might prevent the owner from building, or previous construction might have been undertaken at too low a density; extending an existing building is much more expensive than greenfield construction. The reverse statement does not hold: the value of construction can be lower than its cost, even if the owner is free to decide whether and what to build. This happens when the user value of the plot decreases after the moment of construction, for example because the nearby city declined. Then the user value goes down, but since construction is irreversible, the owner cannot reclaim the cost of construction by tearing down the building. Similarly, the value of construction in built areas can exceed the cost of construction, since the cost of extending an existing building exceeds the cost of construction in empty land by a wide margin.

Figure 4.7 plots this value of construction per square meter for all ZIP codes outside the built area in city centers. For most of the country, in particular in the

80 Coen N. Teulings

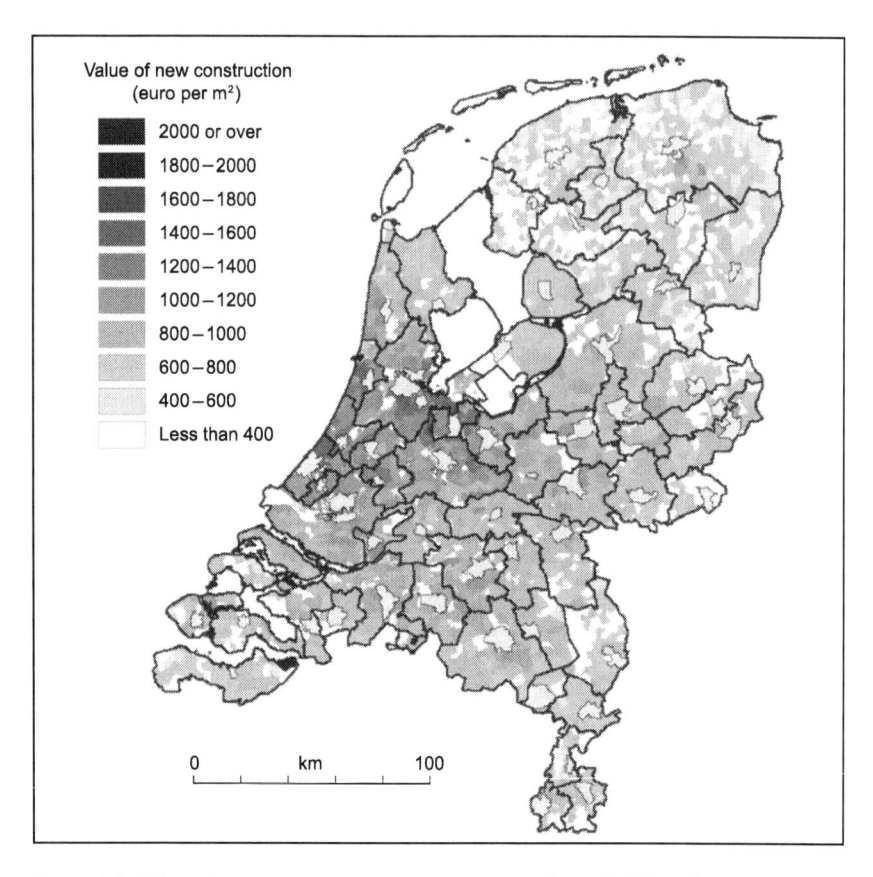

Figure 4.7 Value of new construction per square meter for each ZIP code.
Source: Vermeulen et al. (2016).

eastern part, the value of construction is far below its cost, which is about 1,080 euro per square meter for homes built on empty land. In many of these locations, the value of construction is less than 50% of the cost. On the other side of the distribution, there are some locations where the value of construction is far above its cost. These are the villa villages inhabited by the well-to-do (for those familiar with the Netherlands: Wassenaar, Aerdenhout/Bloemendaal, and 't Gooi). The explanation for this phenomenon is that these people value open space highly. Hence, a larger construction on one's own plot has a negative externality to the value of neighboring plots. Hence, local spatial planning policies that make this impossible have broad local political support.

Other than these villa resorts, the value of construction exceeds its cost almost only in the direct neighborhood of Amsterdam. As we have seen, the Amsterdam region has one of the highest growth rates of its population in the country. Hence, there is a vibrant market for new real estate. In accordance with the monocentric

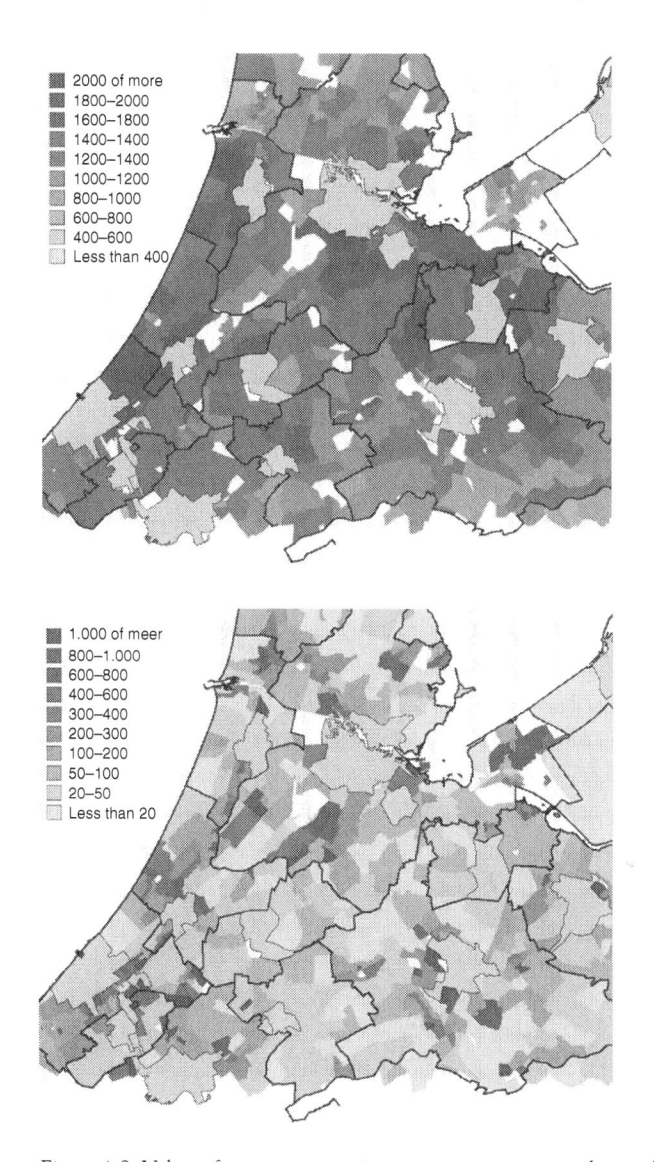

Figure 4.8 Value of new construction per square meter and actual new construction per square kilometer in the Amsterdam region.
Source: Vermeulen et al. (2016).

model of the city (see Figure 4.6 above), the demand for new houses is highest at the city edge. In order to analyze the situation more closely, it is useful to look at Figure 4.8, which is identical to Figure 4.7 but focused on the Amsterdam region. Moreover, Figure 4.8 provides a companion panel showing where houses were actually built. The graph reveals that there is a large demand (= a high value of

82 Coen N. Teulings

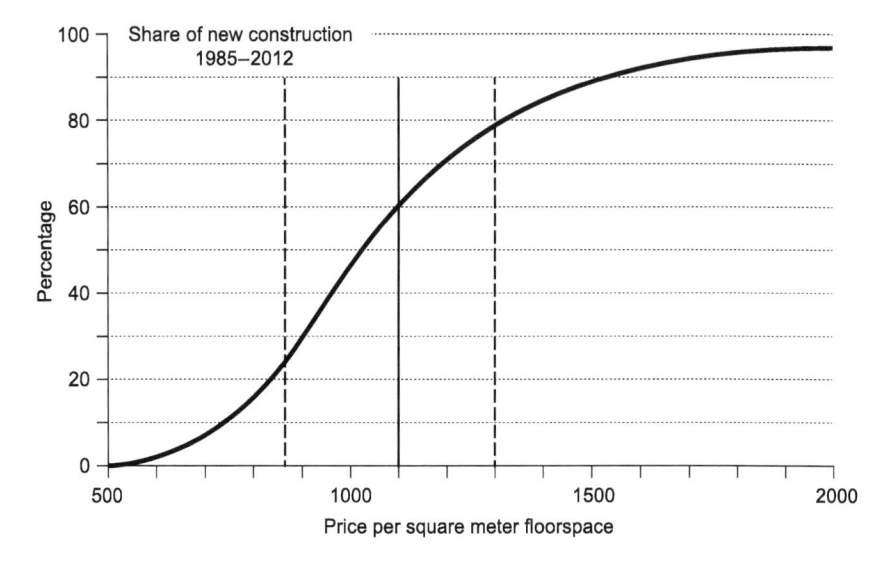

Figure 4.9 New construction ranked by the value of new construction at the location.
Source: Vermeulen et al. (2016).

construction) at locations at the city edge. Nevertheless, almost all new construction took place in the new town of Almere, just east of the city, that was started in a new polder (a tract of low land) that was pumped out only a couple of decades ago. Clearly, spatial planners have located the main construction site at a place where people do not want to live.

Figure 4.9 ranks all newly constructed housing since 1985 by the value of the marginal unit of construction. Roughly 50% of the new construction is built at locations where its value exceeds the cost of construction by 20% or more. Hence, the market operates reasonably efficiently in these locations. However, 20% of the newly built houses are constructed at locations where the value of new construction is more than 20% lower than its cost, as in Almere.

The simple rule discussed above—build as soon as the market value of construction exceeds its cost—is optimal from the point of view of the owner of the land. The study of Ahlfeldt, Redding, Sturm, and Wolf (2015), on the division and subsequent reunification of Berlin, reveals substantial agglomeration benefits of about 7%. As discussed in Section 2, what is optimal for the landowner is not optimal for society as a whole in that case; see Conclusion 4. The city is too small and the density of construction is too low. This conclusion for the static analysis in Section 2 extends naturally to an analogous conclusion for the dynamic framework in this section. Private landowners do not take into account the benefit of the city as a whole from high-density development at the edge of the city, because they do not consider the agglomeration benefit of the new inhabitants of the city for the existing population. Hence we reach another conclusion:

Conclusion 3: Private developers have an incentive to build at too-low densities.

City planners should enforce construction at a rather high density, since private landowners do not take into account the extra agglomeration benefits that fall upon the incumbent population of the city by having more people working in the CBD; see Conclusion 4 in Section 2.

In theory, all surplus value of the real estate—i.e., the difference between the market value of the real estate minus the value of agricultural land minus the cost of construction, or alternatively, the sum of the current location premium and its expected future growth—is a reward for the public goods provided by others than the landowner. Take our example of a city that becomes more attractive due to investment in its CBD. Then, the growth of the location premium for residential real estate is due to investments in the CBD, not to any efforts made by its owner. Hence, since for an efficient outcome all gains from an investment should accrue to the investor so that he has the right incentives to invest, the increase in the location premium should be attributed to the investor.

The growth of the attractiveness of a CBD can have several causes:

1. The increase in the value of a fixed location advantage at the CBD, e.g., the presence of a particular type of expertise at that location for historical reasons, which happens to becomes more valuable over time;
2. Local public goods provided by the city council, like investment in public transport that reduces the commuting cost to the CBD; and
3. Local public goods provided by the inhabitants of the city working in the CBD, like knowledge spillovers to the other workers.

Who will seize the rents from the location premiums associated with these gains in the attractiveness of the CBD? To the extent that the premium is derived from fixed location advantages, as in the first case, the gains arrive as manna from heaven. In that case, it is just a matter of distributing the rents from the happy coincidence. The distribution of the rents will not affect the size of the location advantages. However, if the location premium follows from effort provided by the city council or the city's inhabitants, the second and the third case respectively, these parties should be compensated for that effort if an optimal allocation is to be achieved. One can show that the required compensation is exactly equal to the surplus value of the land; compare Conclusion 5 on Henry George taxation discussed in Section 2. This yields our final conclusion:

Conclusion 4: City planners should have the right legal instruments to extract the gains for private developers of new construction at the edge of the city.

If city planners have insufficient instruments, they cannot generate enough funds to create the local public goods that are required for efficient organization of the city.

Judged by the Dutch experience, it is a major challenge to achieve a workable solution that strikes a proper balance between the necessity to raise tax revenues for the city planner to pay for critical public infrastructure on the one hand, and the preservation of existing networks of agglomeration benefits in historical neighborhoods on the other. In China, city planners seem to have plenty of legal instruments to extract surpluses from landowners. One might wonder whether this wide availability of legal instruments on the side of the city planner might hamper the private incentives for the creation of public goods.

Appendix

The Construction of Land Rents

We apply a two-step approach:

1. We estimate an OLS regression whereby we regress the log transaction price $\ln P$ on time-fixed effects to account for the general increase in house prices and the characteristics of the building, say the vector x and the log of the land surface $\ln A$. Let b be the parameter vector associated with these characteristics. Table 4.4 provides an overview of the coefficients b.
2. We calculate an index for the log physical construction $h = b'x$. Hence, we aggregate the various characteristics of the construction into a single index for the construction using the regression coefficients of the first step as weights. Then, we run a second regression in which we regress the log transaction price per square meter of land, $\ln P - \ln A$, on time-fixed effects, fixed effects a_z for each ZIP code z, and the log physical construction index per square meter of land, $h - \ln A$, crossed with dummies for each ZIP code, c_z.

Table 4.4 Contribution of various characteristics of the log construction index h.

Dwelling's characteristic	Effect
Log living surface (in m²)	0.53
No central heating	−0.14
Semi-detached	−0.12
End house	−0.21
Terraced house	−0.21
Own parking space	0.02
Carport	0.05
Garage	0.08
Carport and garage	0.10
Built before 1906	−0.11
Built between 1906 and 1970	−0.16
Built between 1971 and 1980	−0.13
Built between 1981 and 1990	−0.08

Source: De Groot et al., 2015.

House prices in the Netherlands 85

The second step is equivalent to estimating a Cobb Douglas function for the transaction value with constant returns to scale and with log land ln A and log construction h as its inputs, separately for each ZIP code. The parameter c_z measures the share of construction in ZIP code z, while fixed effect a_z measures the overall price. Hence, the land rent in a ZIP code can be constructed as total price per square meter multiplied by the land share (which is one minus the share of construction), or in logs:[2]

$$\ln R_z = a_z + \ln(1-c_z)$$

Due to the large number of transactions, all coefficients are measured very precisely. The sum of the coefficients on the log land surface ln A and the log square meters of construction is close to unity, suggesting that the constant returns to scale assumption does a reasonable job. All coefficients have the expected sign and reasonable magnitudes.

Measuring the persistence in the growth rate of cities

We hypothesize that the log size of a city n_t follows the following law of motion:

$$n_t = n_{t-1} + g_t + e_t$$

$$g_t = d + r \cdot g_{t-1} + v_t$$

where g_t is the structural growth rate of the city at time t; e_t are sudden changes in the population due to administrative reclassifications; v_t are changes in the growth; the mean value of e_t and v_t is zero; and r, $0 < r < 1$ is the mean reversion parameter in the growth rate. These equations say that the long-term equilibrium growth rate is $d/(1-r)$ per year; this number is obtained by setting $g_t = g_{t-1}$ and setting v_t at its mean value. The parameter r is the rate of persistence in the growth rate; if $r = 0$, past growth rates have no effect on the current growth rate; if $r = 1$, past shocks v_t to the growth rate will last forever. If the administrative changes in the boundaries were not accounted for, and if researchers would use the observed growth rate $n_t - n_{t-1}$ as an exact measurement of the structural growth rate g_t, one would strongly overestimate the degree of mean reversion because this year's shock e_t would add to the growth rate, while next year the same shock would reduce the growth rate, suggesting that there is strong mean reversion. A proper estimation method should correct for that. We use the correlation between successive observed growth rates at different lag lengths k for $k > 1$. It is easy to show that for any $k > 1$ these correlations can be written as follows:

$$\text{Cor}[n_t - n_{t-1}, n_{t-k} - n_{t-k-1}] = c\, r^k$$

where c is a constant. The administrative shocks e_t do not enter these correlations because they are uncorrelated for all lag lengths $k > 1$. There are two simple ways

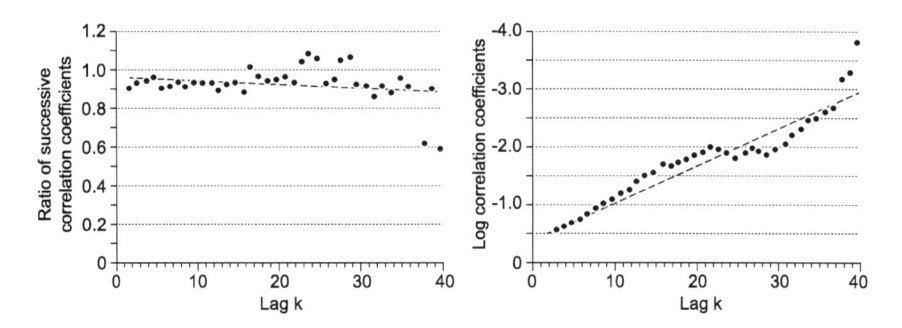

Figure 4.10 Persistence in the growth of regions.
Source: Vermeulen et al. (2016).

to estimate the parameter r. First, one can plot the log of these correlations for all $k > 1$. If the model is correct, the result should be a downward sloping line with a slope approximately equal to the mean reversion parameter $\ln r \approx 1 - r$ for small r. Second, one can plot the ratio of the correlations for k and $k + 1$. This ratio should be constant r for all lag lengths k.

Figure 4.10 reports the results of both exercises. The ratio between two successive correlation coefficients is approximately constant indeed, about 0.95. The slope of the log of correlation coefficients for various lag length is indeed approximately a straight line with a slope of -0.05. Hence, $r = 0.95$, implying an "average" persistence of a shock in the structural growth rate of $1/(1-r) = 20$ years.

Notes

1 See Lange and Teulings (2018).
2 In practice, somewhat different methods are applied across various studies. Land rent presented in various figures might therefore be based on slightly different normalizations than discussed in the text.

References

Ahlfeldt, G. M., Redding, S. J., Sturm, D. M., and Wolf, N. (2015). The Economies of Density: Evidence from the Berlin Wall. *Econometrica*, 83(6), 2127–2189.
Capozza, D. R. and Helsley, R. W. (1990). The Stochastic City, *Journal of urban Economics*, 28(2), 187–203.
De Groot, H. L. F., Marlet, G., Teulings, C. N., and Vermeulen, W. (2015). *Cities and the Urban Land Premium*. Edward Elgar, Cheltenham.
Desmet, K. and Rappaport, J. (2017). The Settlement of the United States, 1800–2000: The Long Transition Towards Gibrat's Law. *Journal of Urban Economics*, 98, 50–68.
Gautier, P. A., Svarer, M., and Teulings, C. N. (2010). Marriage and the City: Search Frictions and Sorting of Singles, *Journal of Urban Economics*, 67(2), 206–218.

Glaeser, E. L., Kahn, M. E., and Rappaport, J. (2000). *Why Do the Poor Live in Cities?* No. w7636, National Bureau of Economic Research.

Glaeser, E. L., Saiz, A., Burtless, G., and Strange., W. C. (2004). The Rise of the Skilled City. *Brookings-Wharton Papers on Urban Affairs*, 47–105. Retrieved from www.jstor.org/stable/25067406.

Jacobs, J. (1961). *The Death and Life of Great American Cities*. New York: Random House.

Lucas, R. E. and Rossi-Hansberg, E. (2002). On the Internal Structure of Cities, *Econometrica*, 70(4), 1445–1476. Retrieved from www.jstor.org/stable/3082004.

Lange, R.J. and Teulings, C.N. (2018). The Option Value of Vacant Land and the Optimal Timing of City Extensions. www.coenteulings.com/option-value-vacant-land-optimal-timing-city-extensions/

Marlet, G. and van Woerkens, C. (2014). *De Nieuwe Gemeentekaart van Nederland (New Municipality Map for the Netherlands)*. Nijmegen: VOC uitgevers (in Dutch).

Ossokina, I., Teulings, C. N., and. de Groot, H. L. F. (2014). *Welfare Benefits of Agglomeration and Worker Heterogeneity* (No. 289). CPB Netherlands Bureau for Economic Policy Analysis.

Rossi-Hansberg, E. (2004). Optimal Urban Land Use and Zoning Policy. *Review of Economic Dynamics*, 7(1), 69–106.

Vermeulen, W., Teulings, C. N., Marlet, G., and De Groot, H. L. F. (2016). *Groei & Krimp* (Growth & Decline), Nijmegen: VOC Publish (in Dutch).

5 Green Belts, edgelands, and urban sprawl
Reconsiderations of Green Belt concepts

Guy M. Robinson
UNIVERSITY OF ADELAIDE

1. Introduction

A key challenge for planners and policy makers is how to manage the interface between the urban and the rural, often termed the peri-urban fringe, as urban growth continues to supplant productive rural land. Planning has focused on how to restrict urban sprawl, which consumes both agricultural land and areas of conservation and recreation value, while still finding land for housing development sufficiently close to centers of employment. Various zoning arrangements have been implemented, sometimes favoring special protection for prime farmland or allocating land specifically to conservation and/or recreational use. Alternatives to sprawl have also been proposed, including a focus on so-called "brownfield" sites in the heart of long-established but expanding cities, and building new towns or satellite towns to serve a major city. These various measures can be seen in both developed and developing countries, but often with little direct focus on what resultant landscapes and land-use patterns are being created in peri-urban fringes. This paper directs its attention to these fringe areas, arguing that many have become the "edgelands" of planning systems, or neglected landscapes that merit more attention if they are to fulfill their potential to supply multiple needs for growing urban populations.

2. The peri-urban fringe

The term rural-urban fringe or peri-urban fringe first appeared in the UK in the 1930s, though various terms for this area have been used subsequently in the developed world, including the "outer city" (Herington, 1984), the "regional city" (Bryant et al., 1982), "the city beyond the city" (Robinson, 1991), the "urban hinterland" (Furuseth and Lapping, 1999), the "peri-urban zone" (Cadène, 1990), and the "edgelands" (Shoard, 2002). It is described as "that area of interpenetrating rural and urban land uses peripheral to the modern city" (Martin, 1953), characterized by distinctive processes, themselves given various terms, e.g., "suburbanization" (Jackson, 1985), "peri-urbanization" (Liu et al., 2004), "exurbanization" (Nelson and Dueker, 1990), and "rurbanization" (Balk, 1945).

One solution to controlling the explosive sprawl of cities worldwide has been to designate parts of the peri-urban fringe as a "Green Belt" around the city to stop new urban development from consuming agricultural land. Often accompanied by other measures, such as protecting land for conservation purposes, designating land for recreation or space-using services, or creating new settlements for urban residents displaced from the city, Green Belts and green "wedges" have been adopted in cities as varied as London, Tokyo, Hong Kong, Melbourne, and Vienna (Amati, 2008; Buxton and Goodman, 2003; Frey, 2000; Lemes de Oliveira, 2014). It has also been argued that it is an idea with potential for greater implementation in China (Li et al., 2005), though its applicability may be questioned in the light of mounting criticisms about the potential effect on housing provision and affordability.

In historical context, initial zoning schemes for cities were largely the product of government regulation dominated by state-led, welfare-focused, macro-economic management. A rational planning process was introduced in which blueprints for land use (including zoning arrangements, designation of Green Belts, and creation of new towns) were privileged over market processes (Alexander, 2009; Hague, 2002). However, it can be argued that as Western society has moved to a neo-liberal, post-Fordist regime from the 1970s onwards, more market-oriented approaches have been adopted, partially realigning planning systems (Prior and Raemaekers, 2007). This has brought threats to pre-existing planning arrangements such as Green Belts, which have often been deemed to be out of step with contemporary socio-economic needs. In some cases, this has brought effective extinction of the designated Green Belt, as in the case of Dundee, Scotland, or major modifications, as has occurred for Edinburgh (Bramley et al., 2004), while a debate currently rages about the future of the London Metropolitan Green Belt.

3. Green Belts: The UK example

In the UK, precursors to Green Belts were established in the 1930s in the form of the 1935 Restriction on Ribbon Development Act, designed to prevent conurbations from coalescing, and the 1938 London Green Belt Act, which introduced the idea of a "green girdle" around the capital to prevent prime agricultural land from being consumed by urban development. A Green Belt for London was enshrined in the wartime Greater London Plan (Abercrombie, 1944); post-World War II, some of these ideas were enshrined in the 1947 Town and Country Planning Act and were pursued in the 1950s by the Conservative government, with the formal designation of new towns and the statutory establishment of Green Belts around several major cities and conurbations. The intention was to create Green Belts to operate as a mechanism "to preserve the integrity of the built-up areas on one side of it and the countryside on the other" (Shoard, 2002). Around 15 percent of England was designated as Green Belt, effectively placing these areas adjacent to 30 million people today, or around 60 percent of the total population. More than one-fifth of Greater London (22 percent) is classified as Green Belt, while 14 of the 33 London boroughs have a larger area of Green Belt than residential land.

The Green Belt is part of a land-use planning system that has sought to distinguish between definitive sets of "urban" and "rural" land uses, designed to sharpen the interface between urban and rural, and to restrict the extent of the rural-urban fringe. In some places, there is a very sharp edge to the boundary between the city and the adjacent countryside; in other cases, there is a great intermingling of different uses, as urban development pushes into what was previously "green" land use. The presence of urban development and associated infrastructure can contribute to degradation of agricultural land, with farmers able to make money from relatively low levels of input, e.g., "horseyculture" and sod farming, speculative land holding awaiting sales for development (Munton et al., 1988), and lucrative conversions of farmland to golf courses and other recreational uses (Booth, 1996; Zasada et al., 2013). Developers can aid the process of speculation by targeting land for housing or other urban uses as they await the next surge in urban expansion. Meanwhile, local residents frequently protest against potential development and may epitomize the "not in my back yard" (NIMBY) response (Bailoni et al., 2012; von der Dunk et al., 2011). They are part of a strong lobby to retain the Green Belt, despite the gradual urban incursions that frequently have led to a confused and generally unattractive pattern of land uses in parts of these "edgelands."

Marion Shoard (2002) refers to the edgelands as "neglected landscapes" because they have not received the attention of the more definitely urban and rural areas on either side of the fringe. Edgelands are characterized by the negative effects of proximity to urban centers (e.g., pollution, vandalism, trespass), reflecting the influence of urban intrusions that mitigate against efficient farming and which can contribute to underinvestment in agriculture, as already described above. Idle or underused farmland on the fringes of cities can also be attributed to the presence of part-time and hobby farmers who do not wish to or cannot make full agricultural use of their land (Houston, 2005; Zasada, 2011), and the effects of high municipal taxes combined with the uncertainties of farming acting as disincentives to invest or go beyond low-intensity farming.

Land uses in the "edgelands" are a random assemblage of service functions, referred to by Kaika and Swyngedouw (2000) as the "urban dowry," comprising the essential apparatus needed for cities to work, but exiled from more central locations. They include commercial activities; noisy and unsociable uses pushed away from people; transient uses such as casual markets; bulk retail; light manufacturing; warehousing and distribution; some public institutions; degraded farmland; planned recreational areas such as country parks and golf courses; and fragmented residential development (often centered on road junctions) interspersed between areas of unkempt rough or derelict land awaiting reuse (Gallent and Shaw, 2007).

The characteristics of these areas have been captured in recent literature (Kabo, 2015; Rippey, 2012), such as that by Farley and Roberts (2012), which refers to the edgelands as "England's true wilderness"; Papadimitrou's (2013) *Scarp*, detailing life and landscape on London's northern fringes; and Sinclair's (2003) *London Orbital*, showing how the M25 motorway circumscribes London while carving through the Green Belt. Nevertheless, Shoard (2013) emphasizes that "it [edgeland] has important qualities: it is a refuge for wildlife driven out of an increasingly inhospitable countryside; it is a living museum of the workings of contemporary society; and it has considerable, untapped recreational potential."

Landowners in the peri-urban fringe may make use of the opportunity to subdivide their land to sell piecemeal to developers, encouraging the process of urban incursion. This process contributes to the seemingly chaotic set of land uses, which essentially are a product of post-war planning legislation that has partly fossilized some patterns of use while also reflecting more recent dynamic change. In terms of the latter, out-of-town shopping centers have frequently been permitted to develop in edgeland locations, often in conjunction with business parks, perhaps adjacent to relics from the pre-World War II period when agriculture was more dominant. So, despite the attempt to preserve rurality and create a "green lung" for the city, the implication is that the creation of Green Belts has neither created an entirely satisfactory edge to the city, nor is it strictly correct to label this edge "green."

In effect, the Green Belt fulfills the function of a firebreak between more genuinely appreciated landscapes. Yet, despite this lack of appreciation, there has been an emergence of edgelands and growing demands to build more houses on Green Belt land; currently across England Green Belts "contain 30,000 kilometers of footpaths and other rights of way, 220,000 hectares of woodland, 250,000 hectares of top quality farmland, and 89,000 hectares of Sites of Special Scientific Interest" (Bryson, 2015).

The dynamism of the edgelands and the Green Belt can be seen in one simple example that shows land-use change over a century in part of Spelthorne Borough in South West London (Table 5.1). Using historical surveys for 1904, 1939, and 2004, it reveals the growth of housing development pre-World War II, when the area occupied by housing nearly trebled through conversion of agricultural land to residential development. It was largely to restrict this "urban sprawl" that Green Belt designation was introduced for Greater London in 1955, affecting the western part of the borough. However, as Table 5.1 reveals, while this designation has reduced the growth of land converted to housing, significant amounts of agriculture and rough grazing land have been lost to other uses, resulting in a further 70 percent decline in the amount of agricultural land in the area between 1939 and 2004, and the complete elimination of commercial orchards. Substantial building of reservoirs and mineral extraction has been permitted, with this area becoming both an important supplier of water for London, and a home to water-based recreation. New "green" land uses have also appeared—notably golf courses, horseyculture, and other recreational uses (Harrison, 1981; Munton, 1983; Robinson et al., 2011). However, the area has become more urbanized, with public utilities, industry, and schools helping to increase the total urban use by a factor of 4.67 since 1939. The built-up area has more than doubled since 1939. Of course, not all of this represents conversion of agricultural land in the Green Belt, but it symbolizes the dynamism of the edgelands and the difficult task faced by planners in accommodating the pressures of urban development.

Green Belt designation has not necessarily acted as a stimulus to the development or enhancement of particular land uses and landscapes in the

Table 5.1 Land-use change in the edgelands: the example of Spelthorne Borough, South West London.

Land use/Year	Area ha (1904)	Area ha (1939)	Area ha (2004)	Change in area ha (1904–2004)
Residential	32.9	90.0	155.5	+122.6
Public utility	–	–	18.5	+18.5
School/Playground	–	1.4	15.5	+15.5
Recreation/Sports ground	1.9	1.9	7.9	+6.0
Industrial	–	0.9	19.2	+19.2
Nursery	1.9	6.8	1.1	–0.8
Allotment garden	–	1.1	1.1	+1.1
Golf Course	–	–	43.8	+43.8
Equestrian/Stable	–	–	19.2	+19.2
Active gravel pit	–	–	21.9	+21.9
Flooded gravel pit	–	–	31.0	+31.0
Reservoir	–	–	142.5	+142.5
Agriculture/Rough grazing	472.4	416.8	125.2	–347.2
Orchard	93.3	83.5	–	–93.3
Total	602.4	602.4	602.4	–
Total built-up area	32.9	92.3	208.7	+175.8
Total non-built-up area	569.5	510.1	393.7	–175.8
Total urban use	36.7	102.1	477.2	+440.5

Source: Gant et al., 2011.

edgelands. In part, this is because the Green Belt itself has proved to be a highly flexible planning measure, in that controls enshrined in this legislation have been frequently overridden. Green Belt boundaries have often been shifted to enable further building to take place around the edge of the built-up area. The land lost in this fashion to urban development has been "reallocated" by assigning a new piece of countryside to the Green Belt. The overall area protected by Green Belt designation has therefore remained approximately the same, but its location has shifted. Local councils have simply changed land-use designations within the Green Belt, labeling areas as "land for non-conforming uses" or "urban envelopes," and then releasing them for building development. Moreover, the presence of the Green Belt has often led to "leapfrogging," whereby urban development has been deflected further from a major city to small towns and villages beyond the Green Belt, where there may be less stringent planning controls.

In recent decades the issue of Green Belt designation has increasingly become part of the growing concern over the perceived shortage of "affordable" housing in the UK (e.g., Gunn, 2008); the Barker Review, commissioned by central government in the 2000s, recommended substantial reforms to permit more housebuilding in Green Belts, partly to reduce the length of commuting journeys, and the introduction of "more flexible planning approaches," including "green wedges," instead of encircling "belts" (Barker, 2006). The wider context for this debate is the reduction in the provision of new social housing in the UK following major reforms to the housing sector in the 1980s (Whitehead, 2014), and pressures at the lower end of the housing market exerted by growing population numbers (from 49.1 million in England in 2001 to 54.7 million in 2015), and an increasing concentration of the population in London and the South East (Jefferies, 2005; http://ukpopulation2016.com/england).

One argument is that the presence of the Green Belt represents a major constraint on new housing developments close to major cities. It also contributes to increases in house prices by artificially dampening supply, and hence there has been a growth of political pressure to release land in the Green Belt for housing. Meanwhile, the rise in house prices has been substantial, with one commentator noting that if general inflation had risen as fast as housing prices had since 1971, a supermarket chicken would now cost £51 (Carlyon, 2013).

In their "Local Plans," local authorities have possessed the capacity to approve new buildings in the Green Belt under "exceptional circumstances," primarily meaning circumstances in which they can demonstrate that this building will boost economic growth. In response to further lobbying for relaxation of development controls in Green Belts, a further caveat has been added in favor of new builds "where there is a significant local need for housing" (Sawer, 2016). It is argued that such need can often be demonstrated, and hence release of land for housing in Green Belts can contribute to central government's target of building one million new houses in the UK between 2015 and 2020. The number of new homes being approved on Green Belt land in England has increased five-fold in the last five years, with some substantial developments occurring on former Green Belt sites. This finding supports the analysis by Amati (2008) in which he concludes, "substantial operations in the green belt obtain the flexibility to operate where they require it" (Amati, 2008). However, he judged that there already exists a high degree of flexibility in terms of development planning in the Green Belt, reflecting both the existing patchwork of land uses and the diversity of forces supporting and opposing development. Moreover, there are also pressures to "regreen" Green Belts by specifically "reconceptualizing green spaces as areas that provide a variety of functions including river management and climatic amelioration" (Amati and Taylor, 2010), and planning for "green infrastructure" (Kambites and Owen, 2006). The latter may include reconsideration of the type of built environment being constructed in the Green Belt, with more focus on aesthetic concerns in which development ceases to be antithetical to landscape enhancement (Edwards, 2000).

4. Green Belts worldwide

The so-called "father" of the Green Belt in the UK, the planner Sir Patrick Abercrombie, visited Hong Kong in 1947 and championed the concept to the colony. Meanwhile, another British planner, F. J. Osborn, travelled extensively in Southeast Asia in the 1950s, popularizing the British "Garden Cities" idea (Osborn and Whittick, 1963). Seoul and Tokyo were just two of the cities that subsequently adopted Green Belt policies, while Hong Kong also designated a Green Belt, which today covers more than 25 percent of all the land area under statutory land-use zoning plans. Yet, "the actual planning intention of the green belt has been ambivalent and flexible and it is a transition zone rather than a zone for conservation in Hong Kong" (Tang et al., 2007). Indeed, the role of Hong Kong's Green Belt has evolved over time, being used primarily as "passive recreation space" in the 1950s, but being eroded for "residual development space" in the 1960s–1970s, and subsequently taking on more of a conservation role (Tang et al., 2005). In analyzing the operation of this Green Belt, Lai and Ho (2001) show how the inherited colonial system of zoning control enforced the preservation of a Green Belt, not as a continuous encircling belt around a settlement, but rather as a means of preserving green space from development. The principal essence of this zonation has been retained, but with some exceptions permitting "limited developments" such as low-rise residences under the "small-house policy," which endows the right of pre-colonial indigenous residents to build strictly specified houses for their offspring.

The examples described for Hong Kong are good illustrations of the way in which, over time, the reality of Green Belt designation can produce varied outcomes, with different pressures producing relaxations in favor of development or greater emphasis on recreational pursuits and/or nature conservation, as well as more flexible, less restrictive measures. This is well illustrated in the collection of essays edited by Amati (2008a), where the experiences of Green Belts for 11 different cities around the world are recorded, including Tokyo and Seoul. That book explores links between Green Belts and rural preservation aims, as well as how the concept was closely associated with the rise of planning as a profession in different countries.

Green Belts can be seen as a modernist, normative, and rationally determined policy that could be adopted in different contexts, but largely with the common concern of how to stop urban sprawl and channel urban spatial growth in preferred ways. Hence, "the Green Belt was invoked as a universal solution to urban growth" (Amati, 2008b), but its application has extended well beyond the initial thinking by planners post-1945, notably with interest shown in China following the post-1978 reforms (Liu et al., 2004). Yet, as shown above, strong pressures for changes to Green Belt legislation can be seen, not only in England but elsewhere, with Amati (2008b) citing four main arguments favoring reform. These are: resistance by landowners, who wish to profit from urban development and argue that the Green Belt distorts the land and housing market (Amati and Parker, 2007); incompatibility with neo-liberal regimes, which seek to deregulate the planning system (Alterman, 1997); re-evaluations of the functions of the peri-urban fringe that have produced alternatives to a continuous Green Belt (Freestone, 2002); and acceptance of the need for greater flexibility in designating green space in the edgelands, with recognition that in reality Green Belts have become a patchwork of different policies that have frequently eliminated a coherent boundary between "rural" and "urban" (Amati and Yokohari, 2006).

5. Green Belts in China

There has been recognition in China that the use of Green Belts as implemented in the UK, possibly also in conjunction with the designation of new towns, offers potential for managing urban growth (e.g., Jia and Gao, 2005; Tan et al., 2011; Wu et al., 2001). Yet it is difficult to make comparisons between Green Belt policies operating in various countries around the world and the potential for implementation of similar measures in China. Apart from the issue of the incredibly rapid rate and huge scale of urbanization in China in recent years, specific contextual issues in China make direct comparison or potential for transfer of policies difficult. Yet, there have been macro-level policy developments in China that have encouraged implementation of Green Belts and other "green" preservation measures. For example, partly because of growing concerns over food security, in 1994 the State Council of the People's Republic promulgated the Basic Farmland Protection Regulations, stipulating that county-level governments should guarantee the "total quantity dynamic equilibrium of basic farmland" (TQDEBF). The TQDEBF requires implementation of plans for land utilization to ensure that there is no reduction in the total amount of cultivated land within given jurisdictions.

In the succeeding two decades, this policy has had to contend with huge pressures from urban development across the country, seen most sharply for the megacities of Beijing and Shanghai. There, to realize the TQDEBF, new arable land has been created, sometimes in ecologically fragile areas (including wetlands and uplands). Elsewhere, to protect the environment and to promote "harmonious growth" between the economy and ecosystems, the central government has provided subsidies for converting farmland to pasture, grassland, and forest in ecologically vulnerable regions to help protect ecologically fragile areas from further agricultural exploitation (Wang et al., 2007), e.g., under the "grain for green" measures introduced in 1998. This conversion accounted for around half of total farmland loss from 1998–2003, compared with 20 percent due to urban construction (Ministry of Land and Resources, 2006). In contrast, vast land reclamation and consolidation schemes have contributed to substantial improvements in crop yield, enhancing overall food production and modernizing the agricultural sector (Liu et al., 2008; 2014; Long et al., 2010). Hence, "the conflicts between growth of the built-up area, arable land protection, and demand for increase of conservation land are getting increasingly acute" (Tan et al., 2011, p.50). Moreover, the area of farmland has been continuously declining due to rapid urbanization and industrialization in recent decades. The total area of cultivated land was 131.1 million ha in 1996, falling to 127.4 million ha in 2001 and to 121.6 million ha in 2011 (National Bureau of Statistics of China, various years).

In spite of farmland protection measures, the financial incentives to local government that underpin land development are a major driver in urban spatial expansion that has been taking place at an unprecedented pace in recent decades. For example, Beijing's urbanized area increased by nearly 30 percent in the 1990s and has grown by nearly 6 percent per annum since 2000, while comparable figures for Guangzhou were near trebling in the 1990s (7 to 8 km^2 per year) and 9.4 percent per annum since 2000 (Ding, 2007; Wang et al., 2012). Smaller cities have also experienced substantial land area growth: Yantai City in Shandong Province nearly trebled its built-up area from 2001 to 2004 (from 120 km^2 to 340 km^2), and Chongqing's urban built-up area increased from 158 km^2 in 1994 to 175 km^2 in 2000.

In tackling these conflicts, Beijing's Urban Master Plan for 2004–2020 (extending ideas first promulgated in the city's 1958 plan) designated a discontinuous circle of green space at approximately 25 km from the central business district, alongside seven new towns, with a further four new towns designated beyond 60 km. Although on a much larger scale (some of the new towns already have a population well in excess of 1.5 million), this combination of a Green Belt and new towns has distinct echoes of planning for urban growth in the UK post-1945. In the case of Beijing, the Green Belt covers nearly 1,000 km^2 in the form of "nine pieces of wedge-shaped green space which are used to prevent new towns from merging into each other or into Beijing" (Tan et al., 2011). There is also an extension along the Sixth Ring Road to produce a "green strip" along the roadside, and another along the Wenyu River to protect the riparian environment (Qiang et al., 2005; Shengquan, 2001). However, the pressures of urban growth from the new towns and Beijing itself threaten the integrity of these green wedges, especially as the boundaries of the designated wedges have not been firmly established on the ground.

Issues of boundary demarcation are not the only problem limiting the effectiveness of this particular Green Belt. Its intentions are only loosely stated in the Master Plan (in a section on "Green Construction in the Plains"), with a lack of clearly permitted land uses. So, in effect, Beijing's Green Belt is perhaps little more than a planning concept, lacking more concrete contents that would enable constraints on development to be properly implemented. Yet, if fully realized, it could be "an interconnected and integrated network of urban greenspaces," one component of which would be "a green network system of green wedges, parks and green corridors ... to limit future urban expansion, improve urban environmental quality and serve as habitats and migration routes for wildlife" (Li et al., 2005). However, based on recent trends in land-use change, modeling exercises predict further losses of green open space both in and around the city (He et al., 2006).

The ecological principles of creating green spaces within Beijing and on its outskirts have been espoused for other Chinese cities such as Nanjing (Jim and Che, 2003), but the most quoted model is Dalian, a major port city in the northeast, with its initiative to become both a "garden city" and an "environmentally friendly city" (Hoffman, 2011); it uses ideas from the example of Singapore alongside those of pioneer UK urban designer/planner Ebenezer Howard, modern Chinese conceptions of urbanism, and experiences from the city's sister city of Kitakyushu in Japan. The success of this "garden city" can be measured by various international awards, acknowledging the introduction of green spaces within the city, relocation of heavy industry from central areas, introduction of substantial recycling measures, and other pro-environmental approaches to urban development. However, this success has not prevented urban sprawl and further pollution of the city's river.

In terms of more effectively controlling urban sprawl around Chinese cities, there will need to be resolution of the conflicting aims of different levels of government. Local administrations have frequently used land auctions as a means of attracting investment and creating employment opportunities in the local economy (Ding and Lichtenberg, 2011). Hence, they have a vested interest in promoting development, as opposed to protecting green spaces in the peri-urban fringe. This interest runs counter to goals expressed, for example, by Beijing's government, whose Master Plan espouses the need for protection of green space at the city's edge, and where a desire for a clear limit to urban sprawl is expressed. To date, the gains to local government revenue from land sales for development have largely outweighed the need to maintain Green Belts. However, any proposals to provide stronger controls on protecting "green space" in the fringe need careful consideration of the particular context that applies to land transactions and planning in China.

A key contextual factor is the current institutional differentiation between urban and rural land in China. Land in cities and towns belongs to the state, so that city governments act as the representative of the state in land management, and this extends to land requisition and public land leasing. In contrast, land in rural areas is collectively owned and managed by all members of village communes, although the ultimate owner is still the state (Lin et al., 2014). Moreover, land-use rights transactions have been strictly limited to urban land officially, but black-market transactions of rural land rights are rampant and seemingly tolerated by the government. Only in the last few years has central government suggested land reforms that enable the creation of land markets for "construction land" in rural areas and integration with urban land markets. This integration may make it easier to introduce more holistic planning in urban hinterlands. However, local administrations still have monopolistic controls in land requisition and land leasing, enabling them to make significant proceeds from land development. Hence, one of the key issues regarding effective management of the peri-urban fringe for Chinese cities will be the need to resolve conflicts among different levels and arms of government, including the rural collective committees.

Tan et al. (2011) highlight that, for Beijing, important planning documents affecting the fringe have been produced by at least three different government departments. Thus there are conflicts and contradictions between the plans, partly reflecting the different data resources utilized in their preparation, but also their basic aims. In part, it is this tension that underpins the assertion by Yang and Jinxing (2007), with respect to the Green Belts designated for Beijing, that "urban sprawl is hard to contain with an arbitrary boundary such as a greenbelt."

More coordination between departments and different tiers of government would help remove some of the contradictions and provide impetus towards greater efficiencies in planning. Simplified, overarching planning legislation would also help to resolve many of the existing governance and jurisdictional problems.

Indeed, impetus for just such initiatives to both simplify and strengthen urban and regional planning came from the Central Urbanization Work Conference in December 2013, at which President Xi Jinping championed implementation of a standard planning blueprint for cities and counties. This conference led to a multi-authority experiment focusing on "Integration of Multi-Planning" pilot projects in 28 city regions (Zhou et al., 2017). However, the initial results suggest that creating a system that can readily resolve conflicts between central government departments and municipal authorities is difficult, especially as there also needs to be consideration of the functions of local government in the planning process. It should be noted, though, that ongoing experiments in ways to manipulate the land market and interpret national policy goals continue to show that there is both considerable scope for new planning *modi operandi* and a general desire to improve existing systems (Zhang and Wu, 2017).

At present, local government plays two roles: one as administrator and the other as a salesman, whereby it expropriates land cheaply from farmers and sells at a much higher price to real estate developers. It is this ability to gain revenue from sales that encourages local government to convert agricultural land to construction land and to amend existing plans. In examining the allocation of rural-urban construction land, Tang et al. (2015) conclude that local governments arbitrarily expand the implementation area of construction land, ignoring the quality of consolidated farmland. They note that it is the local authorities that have the power to drive economic development through land sales, and hence it is the desire to meet development targets that overrides other considerations (see also Yue et al., 2013). Any reformed planning system would need to address this problem by giving planners powers to control such sales and to prevent development on land designated for protection, as in the case of a Green Belt.

Furthermore, the physical quality of farmland has been a key primary factor in geographically delineating the boundaries of basic farmland districts that are prohibited from land development. Site value for alternative uses is seldom considered. This omission has produced two undesirable consequences. The first is to raise land prices, particularly around the major cities of eastern China, where land is a constraint for growth (Lichtenberg and Ding, 2009). As land prices increase, housing affordability deteriorates, particularly for rural-urban migrants. The second unwanted outcome is that land policy is driving the nature of urban sprawl (Ding et al., 2014). Across the developed world, urban sprawl has been associated with falling transport costs, rising incomes, and consumer preferences for houses with bigger yards/gardens. In contrast, farmland protection is a dominant factor producing spatial "leapfrogging" development in Chinese cities. Sites close to the urban built-up areas often have higher agricultural productivity than remoter sites and thus are designated as basic farmland. This pushes urban growth further away from major central business districts. In some cases, notably for Beijing, this has been accentuated by the designation of new towns on greenfield sites to accommodate millions of residents (Yang et al., 2015).

6. Conclusion

Peri-urban fringes worldwide face multiple pressures to fulfill various roles. Shortages of affordable housing for the urban population continue to place pressure on farmland adjacent to major cities. Yet, this farmland is often part of an area specifically protected to preserve amenity and to provide "green" space for recreation and nature conservation. The need for the peri-urban fringes to fulfill multiple roles has also increased pressure on existing planning measures to be modified or scrapped to meet the new challenges.

In the UK, this pressure has brought about further scrutiny of the long-established Green Belt legislation that has been widely adopted as a measure to restrict urban sprawl and provide green space adjacent to major cities. This paper has suggested that the operation of the Green Belt policy has had positive outcomes in terms of contributing substantially to conservation and recreational objectives, but that Green Belts have become a battleground between competing land uses, with pressures for urban development growing in the face of a national housing shortage. Yet Green Belts have been remarkably flexible in their operation and have not necessarily been as antithetical to urban land uses as might be imagined. Indeed, they have contributed to the complexity of the land-use mosaic in peri-urban fringes that is increasingly being termed the "edgelands." This is a growing feature of land use around major cities, reflecting competing market demands as well as the limitations of planning systems that were largely designed under different circumstances.

Having focused specifically on Green Belt policies in the UK, the chapter has noted that this measure and its variants has been adopted in many cities worldwide, including several developing countries, and that some of the underlying ideals of Green Belt policy have appeared in China in recent decades. There are, though, many reasons why simple translation of this planning measure to the Chinese context is difficult, with its different systems of planning and land management. Nevertheless, the discussion has highlighted problems within planning for peri-urban fringe areas in China that will need further urgent attention across all levels of government if the creation of chaotic patterns of land use is to be avoided and key national priorities compromised, notably the preservation of productive farmland, provision of recreational space for the expanding urban population, and conservation of valuable natural assets.

Having reached this conclusion, it is possible to make some simple recommendations that might be considered when planning for the evolution of China's peri-urban fringes. First, given that the demands of national policy goals are often in conflict, a more coherent policy framework should be considered, perhaps along the lines suggested by Liu et al. (2014), in which the ultimate goal of enhancing sustainable land use is sought by establishing a strategic land-use policy system that links different planning "layers": strategy, policy, and protection. However, within this system, greater consideration should be given to protecting prime farmland from conversion to non-farm use, echoing Huang et al. (2015), and to establishing clearly demarcated protected areas, such as Green

Belts and conservation areas, where development is strongly limited. For this effort to be effective, there is the implication that an overhaul of the planning and local government finance systems may be required, with planning authorities having greater powers to implement agreed plans while local authorities become less reliant on land sales for their finance. This tighter planning control may be able to reduce the currently prevalent land-use fragmentation in urban fringes across China (Tian, 2015), in addition to maintaining more green space and creating additional recreational provision.

References

Abercrombie, P. (1944). Greater London Plan. In *Standing Conference on London Regional Planning*. London: Ministry of Housing and Local Government.

Alterman, R. (1997). The Challenge of Farmland Preservation: Lessons from a Six-Nation Comparison. *Journal of the American Planning Association*, 63(2), 220–244.

Alexander, A. (2009). *Britain's New Towns: Garden cities to sustainable communities*. London: Routledge.

Amati, M. (2007). From a Blanket to a Patchwork: The Practicalities of Reforming the London Green Belt. *Journal of Environmental Planning and Management*, 50(5), 579–594.

Amati, M. (Ed.). (2008a). *Urban Green Belts in the Twenty-first Century*. Aldershot, UK and Burlington, VT: Ashgate Publishing.

Amati, M. (2008b). Green Belts: A Twentieth-Century Planning Experiment. In M. Amati (Ed.). *Urban Green Belts in the Twenty-First Century*. Aldershot, UK and Burlington, VT: Ashgate Publishing, pp. 1–18.

Amati, M. and Taylor, L. (2010). From Green Belts to Green Infrastructure. *Planning, Practice & Research*, 25(2), 143–155.

Amati, M. and Parker, G. (2007). Planned by Farmers for Farmers? Twentieth-Century Land reform and the Impact on Twenty-First Century Japan. In C. Miller and M.M. Roche (Eds.). *Past Matters: Heritage and Planning History Case Studies from the Pacific Rim*. London: Cambridge Scholars Press, pp. 172–194.

Amati, M. and Yokohari, M. (2006). Temporal Changes and Local Variations in the Functions of London's Green Belt. *Landscape and Urban Planning*, 75(1), 125–142.

Bailoni, M., Edelblutte, S., and Tchékémian, A. (2012). Agricultural Landscape, Heritage and Identity in Peri-urban Areas in Western Europe, *European Countryside*, 4(2), 147–161.

Balk, H. H. (1945). Rurbanization of Worcester's Environs. *Economic Geography*, 21(2), 104–116.

Barker, K. (2006). *Barker Review of Land Use Planning, Final Report Recommendations*. London: HM Treasury.

Booth, P. (1996). *Controlling Development: Certainty and Discretion in Europe, the USA and Hong Kong*. London: UCL Press.

Bramley, G., Hague, C., Kirk, K., Prior, A., Raemaekers, J., Smith, H., Robinson, A., and Bushnell, H. (2004). *Review of Green Belt Policy in Scotland. Research Report to Scottish Executive Development Department*. Edinburgh: Scottish Executive Social Research.

Bryant, C. R., Russwurm, L. J., and McLellan, A. G. (1982). *The City's Countryside. Land and its Management in the Rural-Urban Fringe*. London: Longman.

Bryson, B. (2015). *The Road to Little Dribbling: More Notes from a Small Island.* London: Doubleday.
Buxton, M. and Goodman, R. (2003). Protecting Melbourne's Green Belt. *Urban Policy and Research,* 21(2), 205–209.
Cadène, P. (1990). L'usage des espaces péri-urbains: Une géographie régionale des conflits. *Etudes Rurales,* 118–119, 235–267.
Carlyon, T. (2013). *Food for Thought: Applying House Price Inflation to Grocery Prices.* London: Shelter. http://england.shelter.org.uk/__data/assets/pdf_file/0005/625550/Food_for_thought.pdf
Ding, C. (2007). Policy and Praxis of Land Acquisition in China. *Land Use Policy,* 24(1), 1–13.
Ding, C. and Lichtenberg, E. (2011). Land and Urban Economic Growth in China. *Journal of Regional Science,* 51(2), 299–317.
Ding, C., Niu, Y., and Lichtenberg, E. (2014). Spending Preferences of Local Officials with Off-Budget Land Revenues of Chinese Cities. *China Economic Review,* 31, 265–276.
von der Dunk, A., Grêt-Regamey, A., Dalang, T., and Hersperger, A. M. (2011). Defining a Typology of Peri-urban Land-use Conflicts – A Case Study from Switzerland. *Landscape and Urban Planning,* 101(2), 149–156.
Edwards, M. (2000). Sacred Cow or Sacrificial Lamb? Will London's Green Belt Have to Go? *City,* 4(1), 106–112.
Farley, P. and Roberts, M. S. (2012). *Edgelands: Journeys into England's True Wilderness.* London and New York: Vintage.
Freestone, R. (2002). Greenbelts in City and Regional Planning. In K. Parsons and D. Schuyler (Eds.). *From Garden City to Green City: The legacy of Ebenezer Howard.* John Hopkins Press, Baltimore, pp. 67–98.
Frey, H. W. (2000). Not Green Belts but Green Wedges: The Precarious Relationship between City and Country. *Urban Design International,* 5(1), 13–25.
Furuseth, O. J. and Lapping, M. B. (1999). *Contested Countryside: The Rural Urban Fringe in North America.* Ashgate Publishing Ltd.
Gallent, N. and Shaw, D. (2007). Spatial Planning, Area Action Plans and the Rural-Urban Fringe. *Journal of Environmental Planning and Management,* 50(5), 617–638.
Gant, R. L., Robinson, G. M., and Fazal, S. (2011). Land Use Change in the "Edgelands": Policies and Pressures in London's Rural-Urban Fringe. *Land Use Policy,* 28(1), 266–279.
Gunn, S.C. (2008). Green belts: A review of the regions' responses to a changing housing agenda. *Journal of Environmental Planning and Management,* 50(5), 595–616.
Hague, C. (2002). Response to Roberts, T., The Seven Lamps of Planning. *Town Planning Review,* 73(1), 1–15.
Harrison, C. (1981). A Playground for Whom? Informal Recreation in London's Green Belt. *Area,* 13, 109 114.
He, C., Okada, N., Zhang, Q., Shi, P., and Zhang, J. (2006). Modeling Urban Expansion Scenarios by Coupling Cellular Automata Model and System Dynamic Model in Beijing, China. *Applied Geography,* 26(3), 323–345.
Herington, J. (1984). *The Outer City.* London and New York: Harper Collins.
Hoffman, L. (2011). Urban Modeling and Contemporary Technologies of City-Building in China: The Production of Regimes of Green Urbanisms. In A. Roy and A. Ong, (Eds.). *Worlding Cities: Asian Experiments at the Art of Being Global.* Chichester and Maldon, MA: Wiley Blackwell, pp. 55–76.

Houston, P. (2005). Re-valuing the Fringe: Some Findings on the Value of Agricultural Production in Australia's Peri-urban Regions. *Geographical Research*, 43(2), 209–223.

Huang, D., Jin, H., Zhao, X., and Liu, S. (2015). Factors Influencing the Conversion of Arable Land to Urban Use and Policy Implications in Beijing, China. *Sustainability*, 7(1), 180–194.

Jackson, K. T. (1985). *Crabgrass Frontier: The Suburbanization of the United States*. New York: Oxford University Press.

Jefferies, J. (2005). The UK Population: Past, Present and Future. In Office for National Statistics, *Focus on People and Migration, 2005*. London: HM Government.

Jia, J. and Gao, J. (2005). The Origination, Development and Challenge of Green Belt Policy in the UK. *Journal of Chinese Landscape Architecture*, 3, 69–72 (in Chinese).

Jim, C. Y. and Chen, S. S. (2003). Comprehensive Greenspace Planning Based on Landscape Ecology Principles in Compact Nanjing City, China. *Landscape and Urban Planning*, 65(3), 95–116.

Kabo, R. (2015). Towards a Taxonomy of Edgelands Literature. *Alluvium*, 4(3), www.alluvium-journal.org/2015/06/26/towards-a-taxonomy-of-edgelands-literature/

Kaika, M. and Swyngedouw, E. (2000). Fetishizing the Modern City: The Phantasmagoria of Urban Technological Networks. *International Journal of Urban and Regional Research*, 24(1), 120–138.

Kambites, C. and Owen, S. (2006). Renewed Prospects for Green Infrastructure Planning in the UK. *Planning Practice and Research*, 21(4), 483–496.

Lai, L. W. C. and Ho, W. K. (2001). Low-rise Residential Developments in Green Belts: A Hong Kong Empirical Study of Planning Applications. *Planning Practice and Research*, 16(3–4), 321–335.

Lemes de Oliveira, F. (2014). Green Wedges: Origins and Development in Britain. *Planning Perspectives*, 29(3), 357–379.

Li, F., Wang, R., Paulussen, J., and Liu, X. (2005). Comprehensive Concept Planning of Urban Greening Based on Ecological Principles: A Case Study in Beijing, China. *Landscape and Urban Planning*, 72(4), 325–336.

Lichtenberg, E., and Ding, C. (2009). Local Officials as Land Developers: Urban Spatial Expansion in China. *Journal of Urban Economics*, 66(1), 57–64.

Lin, G. C., Li, X., Yang, F. F., and Hu, F. Z. (2014). Strategizing Urbanism in the Era of Neo-liberalization: State Power Reshuffling, Land Development and Municipal Finance in Urbanizing China. *Urban Studies*, 20(10), 1–21.

Liu, S. H., Chen, T., and Cai, J. M. (2004). Peri-urbanization in China and its Major Research Issues. *Acta Geographica Sinica*, 59(S1), 101–108.

Liu, Y., Fang, F., and Li, Y. (2014). Key Issue of Land Use in China and Implications for Policy Making. *Land Use Policy*, 40, 6–12.

Liu, Y., Wang, L., and Long, H. (2008). Spatio-temporal Analysis of Land-use Conversion in Eastern Coastal China During 1996–2005. *Journal of Geographical Sciences*, 18(3), 274–282.

Long, H., Liu, Y., Li, X., and Chen, Y. (2010). Building New Countryside in China: A Geographical Perspective. *Land Use Policy*, 27(2), 457–470.

Martin, W. T. (1953). *The Rural-Urban Fringe: A Study of Adjustment to Residence Location*. Eugene: University of Oregon Press.

Ministry of Land and Resources (2006). *China Land and Resources. Statistical Yearbook, 2006*. Beijing: Ministry of Land and Resources of the People's Republic of China.

Munton, R. J. C. (1983). *London's Green Belt: Containment in Practice*. Hemel Hempstead: Allen & Unwin.

Munton, R., Whatmore, S., and Marsden, T. (1988). Reconsidering Urban-fringe Agriculture: A Longitudinal Analysis of Capital Restructuring on Farms in the Metropolitan Green Belt. *Transactions of the Institute of British Geographers*, 13(3), 324–336.

National Bureau of Statistics of China (various years). *China Statistical Yearbooks*. Beijing: China Statistics Press.

Nelson, A. C. and Dueker, K. J. (1990). The Exurbanization of America and its Planning Policy Implications. *Journal of Planning Education and Research*, 9(2), 91–100.

Osborn, F. J. and Whittick, A. (1963). *The New Towns: The Answer to Megalopolis*. London: Leonard Hill Books.

Papadimitrou, N. (2013). *Scarp: In Search of London's Outer Limits*. London: Sceptre.

Prior, A. and Raemaekers, J. (2007). Is Green Belt Fit for Purpose in a Post-Fordist Landscape? *Planning, Practice & Research*, 22(4), 579–599.

Qiang, Z. H. U., Kong-Jian, Y. U., and Di-Hua, L. I. (2005). The Width of Ecological Corridors in Landscape Planning. *Acta Ecologica Sinica*, 25(9), 2406–2412.

Rippey, J. (2012). Wild at heart: Exploring the edgelands of contemporary landscape literature', 紀要 44, 17–31 (in Chinese).

Robinson, G. M. (1991). The City Beyond the City. In G. M. Robinson (Ed.). *A Social Geography of Canada*. Toronto and Oxford: Dundurn Press, pp. 302–329.

Robinson, G. M., Gant, R. L., and Fazal, S. (2011). Neglected Landscapes: A Study of Land-use Change in the Rural-Urban "Edgelands." In G. M. Robinson and F. Molinero (Eds.). *Changing European Rural and Agricultural Landscapes: Papers from the Anglo-Spanish Rural Geography Conference, Canterbury Christ Church University, June 2009*. Valladolid, Spain: Universidad de Valladolid.

Sawer, P. (2016). Plans for Thousands of New Homes Threaten Green Belt Areas. www.telegraph.co.uk/news/2016/04/16/plans-for-thousands-of-new-homes-threaten-green-belt-areas/

Shengquan, C. H. E. (2001). Study on the Green Corridors in Urbanized Areas. *Urban Ecological Study*, 25(11), 44–48 (in Chinese).

Shoard, M. (2002). Edgelands. In J. Jenkins (Ed.). *Remaking the Landscape: The Changing Face of Britain*. London: Profile Books, pp. 117–146.

Shoard, M. (2013). Edgelands of Promise. *Landscapes*, 1(2), 74–93.

Sinclair, I. (2003). *London Orbital*. Harmondsworth: Penguin.

Tan, M., Robinson, G. M., and Li, X. (2011). Urban Spatial Development and Land Use in Beijing: Implications from London's Experiences. *Journal of Geographical Sciences*, 21(1), 49–64.

Tang, B. S., Wong, S. W., and Lee, A. K. (2005). Green Belt, Countryside Conservation and Local Politics: a Hong Kong Case Study. *Review of Urban & Regional Development Studies*, 17(3), 230–247.

Tang, B. S., Wong, S. W., and Lee, A. K. W. (2007). Green Belt in a Compact City: A Zone for Conservation or Transition? *Landscape and Urban Planning*, 79(3), 358–373.

Tang, Y., Mason, R. J., and Wang, Y. (2015). Governments' Functions in the Process of Integrated Consolidation and Allocation of Rural-Urban Construction Land in China. *Journal of Rural Studies*, 42, 43–51.

Tian, L. (2015). Land Use Dynamics Driven by Rural Industrialization and Land Finance in the Peri-urban Areas of China: The Examples of Jiangyin and Shunde. *Land Use Policy*, 45, 117–127.

Wang, L., Li, C. C., Ying, Q., Cheng, X., Wang, X. Y., Li, X. Y., Hu, L. Y., Liang, L., Yu, L., Huang, H. B., and Gong, P. (2012). China's Urban Expansion from 1990 to

2010 Determined with Satellite Remote Sensing. *Chinese Science Bulletin*, 57(22), 2802–2812.

Wang, X., Lu, C., and Fang, J. (2007). Implications for Development of Grain-for-green Policy Based on Cropland Suitability Evaluation in Desertification-affected North China. *Land Use Policy*, 24, 417–424.

Whitehead, C. (2014). Social Housing in England. In K. Scanlon, C. Whitehead, and M. Fernández Arrigoitia (Eds.). *Social Housing in Europe*. London: Wiley-Blackwell, pp. 105–120.

Wu, G., Yu, S., and Wang, Z. (2001). The Discussion on the Concept of the Planning for the Green Belt around Shanghai. *City Planning Review*, 4 (in Chinese).

Yang, J. and Jinxing, Z. (2007). The Failure and Success of Greenbelt Program in Beijing. *Urban Forestry & Urban Greening*, 6(4), 287–296.

Yang, X., Day, J., and Han, S. S. (2015). Urban Peripheries as Growth and Conflict Spaces: The Development of New Towns in China. In T-C. Wong, S. S. Han, and H. Zhang (Eds.). *Population Mobility, Urban Planning and Management in China*. Cham, Switzerland: Springer International, pp. 95–112.

Yue, W., Liu, Y., and Fan, P. (2013). Measuring Urban Sprawl and its Drivers in Large Chinese Cities: The Case of Hangzhou. *Land Use Policy*, 31, 358–370.

Zasada, I. (2011). Multifunctional Peri-urban Agriculture – A Review of Societal Demands and the Provision of Goods and Services by Farming. *Land Use Policy*, 28(4), 639–648

Zasada, I., Berges, R., Hilgendorf, J., and Piorr, A. (2013). Horsekeeping and the Peri-urban Development in the Berlin Metropolitan Region. *Journal of Land Use Science*, 8(2), 199–214.

Zhang, Q. F. and Wu, J. (2017). Political Dynamics in Land Commodification: Commodifying Rural Land Development Rights in Chengdu, China. *Geoforum*, 78, 98–109.

Zhou, X., Lu, X., Lian, H., Chen, Y., and Wu, Y. (2017). Construction of a Spatial Planning System at City-level: Case Study of "Integration of Multi-planning" in Yulin City, China. *Habitat International*, 65, 32–48.

6 Land and housing market dynamics and housing policy in Japanese metropolitan cities

*Eiji Oizumi**

WAKAYAMA UNIVERSITY & REGIONAL ECONOMICS OTEMON GAKUIN UNIVERSITY

1. Introduction: Overview of the post-bubble Japanese economy

Higher market volatility and a greater housing divide constitute common features of many housing markets around the world. These tendencies can be partly explained by the dynamics of capital movements in global financial markets, the ongoing deregulation in housing and housing-finance markets, and worsening income inequalities. This chapter provides some theoretical and empirical explanations for the distinctive dynamics and social, economic, and spatial relationships in Japanese housing markets. To begin, the introductory section presents an overview of the major developments of the Japanese economy during the period from the 1990s to the mid-2010s.

1.1 Prolonged deflationary cycle

As a consequence of the asset bubble in the latter half of the 1980s, and in the aftermath of its burst at the very beginning of the 1990s, the Japanese economy has experienced profound changes including: (1) a recession since 1998 partly explained by "bad debt" problems within Japanese financial institutions, the onset of the Asian currency crisis, and the hike in the rate of the consumption tax in April 1997; (2) an economic recovery during 2003 and 2007 attributed to the prominent expansion of exports; (3) a reversion to recession in 2009 explained by the "Lehman shock" in September 2008 and the world financial crisis; and (4) a significant drop in household consumption due to another increase in the rate of the consumption tax in April 2014. This sequence of events should be complemented with the outbreaks of enormous disasters: (5) the Hanshin earthquake in January 1995, and (6) the Eastern Japan earthquake and tsunami disaster and Fukushima's serious nuclear accident in March 2011.

Illustrating the fluctuations of the GDP in real terms and its demand-side indices, Figure 6.1 provides a clear perspective on the major shocks and developments experienced by the Japanese economy over the last 25 years. Starting with the bust of the asset bubble in the 1990s, the Japanese economy entered a long-lasting stagnation. Given the double constraints of post-bubble over-production capacity and intensive competition from emerging economies,

106 Eiji Oizumi

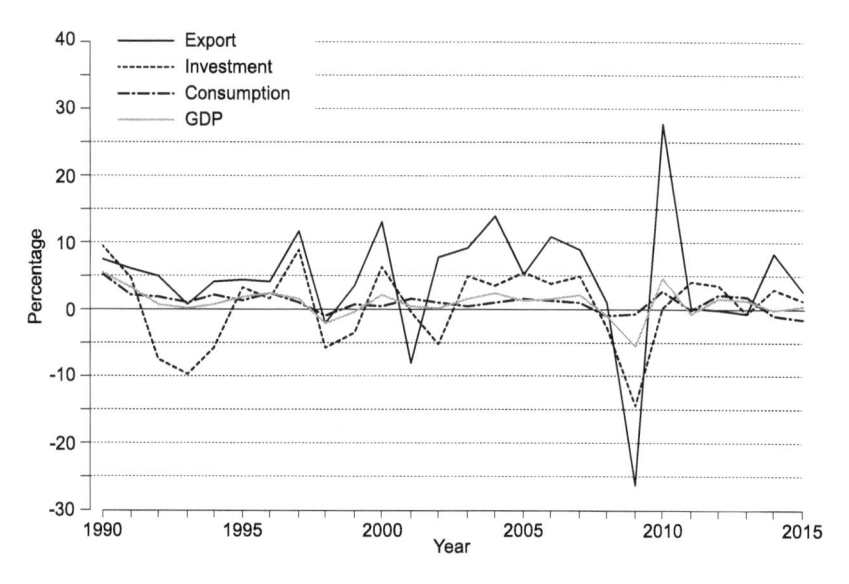

Figure 6.1 Changes in GDP and demand-side indices in Japan, 1990–2015.
Note: The change ratio (percentage) to the previous year of real GDP, household consumption, private capital investment, and exports of goods and services.
Source: Government Cabinet Office, Japan's National Account.

the Japanese manufacturing industry faced difficult strategic decisions. In the context of decreasing exports due to the appreciation of the yen, Japanese manufacturers resorted to the transfer of production bases offshore and to cutbacks in labor costs by substituting long-term regular employment with non-regular employment: part-time, fixed-term, and temporary labor.

With the prominent expansion of exports to the U.S. markets and Asian emerging economies, the Japanese economy underwent a recovery phase during the period from 2003 to 2007. Despite the appellation of "the longest post-war economic recovery," the growth rate of real GDP was confined to only a 1 percent annual average. Due to the rapid increase in exports from 56.7 trillion yen in 2002 to 92.2 trillion yen in 2007, amounting to a 1.6-fold increase, the performance of major manufacturing companies remarkably improved. Capital investment increased by 24 percent, from 65.1 trillion yen to 81 trillion yen over the same period. However, the relative share of labor in added value for major manufacturing companies—composed of 2,500 companies with capital of at least 1 billion yen—diminished significantly from 66.7 percent in 2002 to 56.9 percent in 2007. At the same time, retained earnings in the companies recorded, in contrast, an increase of 38.7 percent from 55.1 trillion yen to 76.4 trillion yen.

There is a clear correlation between the fluctuations of exports and real investment. However, an increase of investment is not necessarily conducive to rising employment and wages. Because non-regular employment with low wages was expanding across all industries, household expenditures were constrained, and household consumption increased by just 3.6 percent, from 277.6 trillion yen in 2002 to 287.5 trillion yen in 2007. Hence, the growth of the GDP remained at a low level as well (Oizumi, 2013).

This export-led economic growth was further weakened by the diminishing volume of international trade after the "Lehman shock." A rapid recovery of exports in 2010, coupled with expansionary fiscal spending on infrastructures for post-disaster reconstruction purposes, enabled the economy to recover from recession. However, negative growth occurred again following the increase in the rate of the consumption tax in 2014 that led to a contraction in household consumption.

Thus, the Japanese economy has lapsed into a long-lasting deflationary cycle involving falling prices in the goods and services, stocks, and real estate markets. It can be argued that the deflationary cycle is partly caused by a complex mixture of macro-economic factors. First, the drastic elimination of manufacturers in regional economies within Japan results from the ongoing external shifts of factories across borders and keener competition among companies in domestic markets under declining national demand. Second, there is a significant transfer of income from the household sector to the financial sector caused by the ultra-low interest-rate policy adopted by the Bank of Japan. Third, the substantially curtailed employment and wage levels reflect the business downsizing strategies pursued by companies. These factors bring about a severe decline in household consumption and, naturally, a lasting economic stagnation. In addition, other aggravating factors include the demographical and income disparity problems in the Japanese economy, which are discussed below.

1.2 Decreasing and ageing population

The Japanese economy faces the full-scale compounded effects of an ageing society and a shrinking population. These demographic changes are the direct result of lower birthrates and long life expectancy. Presently, Japan has one of the lowest birthrates in the world: the total fertility rate dipped as low as 1.42 in 2014. This low rate is, in the long term, conducive to a significant decrease in the size of the population and an increase in the proportion of elderly people. While Japan's population in 2010 was 128 million, it is expected to decline to less than 117 million by 2030, representing a shrinkage of 8.6 percent over just two decades. Over the same period, the proportion of people aged over 65 years is expected to increase from 23 percent to 31.6 percent (IPSS, 2012).

Despite the shrinking population, the number of households continues to increase. In particular, the increase in the number of one-person households is rather conspicuous. This number amounted to 16.8 million and accounted for 32.4 percent of the total households in 2010. The aggregate number of households is estimated to reach a peak of 53.1 million in 2019 and decrease afterwards to 49.6 million in 2035. These alarming trends will certainly create a long-term decrease in housing demand. It is estimated that housing demand in 2035 will be as much as 3 million housing units smaller than its expected peak in 2019.

In addition, there is the problem of population movement towards large metropolitan regions. Only a few prefectures, including Tokyo, Kanagawa, Saitama, Aichi, and Fukuoka, are still experiencing an increasing population. The gravity of problems related to an ageing society extends beyond depopulated local areas to include metropolitan cities as well. As many as 2.97 million people aged over 65 years reside in metropolitan Tokyo alone as of 2015, representing 22.9 percent of its total population. Among the elderly population of Tokyo, the proportion of singleton households amounted to 38 percent in 2010, and singleton households constitute the highest proportion among all prefectures. It is also estimated that the number of elderly singleton households might exceed the 1 million mark in Tokyo, for a remarkable proportion of 44 percent in 2035.

1.3 Inequalities in income distribution and poverty

In light of the structural changes described above, there is indisputable evidence that economic and social polarization is growing. Job security has been substantially eroded for many classes of working people. The drastic changes in the labor market are conducive to increasing disparities of incomes and living conditions, and greater economic uncertainty and social insecurity. These changes affect not only lower income groups but the middle classes as well.

The World Wealth and Income Database (WID), organized by Thomas Piketty and other researchers, provides comprehensive evidence about the concentration of wealth into a handful of ultra-affluent classes in many countries. According to WID statistics for 2010, the share of the top 1 percent in the whole income structure in Japan, Korea, the UK, and the USA amounts to 9.5 percent, 11.8 percent, 12.6 percent, and 17.5 percent respectively. This statistical evidence may suggest that the extent of income disparity is relatively low for Japan. However, the share of the top 10 percent, amounting to 40.5 percent for Japan, is still comparable to the rates reported for the other three countries, which reach equally high levels of 43.3 percent, 38.1 percent, and 46.4 percent respectively.

The main concerns arise rather with respect to the conditions of the remaining 90 percent. As shown by Figure 6.2, it is important to note that the average annual income of the bottom 90 percent in Japan decreased by 30.2 percent, falling from 2.14 million yen in 1990 to 1.49 million yen in 2010. Due to continuous deflation, the GDP deflator decreased by 9.6 percent through the same period, and hence the reduction rate of the average income in real terms is counted as 23.7 percent. However, important issues are not only a fall in income level but also the changes in income distribution.

The Organisation for Economic Cooperation and Development (OECD) defines the relative poverty rate as the proportion of people with income less than half the median of income distribution. In the case of Japan, the poverty rate in 2012 reached 16.1 percent, and the poverty line is associated with an annual income of 1.22 million yen (see Figure 6.3), which is nearly equal to the average annual income of the bottom 90 percent.

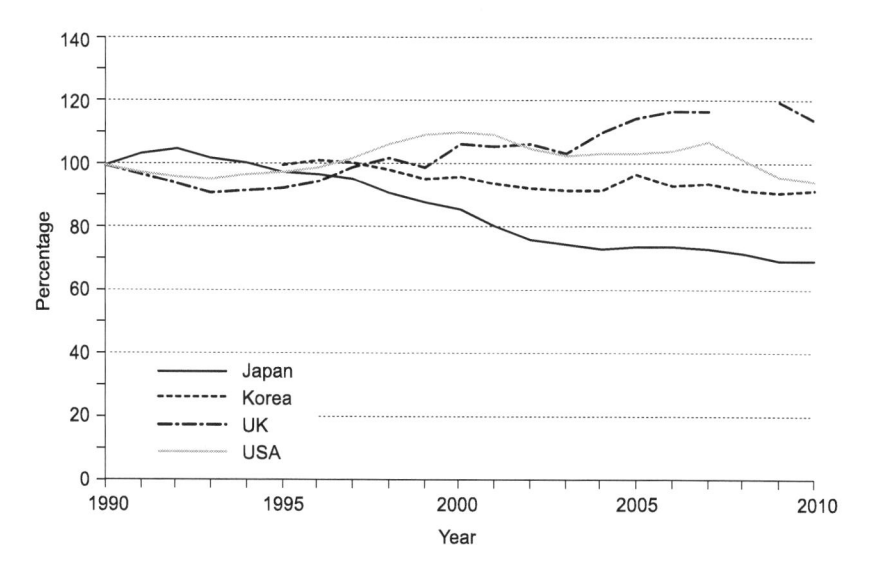

Figure 6.2 Changes in average income of the bottom 90 percent, 1990–2010. Comparison of Japan with Korea, the UK, and the USA.

Note: Each index is calculated on the standard of Year 1990=100. Only Korea's index is calculated on Year 1995=100, as the data dring 1990 and 1994 are not givne. Data of the UK in 2008 is not given.

Source: World Wealth and Incomes Database.

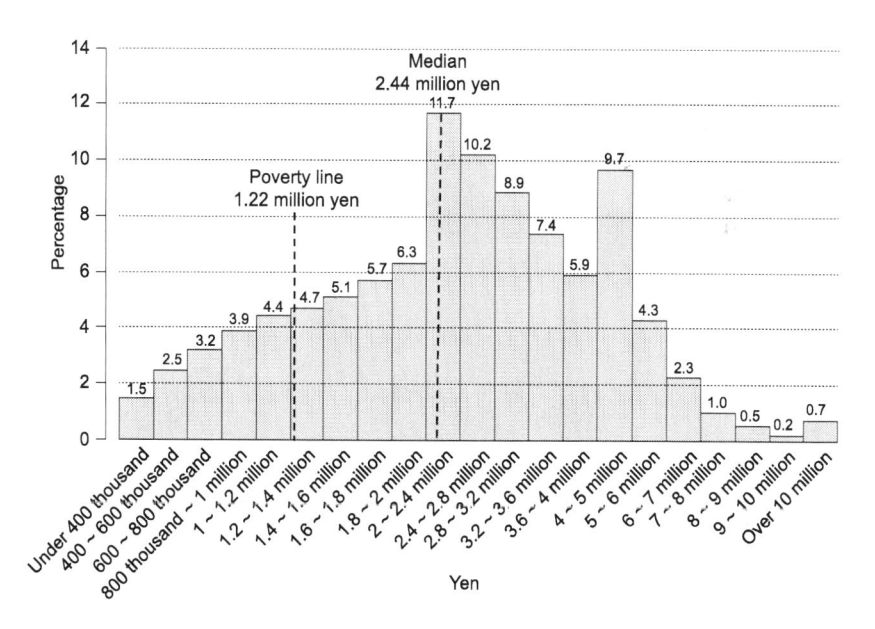

Figure 6.3 Distribution of relative frequency (percentage) of the household members with equivalent disposable income in Japan, 2012.

Note: Equivalent disposable income means the household's disposable income divided by the square root of the number of household member. The median of equivalent disposable income amounts to 2.44.

Source: MHLW, Comprehensive Survey of Living Conditions.

The WID does not consider the stratification of income within the bottom 90 percent, but referring to wage statistics in private enterprises published by the National Tax Administration Agency can be useful in providing proxies for these strata. The reported statistics for the year 2010 suggest that workers with annual wages less than 6 million yen, representing 82.3 percent of all salaried workers, were employed by private enterprises. This income group can be considered within the bottom 90 percent because WID estimates the threshold value of the top 10 percent to be at 5.76 million yen. Within the lower-income group, the largest parts are represented by the income classes of under 2 million yen and 2 to 3 million yen, which are associated with shares of 27.9 percent and 21.4 percent respectively. Meanwhile, the class of income above 4 million and under 6 million yen is associated with a diminishing share from 30.2 percent in 1990 to 28.8 percent in 2010. Thus, there is no doubt that the intensive replacement of regular employees with non-regular employees has depleted the middle class with annual incomes of 4 million to 6 million yen, and expanded the "working poor" with annual incomes of less than 2 million yen.

We expect the powerful socio-economic changes described above to shape the present dynamics of land and housing markets. The next section elucidates the distinctive developments in land policy and land markets in Japanese cities.

2. Deregulation in land policy and urban planning

2.1 Land policy and urban planning in Japanese cities

The City Planning Act of 1968, replacing the former City Planning Act of 1919, introduced various regulative measures about urban land use: (1) designation of the city planning area composed of both the urbanization promotion area (UPA) and the urbanization control area (UCA), (2) initiation of the development permission by planning authorities at the prefectural level, and (3) new subdivision of the land-use zonings composed of eight categories related to residential, commercial, and manufacturing industrial land uses.

Although the Japanese planning system bears some resemblance to the planning system in Western countries, there are exemptions in the regulation of UPA and UCA planning areas. In the UCA, where development should be restrained to prevent sprawl, the development of housing complexes and large shop-buildings was permitted on an exceptional basis. The scope of planning permissions was not exhaustive given the exemption of small-scale development in the UPA from this regulatory requirement. Further, the major pitfall in the regulation UPA land use was the permission for shops and offices to be built anywhere in these areas except in two categories of districts designated exclusively for housing and manufacturing purposes. Although appropriately controlled land-use mix could lead to the formation of favorable urban space, the Japanese urban planning system at that time lacked measures of this kind. These flaws were a natural consequence of governmental and municipal land policies with a strong development orientation.

These regulatory conditions had critical implications for land markets as well as housing markets. Urban land markets actually reflect a mixture of competitive forms of land use due to relatively loose regulation. Residential development has advanced since the 1970s, not just in suburbs but also in city centers and neighboring areas, where the construction of condominiums went into full swing. During this expansion, the commercial estate development towards suburbs and the residential development towards city centers mingled with each other, thereby intensifying competition among the forms of land use in suburbs as well as city centers. Therefore, the dichotomy between centers and suburbs in land use, which has been a key concept of the functionalist urban planning in Western countries since the last century, changed into a complicated and sprawling differentiation. This competitive mixture of land-use forms caused a continuous rise in land prices. In addition, loose constraints on conversion of land use furthered speculative land acquisition. The momentum towards higher pricing reached its climax during the property boom in the second half of the 1980s (Oizumi, 1994 and 2002). An inquiry into the Japanese asset bubble at that time needs a multi-dimensional approach to institutional, administrative, and social factors. However, we focus on the distinctive performance of land markets.

Figure 6.4 clearly presents the propagation process that underlay the behavior of residential land prices during the property boom. The boom started in central

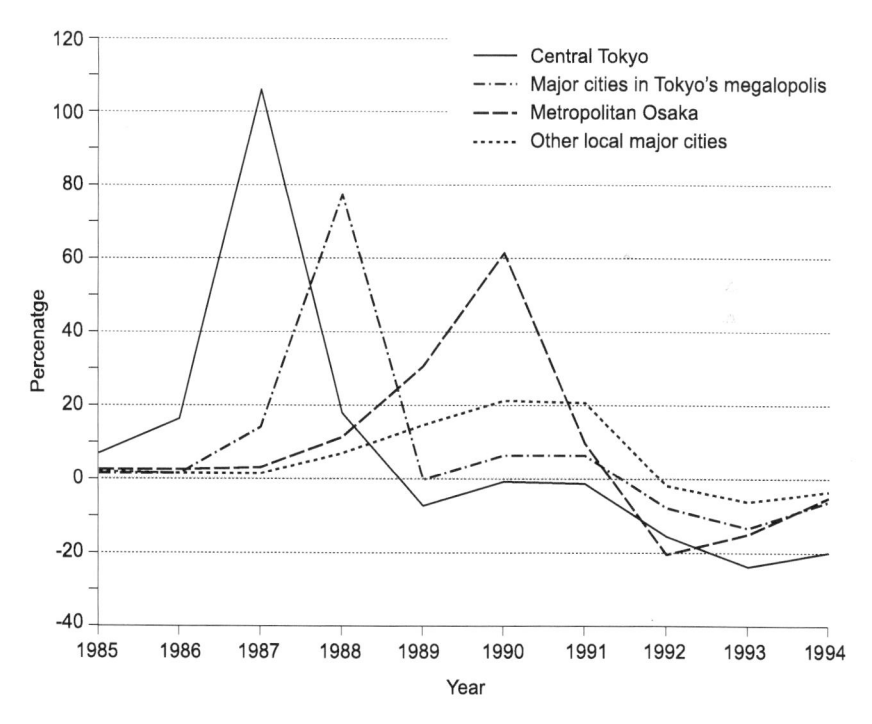

Figure 6.4 Changes in residential land prices of metropolitan Tokyo and Osaka and other cities, 1985–1994.

Note: The change ratio (percentage) to the previous year of residential land prices in each area.

Source: Tokyo Metropolitan Government Office, 1999.

Tokyo and extended to major cities in the Tokyo megalopolis. It subsequently spilled over to Osaka's metropolis and other local major cities, with different time lags and maximum rates of land appreciation. A similar pattern of propagation could be observed in commercial land prices as well. The speculative land acquisitions by many firms provided a strong impetus for this propagation process. Indeed, the momentum was not only supported by real estate companies, as many firms in virtually every sector competed in the launching of projects in the real estate business. Their criteria for land acquisition depended not on zoned land uses, but rather on larger floor-area ratios.

Upon the collapse of the asset bubble, a deregulation policy was adopted with the purpose of reviving real estate markets from stagnation. Its core instrument was the relaxation of floor-area restrictions. The revised City Planning Act of 1992 subdivided three former categories of land use in residential districts into seven categories. In the aggregate, eight former categories were segmented into 12. This meant, in fact, raising the upper limit of the floor-area ratio and promoting mixed land uses including commercial use in residential districts. The revised City Planning Act of 1997 introduced an even higher ceiling in floor-area ratios in newly designated districts to facilitate the inclusion of high-rise dwellings.

2.2 Polarization of land markets and land policies

Figure 6.5 shows the trends of residential land prices in central and metropolitan Tokyo, metropolitan Osaka, and other local cities after the asset bubble burst. There was a long protracted period of declining land prices, and there were also signs of divergence in land markets at the beginning of the 2000s. While there was pervasive stagnation in land markets throughout the country, several "hot spots" in central Tokyo emerged with renewed activities in real estate investment. Residential land prices in central Tokyo seemed to reach bottom levels around 2003 and 2004 and exhibited a rapid increase from 2005 through 2007. Despite the sudden drop in association with the "Lehman shock" and the U.S. financial crisis in 2007–2008, residential land prices seem to have resumed their upward momentum over more recent years. As of January 2015, residential land prices in metropolitan Tokyo and Osaka showed a 1.3 percent and 0.5 percent increase respectively from the previous year, while those in other local areas decreased by 0.7 percent.

2.2 Metropolitan cities: booming construction under deregulation policy

Changes in land market conditions were strongly influenced by the urban development policy as part of the government's business stimulation package. A new stage of deregulation was indeed started by the government's urban regeneration policy. The Headquarters for Urban Regeneration in the Government Cabinet Office was established in 2001, and the Special Measures Law of Urban Regeneration was enacted in the next year. The objective of this policy has been

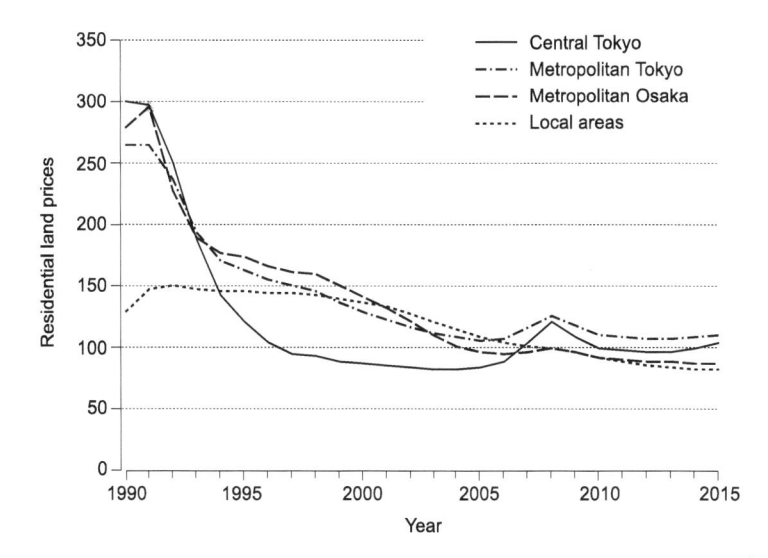

Figure 6.5 Cumulative changes in residential land prices in metropolitan Tokyo and Osaka and local areas, 1990–2015.
Note: Each index is calculated on the standard of Year 1983=100.
Source: Tokyo Metropolitan Government Office, 2014; MLIT, Official llisting of land prices.

the redevelopment of the centers of metropolitan cities, through the promotion of intensive land-use projects in city centers. The government designated a selection of sites within the centers of major cities as Urgent Redevelopment Zones (URZs). Among the 34 URZs covering 4,749 hectares, there were seven sites amounting to 2,375 hectares in central Tokyo, and 12 sites covering 1,072 hectares in Osaka City. That is, both cities occupied 72.6 percent of the total area of URZs. Thereafter, the URZs were expanded to 62 areas covering 8,037 hectares as of September 2014, and 49.3 percent of the total area is occupied by metropolitan Tokyo.

The concept of URZs represented the culmination of the deregulation process in urban planning. Standard regulations concerning land use and buildings were not applicable to these exceptional zones, and developers were allowed to propose their own planning schemes without restrictions. Stimulated by the government's policy, the construction of high-rise office buildings and condominiums has subsequently boomed in central Tokyo, and with some time lag in central Osaka. These market dynamics are discussed in further detail in the next section.

2.3 Local cities: inception of the "compact city" policy

Apart from metropolises, local cities represent the opposite side of the polarization in the real estate markets. Most local cities face twin problems: the decline of commercial estate developments, which are composed mainly of small, privately run stores in city centers, and an ageing-cum-decreasing population. In stark contrast to a once-strong trend towards suburban sprawl, there is a "reverse sprawl" reflected by a disorderly increment of vacant houses and lots in both urban areas and suburbs. Similar conditions can be observed also in the satellite cities of a megalopolis.

In searching for viable solutions to the demographic and urban impasse, many municipalities are eager to introduce the "compact city" policy into their master plans. The "compact city" model can be interpreted as a variant of the European "sustainable city" model. The "compact city" model involves a number of interrelated policy measures: (1) restraint on suburb development; (2) preservation of landscape and environment; (3) control of motorization and advancement of public transport systems; (4) development of renewable energy; (5) improvement and renovation of existing houses; (6) induction of population to city centers; (7) reallocation of communal facilities to city centers; and (8) recovery of good turnout on shopping streets in city centers (Suzuki, 2007; Oizumi, 2013).

Significant challenges exist in pursuing the objective of sustainable communities. Various movements with administrative/residents' initiatives are growing all over Japan. However, permissions are still granted for the development of housing and setting up of large retail stores in suburbs, measures which stand in sharp contrast with the "compact city" policy. These conditions increase the potential for an accelerated "reverse sprawl" in the near future. It is clearly difficult for many municipal authorities to slow down the momentum of development-oriented urban policy.

3. Housing stock and housing market dynamics

3.1 Housing stock: overabundance and differentials in quality

As of 2013, the total number of dwellings in the country amounted to 60.6 million units, while that of households amounted to 52.5 million. The number of vacant dwellings reached 8.2 million, including 412,000 units of seasonal second houses, and its share in the total number of dwellings amounted to 13.5 percent. Among these vacant houses, the proportion of houses for sale or rent equalled 56.1 percent. During the period from 1998 to 2013, while the total number of houses increased by 10.4 million units, with a growth rate of 20.7 percent, the number of vacant houses increased by 2.4 million, and its rate of increase was 42.2 percent (MIC, 2013). Some inferences based on the underlying trends suggest that, unless housing construction unexpectedly decreases or vacant houses are extensively eliminated, the number of vacant houses would exceed 13.9 million units, and their share in the total stock would reach 21 percent by 2023 (NRI, 2015). The vacancy rate in local areas presently amounts to 14.8 percent, while that in metropolitan areas amounts to 12.2 percent. Many provincial cities and rural communities face the critical situation of future disappearance due to population flight.

Although the vacancy rates in the metropolitan cities are generally lower than those in local cities, the incremental increase in vacant houses has strong implications for the housing markets in the metropolis as well. For instance, in central Tokyo, the vacancy rate of non-wooden apartment houses is estimated at 14.5 percent on average (as of 2013 in Chiyoda, Chuo, Minato, Shinjuku, and Shibuya Wards). While the construction of towering condominiums is booming in the central area, vacant dwellings continue to increase in the existing apartment stock located in the surrounding areas. In particular, the Minato Ward has a vacancy rate as high as 15.6 percent, despite a booming bayside housing development. Faced with the advent of the era of housing overabundance, the Ministry of Land, Infrastructure, Transportation, and Tourism (MLIT) regards the solutions associated with the redistribution of vacant houses as an important policy issue.

Another area of concern in the housing stock is the disparity in quality between owner-occupied and rental houses. An international comparison of the average floor space of houses helps to put this problem into perspective. As shown by Figure 6.6, the average size of owner-occupied houses in Japan (estimated at 122.3 square meters as of 2013) approaches or exceeds those in Britain, France, and Germany. On the other hand, that of rental houses (estimated at 46 square meters) is considerably smaller than comparable figures in other countries. There is a long-term significant improvement in the size of owner-occupied houses, which evolved from a historical estimate of 91.3 square meters in 1963. In

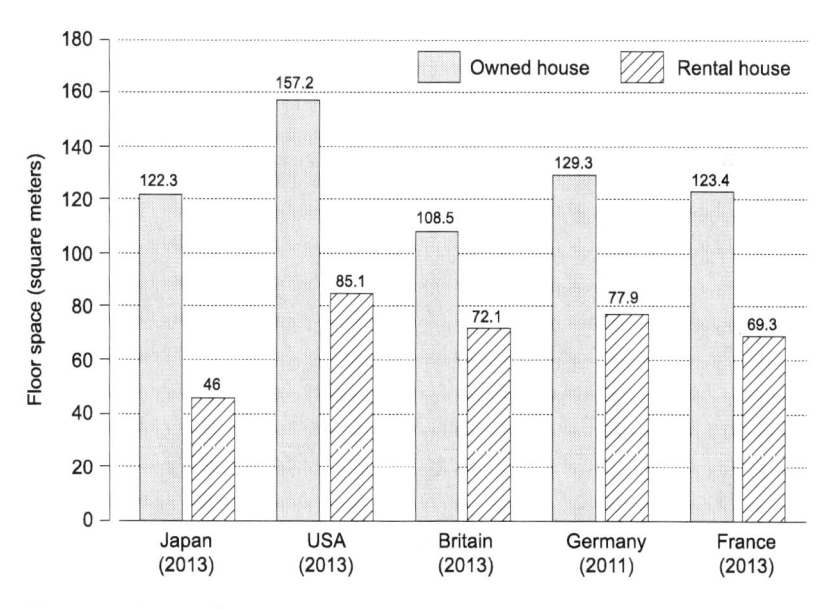

Figure 6.6 Average floor space per unit of owner-occupied and rental houses. Comparison of Japan with Britain, France, Germany, and the USA.
Source: MLIT, 2015a.

contrast, the size of rental houses remains almost unchanged, compared to its estimate of 38.8 square meters a half-century ago.

The problem of qualitative differentials is conspicuous for private rental houses in particular. In 1976, the former Ministry of Construction introduced minimum housing standards with the policy objective of improving the quality of houses. The housing standards were conceived as the basis and essential conditions for wholesome and decent residences, to be achieved for every household by 1985. However, this goal has not been attained, and few signs suggest, 30 years later, that it will be achieved in the near future.

While the specific figures of these housing standards have often been revised, the present norms determine, for instance, the minima of 25 square meters of dwelling floor area for a one-person household, and 40 square meters for a family household composed of parents and two children aged under six years old. The number of households residing in below-standard houses amounts to 3.7 million, or 7.1 percent of the total households. Among the households residing in private rental houses, the below-standard proportion reaches 20.6 percent. In particular, 29.1 percent (535,000 households) of the total households resides in such substandard private rental houses in Tokyo metropolis, and 30.8 percent (152,000 households) in Osaka City.

In fact, the problem of disparities in the quality of housing stock lies principally in the private rental houses located in metropolitan cities. The MLIT is pursuing policies to promote exchanging houses among residents and thereby resolve the twin problems of overabundance and differentials in quality. To facilitate the exchange of housing, market mobility is therefore ranked as an important policy issue. However, the redistribution of the housing stock through market solutions is not advancing as well as expected. On the contrary, the housing market dynamics seem to be aggravating the existing problems of the housing stock.

3.2 Housing market: polarization and high volatility

Housing construction can be considered a barometer of housing dynamics. Figure 6.7 demonstrates the behavior of housing construction classified as owner-occupied and rental houses. After the collapse of the asset bubble, housing construction amounted to 1.5 million units per year by the end of the 1990s, compared to 1.7 million units per year at the height of bubble from 1987 to 1990. During the 2000s, housing construction recorded 1.2 million units per year by 2006, but in association with the onset of the U.S. financial crisis in 2008, it sharply reduced to 780,000 units in 2009, and then hovered around 800,000 to 900,000 units per year. A short-lived increase appeared in 2013, due to a "last-minute" surge in housing demand precipitated by the imminent increase in the rate of the consumption tax in 2014.

Widening disparities across regional economies and across household incomes now cause uneven distribution of housing demand among various cities. Under these conditions, housing markets have been bifurcating into growth segments in the metropolitan cities and stagnation segments in the local cities. The booming

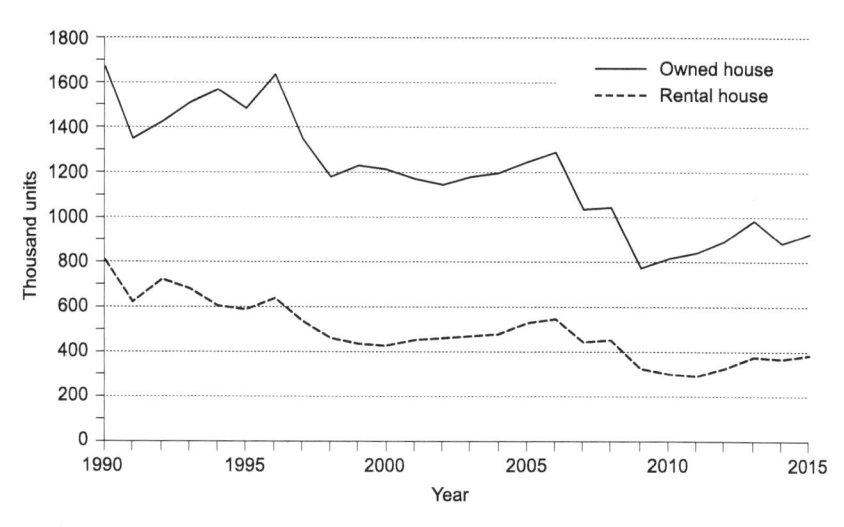

Figure 6.7 Construction of owner-occupied and rental houses in Japan, 1990–2015.
Note: Data (thousand unit) are shown stacked by sector. Rental houses include public rental and company housing.
Source: MLIT, Housing Starts Statistics.

construction of condominiums is representative of these particular developments. Since the beginning of the 2000s, the construction of this type of housing has been brisk in metropolitan areas, particularly in metropolitan Tokyo. Although the construction boom of high-rise condominiums in 2006 and 2007 was disrupted by the onset of the U.S. financial crisis, it is emerging again with anticipations of greater prosperity in association with the 2020 Tokyo Olympics. During the period from 2010 to 2014, the total number of condominiums under construction amounted to 576,000 units, with 56.7 percent concentrated in the metropolitan Tokyo area.

As already stated in the second section, the booming markets were the result of the government's urban redevelopment policy targeting the centers of metropolitan cities. The revised City Planning Law of 1997 introduced a new zoning district for high-rise dwellings, where the upper limit of the floor-area ratio of buildings was raised to 600 percent, and the offsite shadow control was relaxed. From 2002, URZs were designated at the central areas of metropolitan cities, where the maximum floor-area ratio was raised from 800 to 1,200 percent. Thus a series of deregulatory measures has boosted the construction of towering office buildings and condominiums in metropolitan cities. During the period from 2004 to 2014, the number of towering condominiums newly provided in Tokyo and adjacent prefectures (Kanagawa, Saitama, and Chiba) exceeded 525 blocks, involving 158,000 units. It was equivalent to a quarter of the total number of apartment houses newly provided in these districts over the same period.

118 *Eiji Oizumi*

However, these booming markets are also more volatile. Because the supply of land exhibits strong rigidity against changes in price and demand, land and housing markets are imperfectly competitive markets in essence and hence contribute to cumulative disequilibrium. Under the conditions of diminishing housing demand, only a few market segments show signs of growth in investment and financing. Though this process is conducive to reducing disequilibrium, it cannot persist. Because only isolated booming markets exist, there is limited room for further investment and construction, and the increase in demand is further repressed by soaring house prices due to rising costs of construction.

In fact, the sales of condominiums in the whole country were consecutively reduced in 2014 and 2015. The sales in 2015 amounted to 78,000 units, representing a drop of 6.1 percent from the previous year's sales. There is, in particular, a decrease of 11.1 percent in metropolitan Tokyo, as shown by Figure 6.8. Due to a surge of building materials and labor costs, the average sales price of condominiums in the whole country reached 46 million yen per unit, a rise of 7.2 percent. In metropolitan Tokyo, the average sales price reached 67 million yen, a rise of 12.3 percent. The market faces a depressed demand caused by aggravated conditions in housing affordability.

Figure 6.7 also indicates that the construction of rental houses, a large portion of which is represented by private houses to let, has maintained historical high levels. Nevertheless, as shown in Figure 6.6, the problem with the quality of private rental housing stock remains unsolved. Figure 6.9 shows that a large share

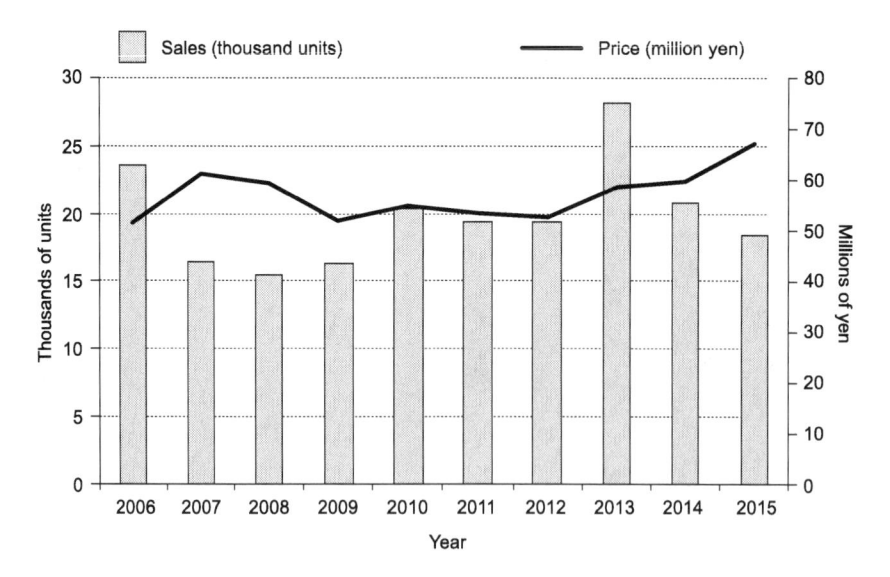

Figure 6.8 Changes in sales and average prices of condominiums in metropolitan Tokyo, 2006–2015.

Note: Data of 23 Wards of metropolitan Tokyo.

Source: REEI, 2016.

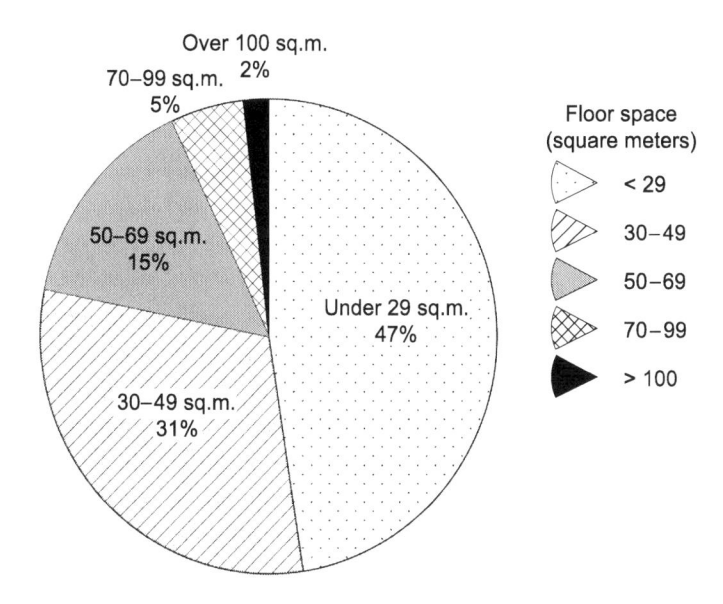

Figure 6.9 Component ratio of the volume of private rental apartment houses classified by floor space, in 23 wards of metropolitan Tokyo, 2013.

Note: Total number of the houses amounts 1.4 million units.

Source: MIC, 2013.

of 47.6 percent of the private rental apartment houses located in 23 wards of metropolitan Tokyo is occupied by cramped dwellings with fewer than 29 square meters. In fact, a larger part of the rental housing construction has been devoted to the speculative construction of small studio apartments with 20–30 square meters, which are associated with anticipation of high profitability.

The housing provision through volatile markets thus furthers the problems of overabundance as well as disparity in housing stock. This fact adds to the aggravated problem of affordability, which represents an obstacle to the appropriate distribution of houses. The focus hereafter falls on this important issue of housing affordability.

3.3 Housing affordability problems: homeowners and tenants

Figure 6.10 shows the income group composition of homeowning households. The proportion of homeowners with annual incomes of less than 4.9 million yen increased considerably from 32.8 percent in 1998 to 46.3 percent in 2013. In contrast, the proportion of households with annual incomes of 5–7 million yen decreased from 19 percent to 17.4 percent, and that of households with annual incomes of 7–10 million yen decreased from 20 percent to 13.2 percent. This middle-income class decreased not just on a relative basis but in absolute terms as

120 Eiji Oizumi

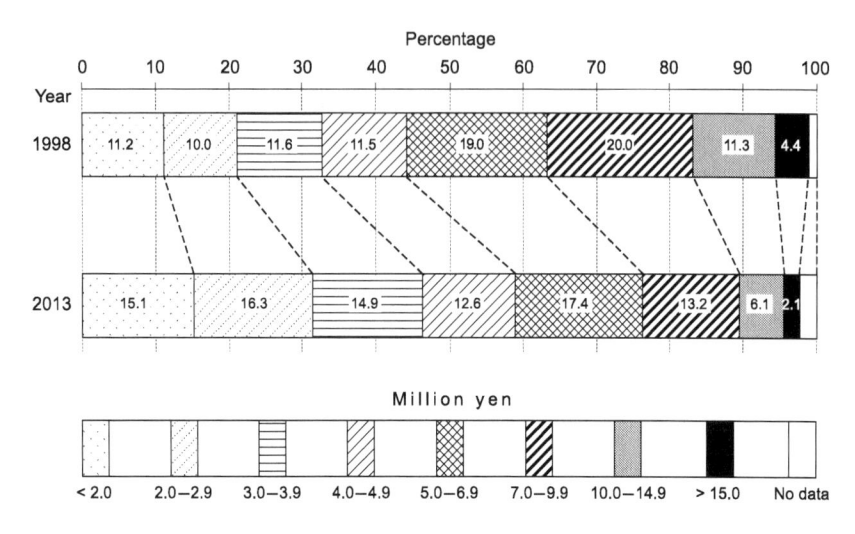

Figure 6.10 Component ratio of each income group in homeowning households in Japan. Comparison between years of 1998 and 2013.

Note: The number of houseowning households amounted to 26.5 million in 1998, and 32.2 million in 2013.

Source: MIC, 2013 and various years.

Table 6.1 Average income and loan repayment of worker households borrowing housing loans in Japan. Comparison between years of 2000 and 2013.

Item description	Year 2000	Year 2013	Rate of change
Number of persons in household	3.77	3.67	-
Householder's age	46	45.9	-
Total income (yen per month)	653,760	601,951	−7.90%
Disposable income (yen per month)	545,334	485,331	−11.00%
Loan repayment (yen per month)	101,770	99,867	−1.90%
Loan-to-income ratio (%)	19	21	-

Note: Loan-to-income ratio = loan repayment/disposable income.
Source: MIC, Household Accounts Survey.

well. Hence, it is evident that income distribution among homeowners is increasingly skewed towards lower-income groups.

One of the reasons for this worsening skew towards lower-income homeowners may have been the increase in aged homeowners reliant on pension benefits. But there are other important reasons behind the increase in lower-income homebuyers, including the decrease in land and house prices and historically low interest rates on housing loans. However, the burdens of loan repayment are putting stronger pressure on household accounts despite the very accommodative ultra-low interest rate policies. Table 6.1 shows the average income and

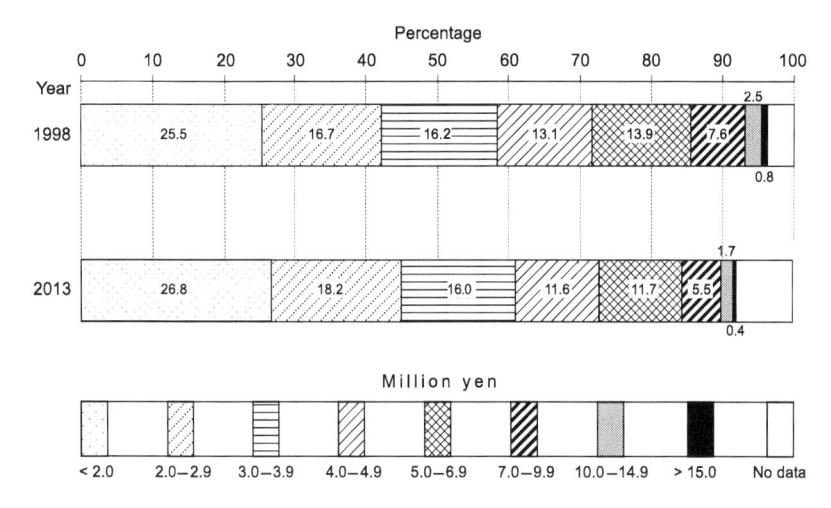

Figure 6.11 Component ratio of each income group in households residing in private rental houses in Japan. Comparison between years of 1998 and 2013.

Note: The number of households residing at private rental houses amounted to 12.1 million in 1998, and 14.6 million in 2013.

Source: MIC, 2013 and various years.

Table 6.2 Average income and rent payment of worker households residing in private rental houses in Japan. Comparison between years of 2000 and 2013.

Item description	Year 2000	Year 2013	Rate of change
Number of persons in household	1.47	1.28	-
Householder's age	38.7	40.6	-
Total income (yen per month)	459,826	369,979	−19.50%
Disposable income (yen per month)	397,256	305,674	−23.10%
Land and house rent (yen per month)	62,693	57,296	−8.60%
Rent-to-income ratio (%)	16	19	-

Note: Rent-to-income ratio = rent/disposable income.
Source: MIC, Household Accounts Survey.

loan repayment of worker households with housing loan borrowings. Comparison between data observations from 2000 and 2013 shows that a remarkable decrease in disposable income (a reduction rate of 11 percent) results in a heavier burden of loan repayment.

Next, Figure 6.11 shows the income group composition of households occupying private rental houses. The proportion of tenants with annual incomes of less than 3 million yen increased considerably from 42.2 percent in 1998 to 45 percent in 2013, with the aggregate number reaching 6.6 million. Table 6.2 shows the average income and rent payment of worker households occupying private rental houses. There is also a remarkable decrease in disposable income (indeed, at a faster diminishing rate of 23.1 percent), which resulted in a heavier burden of rent payment.

It must be noted that the data presented above represent average values. Under these overall conditions of worsening job security and declining household income, there is also a proliferation of marginal groups including younger, middle-aged and elderly singles with lower incomes, single-parent households and homeless people.

The National Consumption-state Survey by the government (MIC, various years) reveals the increasing rent burden of young singles. Judging from the expenditure of working singles less than 30 years old, the proportion of rent payments in the monthly average consumption expenditures considerably increased between 1999 and 2009. The rent ratio for a single male increased from 14.6 percent to 21.6 percent, while that for a single female increased from 24.2 percent to 31.1 percent. Such wide differentials in rent expenses between genders may be explained by two principal reasons. First, the proportion of non-regular and low-wage employees is much higher among the population of female employees than that of male employees. Second, there is a stronger tendency for single females to prefer better-located, well-equipped and hence more expensive houses, as security factors remain an important determinant of their housing preferences.

The proliferation of these marginal groups of low-income households provides an economic rationale for the discussions about the "housing poor." Households with lower income levels have to unwillingly accept residences of very poor quality and insecure tenancy. This state of affairs is symptomatic of the serious socio-economic implications of the present housing problems (Inaba, 2009).

At the bottom of the housing markets, there are indeed lucrative opportunities that can benefit some businesses at the expense of the "housing poor." The provision of "illegal rental rooms" is achieved through the partitioning of house floors or buildings into many small rooms that are more akin to "rabbit holes" than habitats for human beings. In fact, those rooms are divided by flammable-panel walls without windows. It is obvious that such floor partitions are made in violation of the Building Standards Law. The MLIT investigated in December 2014 the prevailing floor conditions for the first time and found 1,148 illegal cases in the whole country. Metropolitan Tokyo presented the highest number with 944 cases, accounting for more than 82 percent.

4. Housing problems and housing policy

4.1 Restructuring of the government's housing policy

In post-war Japan, public policy to promote housing provision was served mainly by the three institutions established in the 1950s: the Government Housing Loan Corporation (GHLC), financing the middle-class purchase of homes; the municipality-managed Public Housing System, providing the lower-income groups with rental houses; and the government-sponsored Japan Housing

Corporation (JHC), providing residents of the metropolitan cities with houses for rent and sale. Those institutions played important roles in the mass housing system whereby the acute housing shortage caused by rapid urbanization was to be solved. However, from the mid-1990s on, the government engaged in drastic restructuring of those institutions.

With the revision of the Public Housing Law in 1996, the upper income limit for tenants of public housing was lowered, while the government simultaneously curtailed its subsidies for public housing construction. The former JHC, which was reorganized as the Urban Development Corporation, began to withdraw from housing provision; it was eventually closed in 2004 and replaced by the Urban Renaissance Agency (UR) to provide support for private urban redevelopment business. The GHLC was abolished in 2007 and replaced by the Housing Finance Agency (HFA), dealing exclusively with the securitization of bank housing loans. This series of agency restructurings was in accordance with the principle that housing policy was to be limited to supplement the operation of the market. This reform essentially followed a neo-liberal approach that supported both privatization and deregulation in the housing system. To complete these reforms, the Housing Construction Plan Act of 1966, which provided the legal basis for the post-war housing policy, was abolished and replaced by the Basic Law of Housing Life of 2006.

Presently, the government determines and regularly revises the Housing Life Master Plan that provides the guidelines for each municipality's housing policy. The recent Master Plan sets the policy targets as follows: (1) creating a living environment that supports safe, secure, and affluent residential living; (2) providing proper management and renewal of existing apartment buildings; (3) providing the conditions for a housing market in which the circulation of existing houses grows; and (4) ensuring housing stability for those who require special support for their housing (MLIT, 2015b).

The government's housing policy is challenged by the emerging scale of an ageing society with falling birthrates, declining population, and shrinking household sizes, together with the socio-economic climate of deteriorated employment and decreasing household income. In this new context, the third and fourth targets of the Master Plan are very significant.

Improving the fluidity of the market for existing houses. Within the whole housing market in Japan, the transactions of existing houses represent only 14.7 percent of total housing transactions, which is very low compared to Britain and the USA, where trading of existing housing units is on the order of 88–89 percent. This very low level of housing trading mobility means that, in spite of there being actual housing stock available, this stock is not used for upgrading housing conditions by moving house. Therefore, the MLIT emphasizes the importance of increasing the trading of existing houses through the market.

Enhancement of the housing safety net for low-income earners. The provision of public housing is an important element of the safety net for low-income earners including the unemployed, the elderly, the disabled, and single-parent families.

Table 6.3 Housing stock and component ratio of housing tenures in Japan, 1958–2013.

Year	Total stock (million units)	Self-owned house (%)	Private rental house (%)	Company house (%)	Public rental house (%)	Vacancy ratio (%)
1958	17.9	71.2	18.5	6.7	3.5	2
1963	21.1	64.3	24.1	7	4.6	2.5
1968	25.6	60.3	27	6.9	5.8	4
1973	31.1	59.2	27.5	6.4	6.9	5.5
1978	35.5	60.4	26.1	5.7	7.6	7.6
1983	38.6	62.4	24.5	5.2	7.6	8.6
1988	42	61.3	25.8	4.1	7.5	9.4
1993	45.8	59.8	26.4	5	7.1	9.8
1998	50.2	60.3	27.4	3.9	6.7	11.5
2003	53.9	61.2	26.8	3.2	6.7	12.2
2008	57.6	61.1	28.7	2.8	6.1	13.1
2013	60.6	61.7	28	2.2	5.4	13.5

Note: Public rental houses include JHC/UR housing.
Source: MIC, 2013 and various years.

Nevertheless, under the subsidy reduction policy adopted by the government, many municipalities have reduced the construction and provision of public housing. In line with the "marketization" of housing policy, the measures of government and municipalities are shifting to using private rental housing for their safety net.

4.2 Housing policy and private rental housing market

The changing composition of housing tenures shown in Table 6.3 reflects some features of the Japanese housing policy. While the housing policy has strongly promoted the spread of owner-occupied dwellings, the provision of public rental housing has remained at a low level, and public intervention in the private rental housing market has not been executed. Instead, a large part of the assistance for residence in rental houses remains the corporate housing welfare, providing employees with low-cost rental housing (company housing) or housing allowances. However, the share of corporate rental housing has declined, and the beneficiaries of this system are limited to the regular employees of companies, with no access for part-time employees.

Given the scarcity of public rental housing, many people have to rely upon private rental housing, in the absence of a general scheme of housing allowances. Thus, low-income tenants are faced with a structural gap between supply and demand in the market: the tenants' limited rent-payment capacities cannot meet the supply prices that cover costs for appropriate maintenance and/or replacement of rental dwellings. This dilemma naturally follows the serious affordability problem for many tenants and the provision of rental housing of poor quality, as mentioned in the third section.

Under the present conditions where both the provision of public housing is shrinking and the private companies are withdrawing from housing welfare, the structural imbalances in the private rental housing market and the housing problems for low-income people are increasingly aggravated. A survey recently conducted by the Big Issue Japan Foundation, an advocacy NGO to save the homeless, reveals the housing distress of young people. Among the 1,767 20–39-year-old singles surveyed, with annual incomes less than 2 million yen and who live in metropolitan Tokyo and Osaka, 57.4 percent suffered from a rent-to-income ratio of over 30 percent. Further, 30.1 percent of these singles bear rent burden ratios of more than 50 percent. Among those singles, those with annual incomes of less than 2 million yen meet the official standards of the Livelihood Protection System. Nevertheless, they are left without any public assistance.

The housing situation of the elderly singles is also serious. As stated in the first section, metropolitan Tokyo has the highest proportion of elderly singles among all prefectures, amounting to 38 percent in 2010. For instance, in Minato Ward of central Tokyo, where the construction of high-rise buildings and condominiums is booming and the population is increasing, there were 5,656 elderly single residents in 2011. Almost 16 percent (15.9 percent) of them reside in private rental housing, while 20.8 percent reside in public rental housing. Their income was generally low: 48.6 percent of them had annual incomes of less than 2 million yen. The housing problem of elderly singles includes not only housing affordability but also the risk of social isolation (Kawai, 2015).

The housing market tends to exclude minorities such as the elderly, the handicapped, foreigners, and single parents with low incomes. There are actually many cases where those households are refused by private landlords who are prejudiced against the elderly and attempt to evade troubles with tenants. To overcome this problem of social exclusion, the government attempts to facilitate their access to a home through the Housing Assistance Council, composed of administration, real estate agents, landlords, and citizen organizations, providing housing assistance such as information and consultation services. While there were 61 councils in all prefectures and 14 municipalities as of April 2016, the government aims to establish them in more cities.

Although such assistance is a welcome development, it is obviously insufficient to ensure housing stability for low-income earners in serious need of decent residential living conditions. Given the rapid expansion of low-income people, the provision of good-quality and low-cost rental housing should be the core issue of Japan's housing policy. As stated in the third section, the dominant supply of low-quality rental housing was a result of the lack of housing policy for this sector. Desirable development of private rental housing markets will not be attainable unless production subsidies and housing allowances are extensively provided to landlords and tenants.

4.3 Proposal for housing policy: housing allowance and social housing provision

The rationale of the government's policy is to turn the vacant stock of good-quality privately owned houses into rental housing. However, affordability problems for low-income tenants are a crucial bottleneck that prevents smooth transfer. A gap between demand and supply exists in the private rental housing market. The measures to solve the gap are a combination of housing allowances and social housing provision. To provide low-income people with decent homes, vacant houses meeting standards should be kept in priority as "quasi-public housing."

In the context of a significant increase in low-income households and single households, the need for decent houses with low rent are growing. However, there are striking disparities in the quality of the stock of owner-occupied houses and rental houses, besides the increase in risk-vulnerable tenants and the "housing poor." Therefore, housing security, which leads to human security, is a keystone in rental housing policy.

However, the principle of a secure, decent residence for all is irreconcilable with the housing policy of neo-liberalism. The "safety net" allowable in the neo-liberal approach is a minimal relief measure for the economically disadvantaged, who have difficulties gaining access to the market. In contrast, residential security is a general principle for all people, without distinction between homeowners and tenants. Although the secure residence policy might be irreconcilable with pure neo-liberalism, it is not completely irreconcilable with the operation of a housing market. We thus must consider the relationship between housing security and the housing market.

The present housing market is facing a new structure of housing demand. First, the aggregate amount of housing demand is decreasing because of the ageing population and the drop in household income as a result of deteriorating employment conditions. Second, the disparities in the composition of housing demand from different households are widening in two ways. On one hand, the regional disparities in housing demand across cities and rural areas are widening. On the other hand, the disparities in households' housing demands are widening because of increasing income disparities. These structural changes result in the polarization of housing markets in both metropolitan cities and local cities.

The structure of housing demand is a precondition for the market, and the market is not capable of altering the structure of demand itself. However, the housing system and the policy of residence security combined can change the structure of housing demand through the provision of subsidies to rectify the disparities in household income. Therefore, the provision of secure housing can create a favorable condition to stabilize the growth of housing markets.

Rental housing policy can be designed to provide both rent assistance and social housing. First, the government should establish a comprehensive and permanent system of rent subsidy for low-income tenants. In fact, rent subsidy schemes for private rental housing were conducted by 75 municipalities as of 2009. The tenants who qualified for assistance varied depending on the purpose of each scheme: to help the elderly, young people with low incomes, the handicapped, single-parent households, newly married couples, and couples raising young children. However, the municipalities conducting this scheme are very few; in addition, the duration of benefits is limited to varying degrees. Supported by sufficient government grants, such schemes should be introduced by all municipalities and should continue as long as tenants are eligible for the benefit.

Second, the improvement in the quality of the rental housing stock should be promoted through the increasing provision of social housing. The municipalities should actively participate in the construction of public housing, leasing of private rental houses, and introduction of a social housing system modeled on those of Germany and the Netherlands. The provision of both a rent subsidy and social housing can enable decent and affordable housing in the rental housing market. Such a coherent public policy approach to the rental housing market can be conducive to stable growth of the owner-occupied housing market, and hence sustainable growth of the whole housing market. (For wider policy perspectives, see Japan Housing Council, 2016; Hirayama, 2017).

5. Conclusion: lessons from the Japanese housing experience

Growing divides in housing markets constitute the core of the current housing problems around the world. This chapter reveals the distinctive dynamics and relations involved in housing divides in Japan's case. The final section summarizes several lessons and policy agendas derived from the Japanese housing experiences.

Polarized land and housing markets. The collapse of the asset bubble after 1991 has resulted in the long-term stagnation of land and housing markets. However, since the early 2000s, increases in property and housing investment have gradually resumed in metropolitan Tokyo. Since then, the polarization of land and housing markets has increased, with growth in a few market segments and stagnation in the others. In this context, land policy has been also become bifurcated within both metropolitan cities and local cities.

There is a boom in the construction of condominium towers and office buildings in the centers of metropolitan cities. This expansion is strongly promoted by deregulation in land-use policy. In contrast, there is the ongoing "reverse sprawl," characterized by disorderly increases in vacant houses and sites, in both urban areas and suburban areas of local cities facing increasingly deeper depopulation. Faced with this spatial problem, many local municipalities are eager to introduce a "compact city" policy. This is a significant challenge in the pursuit of sustainable communities.

Housing market dynamics and housing problems. Housing stock has increasing vacancies and widening disparities in the quality of owner-occupied and rental houses. These problems are closely associated with the housing market dynamics in metropolitan cities. Volatile housing construction aggravates further increases in vacant houses, and speculative construction of small apartments causes the incremental differentials in rental housing stock. In fact, disparities in the quality of housing stock center on the private rental stock located in metropolitan cities.

Given that the income distribution among homeowners and tenants reflects the increasing number of lower-income households, the present housing market dynamics worsen housing affordability conditions. Despite a few declines in loan repayment and rent costs, a significant decrease in disposable income implies heavier financial burdens on both holders of mortgage loans and tenants. In particular, marginal groups with lower incomes suffer from excessive housing expenses and deteriorating housing quality.

Proposal for alternative housing policy. In the context of rising vacancies, there is a need to develop the housing markets for a better circulation of existing stock. The provision of affordable housing for low-income earners is a necessary remedy to the problems of the "housing poor." The government aims to use the vacant stock to provide good-quality homes by making the privately owned vacant houses available for rent. However, the affordability problem for low-income tenants constitutes a serious bottleneck that prevents smooth movement into decent housing.

The appropriate measures to address the affordability problem for low-income tenants include a combination of rent assistance and social housing provision. A comprehensive and permanent scheme of rent allowances is crucially needed for low-income tenants. To provide low-income tenants with decent homes, the vacant houses that satisfy the minimum housing requirements and standards should be reserved and prioritized as "quasi-public housing." A coherent public policy approach to the rental housing market can lead to stable growth of the whole housing market.

Note

[*] The author would like to thank the editors and Professor Nabil Maghrebi (Wakayama University, Japan) for their helpful comments and advice on earlier drafts.

References

Alvaredo, F., Atkinson, T., Piketty, T. Saez, E., and Zucman, G. The World Wealth and Income Database. www.wid.world. Accessed July 26, 2016. (This website is now succeeded by the World Inequality Database.)

Big Issue Japan Foundation (2014). Housing Problems for the Young (in Japanese). http://bigissue.or.jp

Government Cabinet Office, Japan's National Account, Various Years. http://esri.cao.go.jp

Hirayama, Y. (2017). Individualisation and Familisation in Japan's Home-owning Democracy. *International Journal of Housing Policy*, 17(2), pp. 296–313.

Inaba, T. (2009). *The Housing Poor: Understanding Housing Poverty*. Tokyo: Yamabuki Shoten (in Japanese).

IPSS (National Institute of Population and Social Security Research) (2012). Population Projection for Japan: 2011–2060. http://ipss.go.jp

Japan Housing Council. (2016). *Citizens' White Paper on Housing 2014–2016*. Tokyo: Domesu Shuppan (in Japanese).

Kawai, K. (2015). *The Elderly in Poverty and Social Isolation*. Tokyo: Kobunsha Shinsho (in Japanese).

MHLW (Ministry of Health, Labor and Welfare). Comprehensive Survey of Living Conditions, Various Years. http://mhlw.go.jp

MIC (Ministry of Internal Affairs and Communications). Household Accounts Survey, Various Years. http://soumu.go.jp

——— (2013). Statistical Survey on Housing and Land 2013, and Various Years. http://soumu.go.jp

MLIT (Ministry of Land, Infrastructure, Transport and Tourism). Housing Starts Statistics; Official Listing of Land Prices, Various Years. http://milt.go.jp

——— (2015a). Collective Data on Housing Economy 2015. http://mlit.go.jp

——— (2015b.) White Paper on Land, Infrastructure, Transport and Tourism in Japan 2015. http://mlit.go.jp

NRI (Nomura Research Institute). (2015). Projection for Japan's total number of houses, vacant houses and vacancy rate in 2018, 2023, 2028 and 2033. http://nri.com

Oizumi, E. (1994). Property Finance in Japan: Expansion and Collapse of the Bubble Economy. *Environment and Planning A*, 26(2), pp. 199–213.

———. (2002). Housing Provision and Marketization in 1980s and 1990s Japan. In G. Dymski and Isenberg, D. (Eds.). *Seeking Shelter on the Pacific Rim*. New York and London: M. E. Sharpe, pp. 169–186.

———. (2007). Transformations in Housing Construction and Finance. In Y. Hirayama and R. Ronald (Eds.). *Housing and Social Transition in Japan*. London and New York: Routledge, pp. 47–72.

———. (2013). *Housing Market at Instability and Polarization*. Tokyo: Hakuto Shobo (in Japanese).

REEI (Real Estate Economic Institute). (2016). Trends of Condominium Market 2015 (in Japanese). www.fudousankeizai.co.jp

Suzuki, H. (2007). *The Japanese Version of Compact City*. Tokyo: Gakuyo Shobo (in Japanese).

Tokyo Metropolitan Government. (1999, 2014). *Land of Tokyo 1999, 2014*. Tokyo: Bureau of Urban Development (in Japanese).

7 The role of government and housing market dynamics in Korea

Seong-Kyu Ha

CHUNG-ANG UNIVERSITY AND KOREA RESEARCH INSTITUTE
OF HOUSING MANAGEMENT

1. Introduction

Housing is a central component of daily life, and it is intimately inter-related with its socio-economic, cultural, and political environment. The housing market and housing policy in South Korea (hereafter Korea) are the inevitable products of the economic development, political system, and housing norms and preferences of the country. Urban housing is perhaps as diverse as the people living within it, and the dynamics of the housing market directly reflect socio-economic status and cultural diversity.

In the past several decades, Korea has made tremendous progress in meeting the housing needs of the poor, while some cities still face great challenges of severe housing shortages and substandard housing. The Korean housing market is tempered by pervasive government intervention; the Korean government has been controlling the whole process governing housing supply (see Cho et al., Chapter 8 in this volume). One major instrument has been the price control on new housing. Meanwhile, the significant improvement of housing conditions can also be largely attributed to the active role of governments in the housing sector.

Despite the expansion of housing supply and government intervention in the housing market, housing remains a social and economic issue that has not been resolved in Korea. Why has this paradox arisen? How have governments played their role in the housing market? And more specifically, do governments achieve their policy goals through housing market intervention?

Understanding the structure of housing markets is crucial to both the government and private agencies. Housing markets, particularly housing submarkets, are usually categorized according to structure and spatial characteristics, as well as the socio-economic and cultural characteristics of their inhabitants. This paper addresses the role of governments in the housing sector and the effects of governmental intervention on the housing market in Korea during the last 40 years.

2. The housing situation and housing market dynamics

Korea's overall housing conditions have improved substantially since the beginning of the 1980s, as can be seen from the key indicators summarized in Table 7.1. Overcrowding, as well as the quality of dwellings and facilities, has improved

Table 7.1 Housing conditions in Korea.

	1980	1990	2000	2015
Housing supply ratio (%)	71.2	72.4	94.1	102.3
Dwellings per 1,000 inhabitants	142.1	169.5	248.7	383.0
Per capita floor space (m^2)	10.1	13.8	20.2	33.5
House price-to-income ratio (Seoul)	N/A	N/A (9.2)	5.0 (7.9)	6.1 (11.0)[1]

Note: [1] Korean real estate, how expensive? (March 1, 2017), www.hani.co.kr/arti/economy/property/784531.html. Accessed July 24, 2018.

Source: Korean Statistical Information Service, Housing Census, various years. http://kosis.kr/eng/search/search01_List.jsp. Accessed May 3, 2016.

Table 7.2 Dwelling stock by housing type in Korea (unit: no. of thousand dwelling units (%)).

	1995	2005	2015
Detached house	4,490 (46.9)	4,264 (32.2)	3,974 (24.3)
Apartment	3,610 (37.7)	6,963 (52.7)	9,806 (59.9)
Terraced house	1,122 (11.7)	1,788 (13.5)	2,383 (14.6)
Etc[1]	348 (3.7)	209 (1.6)	204 (1.2)
Total	9,570 (100.0)	13,224 (100.0)	16,367 (100.0)

Note: [1] Dwelling units in building not intended for human habitation and living quarters other than housing units.

Source: Korean Statistical Information Service, Housing Census, various years. http://kosis.kr/eng/search/search01_List.jsp. Accessed May 3, 2016.

remarkably. With respect to housing types (structural types), traditional single-story family dwellings have been quickly replaced by high-rise apartments (Table 7.2). The substantial increase in apartments is remarkable; this type of housing development represented only 38% of the market in 1995, but its share among all housing types had increased to 60% by 2015, signaling a significant change in housing norms in Korean society.

Korea has a unique and dynamic housing tenure system consisting of owner-occupancy, *Chonsei* (or *Jeonse*), *Bojeongbuwolse*, and others (Ha, 2002) (Table 7.3). Homeownership is a key cultural goal in Korea, widely considered to be an innate desire for all members of the population. However, the proportion of owner-occupancy households in the country as a whole has declined from 63.6% in 1975 to 53.6% in 2015 (Table 7.3).

Chonsei (or *Jeonse*) is a rental system in which the tenant pays a lump sum to the landlord and receives the same sum back when he or she leaves the rental unit. The landlord will usually invest the lump sum, and the interest earned represents the imputed rent. Landlords benefit during prosperous times by investing the deposits, generating good returns. Renters also benefit by not

Table 7.3 Housing tenure in Korea, 1975–2015, (unit: %).

	1975	1995		2005	2010		2015	
Owner-occupied	63.6	53.3		55.6	54.3		53.6	
Chonsei (Jeonse)	17.5	29.7		22.4	21.7		19.6	
Bojeongbuwolse (monthly rent with deposit)	-	10.3	14.5	-	18.2	21.3	21.8	23.9
Wolse (monthly rent without deposit) and Sakwolse	15.6[a]	4.2		19.0[b]	3.1		2.1	
Other (no rent)	3.3	2.5		3.0	2.7		2.9	
Total	100.0	100.0		100.0	100.0		100.0	

Note: [a] and [b] include *Bojeongbuwolse*.
Source: Korean Statistical Information Service, Housing Census, various years. http://kosis.kr/eng/search/search01 List.jsp. Accessed May 3, 2016.

having to make monthly payments for the duration of the contract. The *Chonsei* system was a "win-win method" for both those seeking to buy a house in the future, and landlords. However, the *Chonsei* system no longer seems to be win-win. Homeowners are not able to gain much financial leverage from *Chonsei* leases as they are no longer assured of significant house price increases, as in the old days during the real estate boom. Banks' deposit interest rates, which once reached double-digit highs in the 1980s amid the economic expansion, hover in the low single digits, with a commercial banks' interest rate being below 2%.

Bojeongbuwolse is a kind of rental involving a security deposit and monthly rent. At the beginning of the contract period, the tenant provides a refundable deposit and pays monthly rent. In case of non-payment, the rent is deducted from the deposit.

Sakwolse is a type of declining *Chonsei*. The tenant makes a one-time deposit from which a given amount is deducted. Hence, the total amount of rent is the deducted amount plus interest earned on the remaining deposit. *Wolse* is the monthly rental system found in most countries and is the most common system used in Korea.

Ilse is a new type of rental housing or accommodation system for the lowest-income groups and the homeless. It is a daily rental system in which people can rent a room, bed, or bed hours in a boarding or rooming house or an illegal inn. With respect to the income levels of different tenure types, homeowners generally earn higher incomes than the other tenure groups. The lowest-income households usually have the *Sakwolse* or *Ilse* arrangements (MLTMA, 2012).

Understanding the characteristics and dynamics of the housing market in Korea requires looking at the nature and types of housing submarkets. A number of previous studies have established the idea that the housing market can be better analyzed as inter-related submarkets rather than a single homogeneous market (Wu and Sharma, 2012; Watkins, 2001; Bourassa, Hoseli, and Peng, 2003; Palm, 1978). The types, classification, and dynamics of housing submarkets are based on a combination and application of spatial boundaries, housing characteristics, demographic factors, and socio-economic dimensions (Figure 7.1).

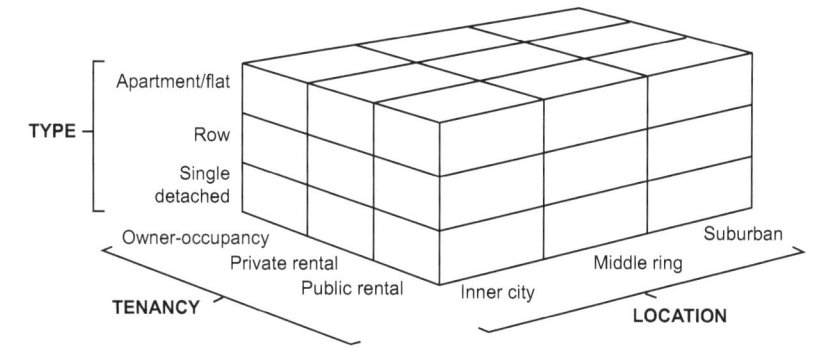

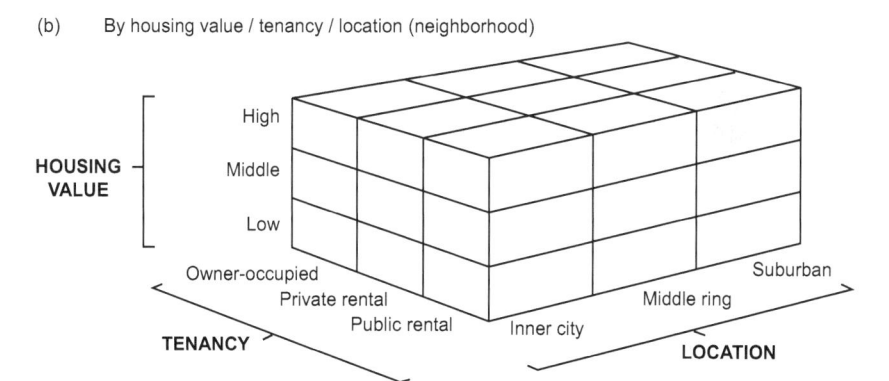

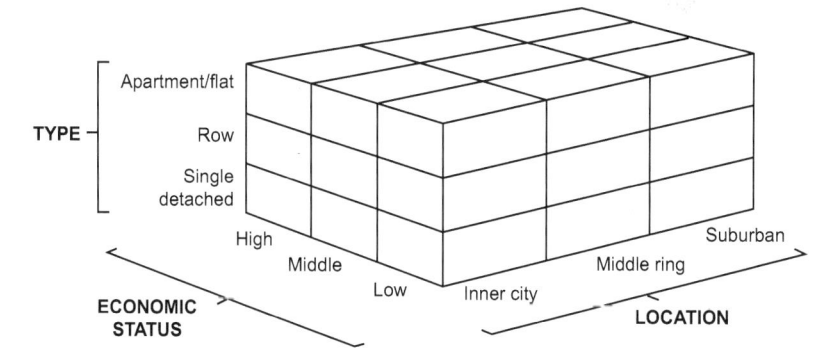

(a) By housing type (stock) / tenancy / location (neighborhood)

(b) By housing value / tenancy / location (neighborhood)

(c) By housing type (stock) / economic status / location (neighborhood)

Figure 7.1 Characteristics and dynamics of urban housing markets.

It is worth looking more closely at the types and dynamics of housing submarkets in urban Korea. Figures 7.1a, b, and c show various combinations and segments of submarket. Figure 7.1 indicates the combination of housing types, tenancy, and location. From a theoretical point of view, there are 27 submarkets being formed (3 raised to the power of 3 is 27). Among 27 submarkets, the most active submarket was the combination of apartment, owner-occupancy and inner city. In the case of Seoul, the traditional owner-occupied single detached houses in inner-city areas are being redeveloped through urban renewal or regeneration projects. The old residential areas have changed into high-rise apartment complexes.

Though an analysis of combinations in housing submarkets, we can recognize the following characteristics and dynamics of Korean housing submarkets: first, location and tenancy are the most crucial factors in urban areas in light of the assessment and significance of the housing submarket. Urban Korea is geographically subdivided into "spatial submarkets" and operates a series of linked submarkets in which location and tenure vary. Second, the apartment submarket plays a pivotal role in the housing sector. In the past, apartments have been located mainly in downtown areas. However, in recent years, apartments have been widely constructed in rural areas. Third, housing submarkets may reflect issues of social tenure – polarization and locational disadvantages. In the Korean context, it is necessary to consider an escalation of residential segregation and social exclusion. Fourth, after the Global Financial Crisis of 2008, the Korean government relaxed most restrictions on the housing market to prevent rapid declines in housing prices. That action led to a slight rebound in housing prices in certain submarkets.

3. The role of government and intervention in housing markets

An analysis of housing markets in Korea inevitably involves a consideration of the role of government in housing provision and consumption. Governments have policies that influence markets. Korea's housing market analysis needs to refer to the government's role in housing supply and consumption. Governments have policies and programs that influence markets. Throughout Korea, for over 40 years, governments have concluded that housing is not something that can be supplied purely according to free-market criteria. Intervention in housing markets has been the norm.

There are two principal mechanisms for housing production and allocation (Spence, 1993; Harvey, 1977; Downs; 1975; Bourne, 1981). One is the traditional private market, which produces and allocates housing to households on a competitive basis in light of a consumer's ability to pay. Second is public sector production and allocation, in which the government distributes according to individual and collective needs and the objectives of the agency involved. From a theoretical point of view, the private market emphasizes efficiency, maximizing output, and profits. For the public sector, the main objective is greater equity or social welfare according to housing needs. In the Korean context, it is possible to rationalize governments' reasons for not leaving housing provision to market forces in terms of a desire to correct for market failure, to redistribute resources on equity grounds, to achieve macro-economic objectives, or to compensate for disequilibrium in the market.

Korean governments have played a large and increasing role in almost all aspects of housing production and consumption. They act as financiers, regulators, administrators, builders, landlords, and frequently destroyers. Governmental intervention that aims to stabilize prices and to control for speculative demands in the housing submarket has been a popular approach. UN Habitat (2008) has argued this point:

> *The Government controls all the major elements of the housing sector which include housing prices, land supply, size distribution of housing units, housing loans with subsidized interest rates; and even the customers of private developers are controlled by the Government. The private sector can only sell housing units to those selected according to the Government's rules. The degree of intervention in South Korea may be higher than in China. The South Korea case demonstrates that attribution of South Korea's success in the housing sector to the free market model is misleading, rather the state can play a more constructive role in meeting the housing demands at the earlier stages of economic development and in conventionally state-dominated countries.*

The basic aim of the government's housing policy is to ensure that, as far as the resources of the economy permit, every family can obtain a dwelling of a good standard, located in an acceptable environment, and at a price or rent they can afford (Housing Act, 2016).[1]

Governmental efforts had concentrated on the expansion of housing production, which was mainly materialized in the Two Million Houses Construction Project (1988–1992). Massive housing construction has played a part in stabilizing home prices and rents.

3.1 Price control on the new apartment submarket

The major instruments have been price control on new apartments, regulations on their size distribution, and rules for selecting purchases of new dwelling units. The ceiling price system was first applied in 1963, to apartment units constructed by the public sector of 85 square meters or less, and was then extended in 1977 to private-sector apartments.[2]

The purpose of the price ceiling system was to meet the demand of prospective homeowners and to control skyrocketing housing prices. The extension of the ceiling price system increased effective demand, resulting in the explosive housing boom of 1978. The price control applies to any new housing project of 20 units or more. The price ceiling was indexed to the cost of production in 1989 to stimulate housing supply. Although the price control applies to all types of housing, in practice it affects only apartments. However, as these regulations cause side effects, the price ceiling has been gradually deregulated.

The price control on the new apartment submarket has negative impacts and distorting consequences. First, a price ceiling discourages the supply of the good it is imposed upon. Since the controlled price is determined regardless of the quality of the floor space supplied, a profit-maximizing developer has an incentive to maximize the total floor space of apartments, and especially commercial premises. The developer tries to fulfill only the minimum safety and other quality standards set by the government, such as the provision of open space.

Second, a crucial housing problem is the increasing polarization between the housing conditions of the lower and upper economic classes, which is the most important consequence of the price control. For several reasons, the main beneficiaries of the price control have been middle- and higher-income households. The difference between the market price and the controlled price of an apartment increases with its size. But a larger sum of money is needed to reap the profit from purchasing a larger apartment, and the amount of deposit required to be eligible to bid for a larger unit is also larger. In the past, the rights to purchase apartments were transferable, so that some moderate-income households were able to cash in a part of the premium by selling off their privileges. However, this practice was made illegal because it was considered an act of speculation (Ha, 2018). Thus, the system favors the relatively well-to-do. The Korean government's intervention in housing submarkets focused on promoting homeownership by middle-income groups. One may wonder how such an inefficient and inequitable system has been upheld for so long. Kim and Kim (2000) argue that more costs than benefits accrue:

> *The system generates its own demand because lots of people are waiting for their turn to enjoy the privileges of buying apartments at discounted prices. Discontinuing the system is a politically dangerous move. Although it can be argued that the current system is inequitable because it favors the middle- and high-income groups more than lower income households, the same fact provides a good excuse for maintaining the system because it helps build a political constituency.*

Therefore, it seems that the Korean government intervention has roiled the housing markets. When the Global Financial Crisis broke out in 2008, housing prices collapsed, and the government wanted to boost the housing sector in an attempt to stimulate economic recovery. Many regulations that had been taken for granted for many years were suddenly removed or amended. In April 2015, the abolition of the price control on new apartments was an example. But the system of price control is still implemented in public-sector housing for sale. A major reason for the delay in lifting the price control on new apartments was the pervasive belief that deregulation would raise the overall prices of housing, at least in the short term.

The housing market has evolved from a seller's market to a buyer's market. Newly built, modern apartment houses were once tremendously popular. But they increasingly remained unsold, and the number of unsold dwellings began to increase after the Global Financial Crisis. The market for owner-occupied housing in Korea was driven by the expected capital gains, but the fall in housing prices since the financial crisis began has changed the perception of many homebuyers. Demand for housing has been shrinking mainly because the pessimistic expectations about housing prices in the near future suppressed the demand for ownership and increased demand for rental housing. Even though interest rates dropped significantly to encourage individuals to purchase houses with lower borrowing costs, rental housing was preferred over purchasing a house, since the modest increase in housing price was not likely to be large enough to reap any capital gains (*The Korea Herald*, 2016). Various governmental regulations on the supply side of the market, as well as direct controls on speculative transactions, have made the housing supply response to changing submarket demand conditions very inelastic. Such an inelastic supply may be an important factor in explaining house prices in Korea.

3.2 Measures to boost the housing market

House prices are strongly steered by the government (Figure 7.2 and Figure 7.3). The peak of Korea's house price boom was reached in 2006, when Seoul prices rose almost 20%. Then the government applied the brakes, imposing controls on housing loans and hiking capital gains taxes on "speculative areas." These cooling measures caused a slight slowdown in 2007. From April to October 2009 there were house price declines, triggered by the Lehman shock[3] and government curbs. Property transactions (but not prices) fell 35.8% year on year to September 2010, in the midst of a cycle of overbuilding. The mini-slump caused the construction industry severe problems (*Korea JoongAng Daily*, 2008). The government began reviving the housing market in 2009, followed by more intensive expansion measures in August 2010, as the government partially eased real estate lending restrictions through the following measures:

- Restrictions on total debt payment ratios in non-speculative areas were abolished;
- Households with annual incomes of KRW 40 million (US$35,787) or lower could receive loans worth up to KRW 200 million (US$178,937) for house purchases;
- The grace period for extra tax on asset transfer income was extended; and
- Housing registration tax exemptions were also extended.

In December 2011 the government announced another new set of policy measures, including the following five measures: (1) abolition of punitive capital gains taxes on owners of more than one property; (2) securitization of up to KRW two trillion (US$1.79 billion) worth of debt owed by construction companies; (3) expansion of eligibility for cheap loans offered to first-time-buying married couples and low-income earners, and the reduction of lending rates by half a percentage point to 4.2%; (4) reduction of the levy to 6–35% on profits obtained

138 Seong-Kyu Ha

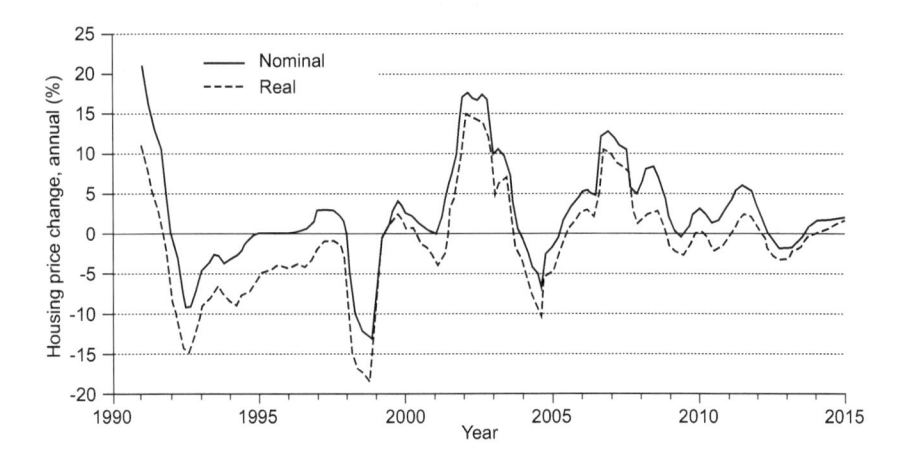

Figure 7.2 Percentage changes 1990–2015.
Source: Korean Statistical Information Service: www.globalpropertyguide.com/real-estate-house-prices/S#south-korea (Retrieved March 29, 2016).

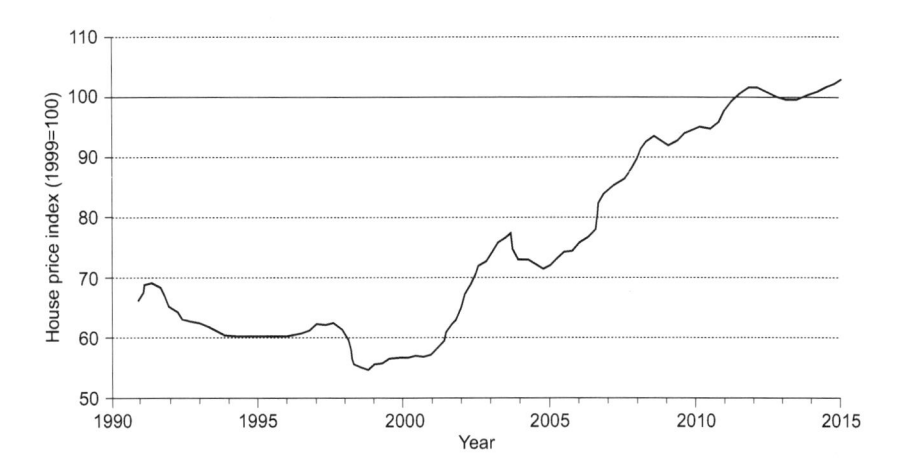

Figure 7.3 Housing purchase price index (June 2011=100).
Source: Korean Statistical Information Service: www.globalpropertyguide.com/real-estate-house-prices/S#south-korea (Retrieved March 29, 2016).

from home sales made by multiple homeowners from the 50–60% rate introduced in 2005; and (5) lifting of the rules preventing the quick sale of properties in Seoul's real estate hot spots.

However, house prices declined from 2012 to 2013. House prices in Seoul fell further than national prices, declining by an average of 2.32% in 2012 and 3.62% in 2013. The house price declines were likely due to the economic uncertainty, a seasonal slowdown, mortgage restrictions, and fears of another housing market recession, among other reasons. Further measures were taken, such as the introduction of tax breaks to induce more demand, as well as other supply control measures to improve prices. The effects of these measures were seen after a year and have been reflected in the gradual improvement of prices for the past two years.

On April 1, 2014 the Korean government unveiled a package of measures to boost the housing market using the loan-to-value ratio (LTV) and the debt-to-income Ratio (DTI). The Korean government reset the LTV limit to a uniform 70%, from between 50% and 85% depending on the region; at the same time, the DTI limit became 60% across the country, whereas before it had varied between 50% and 65%. The constrained households are easy to identify, and the social goals associated with housing may pose important policy challenges in implementing LTV and DTI limits (Lee, 2014).

The Korean national housing purchase price index rose by 3.47% (2.44% inflation adjusted) during the year to end in November 2015. Seoul's house price index rose more, with 4.29% year-on-year growth. The house price increases were therefore due to a combination of low interest rates and economic growth, along with price rises for *Chonsei* houses, leading potential renters to buy their own houses rather than renting. The *Chonsei* system eliminates the likelihood of a tenant defaulting on monthly rents. However, the lump-sum deposit, equivalent to 70–80% of the property value, imposes a huge burden on younger renters and new households.

3.3 Public housing provision

How has the Korean government reacted to low-income urban housing problems? The rationale for government intervention in housing has generated a lengthy debate on the relationship between political structures and the established housing system (Oxley and Smith, 1996; UN Habitat, 2008; Chiu, 2008; Balchin, 2013; Meen and Gibb, 2016). In response to the apparent inability of the private market to produce housing for all, in sufficient quantity and quality and at reasonable prices, the Korean government intervened directly by constructing public housing. It is necessary to examine the role of the government for low-income families, and to explore the public housing (or state-developed housing) sector and the low-income housing policy issues.

In Korea, three categories of policy measures have been adopted for the housing needs group: (1) public housing provision; (2) the control of housing prices and sizes; and (3) financial assistance to homebuyers and tenant households. This section focuses on public housing provision and direct financial assistance.

In the late 1980s, comprehensive housing development planning by the central government of Korea was instituted to determine the extent to which national resources should be allocated to public housing development for the poor. A permanent public rental dwelling program was launched in 1989, and it represented the beginning of a social housing tradition directed to low-income households in

Korea. In the late 1980s, the government formulated a five-year housing supply plan for the purpose of constructing two million dwelling units between 1988 and 1992. The government allocated 900,000 units out of the two million for low-income families. It was in this circumstance that the government drew up the plan to provide 250,000 permanent rental units for the lowest-income families. Although the program for permanent rental housing ended after 200,000 units were constructed by Housing Corporation, it was the first time in the history of Korean housing that the government tried to address the housing problems of different income groups with different packages of subsidies and finance.

Public housing is defined as that constructed with the support of the National Housing Fund in Korea. Public housing comes in two types: public housing for sale and public housing for rent. In terms of public housing for sale, the households eligible for this program are those who have resided in the same administrative area as the place where the new housing is being provided and have been statutorily non-homeowners for one or more years at the time of the first advertisement for the sale of the new houses. Priority is given to those who have saved with the housing subscription savings in the order of the amount and the period of savings.

There are several types of public housing program. At the end of 2014, the proportion of public rental housing (1,709,000 units) to the total housing stock (15,988,800 units) was about 10.7%, but over the long term (more than 20 years) the amount of public housing (820,000 units) to total housing stock has been only 5.1% (Table 7.4). In the case of capital city, 210,000 units of public rental housing were available as of 2013, bringing the percentage of public rentals to 6.1% of the total housing stock (Table 7.5). Recently, more varied approaches have been taken to resolve the shortage of housing sites: studio-type housing on small, city-owned plots; the supply of public rental housing based on a cooperative-type approach; and quasi-public rental housing supplied by remodeling privately owned homes.

Since 1962, the Korea Land and Housing Corporation (LH) has provided a total of 2.3 million public housing units for low-income households. LH was

Table 7.4 Public rental housing stock in 2014.

Type (in Korean)	Permanent rental (Young gu)	Public rental (Gong Gong)	National rental (Kookmin)	Company employee rental (Sawon)	Public and private rental (Keunseol)	Public purchase and rent (Maeip)	Total
Rental period (years)	50	50	30	(50/10/5)	5	3/5	-
No. of housing units (1,000s)	193	106	521	30	501	358	1,709

Source: Public Rental Housing (2016), http://stat.molit.go.kr/portal/stat/yearReport.do.
Ministry of Land and Transportation (2016), http://stat.molit.go.kr/portal/stat/yearReport.do. Accessed April 7, 2016.

Table 7.5 Percentage of public housing for 20-year or longer leases against total housing stock in Seoul.

Year	1995	2000	2004	2010	2012	2013
Percentage	3.5	4.5	5.0	4.6	5.2	6.1

Source: Seoul Metropolitan Government, 2013. *Basic Seoul Urban Plan for 2020*, Seoul: The Seoul Institute, 2013; Jang, Y.-H. (2015), Seoul Housing Policy, www.seoulsolution.kr/en/node/3448. Accessed July 23, 2018.

established under the Korea Land and Housing Corporation Act on October 1, 2009 in a merger of the Korea Land Corporation (KLC) and Korea National Housing Corporation (KNHC). The figure accounts for 15.5% of the total 14.7 million homes that have been constructed in Korea.

The National Public Housing Fund is the heart of South Korea's public housing financing. The fund was first created in 1981 and was funded by government contributions; money from issued National Housing Bonds and housing lottery tickets; deposits from the general financial market and National Public Housing Fund bonds; and housing savings accounts, which give priority for housing to their holders. The fund has various uses: construction of housing for lease or sale, assistance with house purchases or *Chonsei* deposit loans, improvement of seriously deteriorating houses, or purchase of housing sites. The fund has financed approximately 4.5 million housing units, with public lease housing accounting for 48.2% and small housing units for sale making up the remaining 51.8%.

Despite the continuing supply, there is still a serious shortage of public housing. For low-income tenants who cannot move into public rental housing, the *Chonsei* Deposit Loan Program and the Monthly Rent Assistance Program are available, which we could call demand-led services. In addition, from 2015 onwards, the Housing Voucher Program has been implemented at the national level. The number of target households is about 970,000 nationally as a whole, and funds paid were KRW 110,000 (US$96.20) in 2015. The rent assistance system is deemed an effective policy as it utilizes an extensive range of private-lease housing to minimize the number of those who are left behind, and it responds to varied demands for housing assistance.

Under the Park Geun-hae government (2013–2016), there were two new housing programs: the first was "Happiness Housing," which aims to provide public rental housing for people in their 20s and 30s, including newlyweds, college students, and young workers. Happiness Housing is also set at a lower rate than the market prices. The government provided 140,000 units of Happiness Housing until 2017, so that young generations could find houses near their workplaces or schools without too much worry about financial difficulties.

The second program was "New Stay," which was a new type of long-term rental housing program. Unlike existing public rental houses, anyone can apply for these units, without consideration of their income level or whether they own a house. Residents can rent units without having to worry about moving for up to eight years. The annual increase of rental fees is limited to a maximum of 5%. The government planned to supply 18,000 New Stay units in 2016, and more than 60,000 units by 2017.

The real estate policy of the new Moon Jae-in[4] government (from May 2017) is likely to focus on urban renewal and expanding low-cost housing projects. During his election campaign, Moon promised on many occasions to undertake a New Deal to renew urban scenes by investing a total of KRW 50 trillion (US$44.2 billion) for his five-year term until 2022. He also pledged that his government would increase the supply of low-cost housing units through a policy of providing 170,000 homes a year, including 130,000 public rental housing units and 40,000 publicly subsidized rental units. This is 69% higher than that provided by the previous administration of 100,700 units a year.

The housing shortage caused by Korea's continued economic growth and urbanization often precipitated housing price increases and speculation. Finding solutions for this problem has always been a crucial national challenge while the country is enjoying rapid economic growth, and was the reason that housing supply and price control policies were alternately implemented in response to the cycle of housing shortage, followed by rising house prices, followed by speculation, and then control policies.

4. Policy issues and implications

4.1 Disputes over price control

Even if markets work well, they can only be expected to provide an efficient allocation of resources and not necessarily a fair allocation. Governments are concerned not only with efficiency but also with equity. The latter concept involves considerations of fairness in the distribution of resources.

Many housing analysts in Korea have criticized the price control of new housing. Under the price control system, the profitability of house-builders has not been assured. Therefore, the supply of new apartment housing has been reduced in Seoul and large cities. In Seoul, the supply of new residential units has expanded to the outer rings, where land prices are relatively cheap. The expansion of the residential area around Seoul has been much affected by the production of new apartments. This can be explained by the house-builders' strategy under the price control: as far as they are concerned, the cheaper the land they can get, the more profit they can make. They have also built new apartments in other cities around Seoul where the land prices are cheap. The sale price ceiling system has not necessarily contributed to stabilizing the market price; instead, it may have raised the housing price in the long run as it has impaired the financial position of the builders, increased land-use inefficiencies, and in the process lowered the welfare of households driven to live in more distant locations (Kim, 2004; Ha, 2010; Kim, 1997).

There have been many disputes over price control between the government and private house-builders, who have continuously demanded the abolition of the price control since its introduction. In November 1989 the new apartment price ceiling system was replaced by a new type of price control, which is called the "construction cost linkage system," with new apartment prices. The sale price ceiling system was one of the main causes of housing shortages in the Seoul Capital Region. Before the introduction of the construction cost linkage system, the sale price was set and fixed by the government, while since November 1989 the sale prices have varied in accordance with land values. Since the International Monetary Fund (IMF) program and the Global Financial Crisis, the government has removed the price control system (Ha and Lee, 2001).

4.2 Role of the public sector and the allocation process

It is important to ask why the government concentrates on the policy of public housing for sale, rather than on expanding rental dwellings. The policy offering public housing for sale now forms part of a policy package designed to lessen the financial difficulties of public sector housing. This policy aims to invest in public housing for sale and to plow the profits back into low-cost housing for the low-income group and other recipients in urban areas (Yoon, 1994).

The housing authorities seek to encourage homeownership through the sale of public housing. There are huge numbers of households who want to be owner-occupants. In fact, the private sector cannot fulfill all the demand from various income strata, especially low-income households. Thus, the aim of public housing for sale is to meet a demand for owner-occupation, through providing the National Housing Fund loans. Even when the LH Corporation built rental housing, the period during which it was available for renting was short: one year before and five years after 1982. At the end of the rental period, this housing was sold to the tenants whose rental contracts were leases with the option to buy. In 2014, 25% of public rental housing stock was the five-year rental housing type.

The public housing for sale system is administratively simple and easily understood by the public. But it ignores the principle that public housing should be allocated to those households in the most urgent need, considering household size, present living conditions, and health conditions. However, applicants for public housing are required to deposit a considerable amount of money upon their application. So, many low-income people and tenants do not deposit their money in the bank. A common criticism is that public housing nominally targeting low-income people has often been allocated to middle-income households.

In institutional terms, the central government, as the key policy-making agent, plans the aggregate volume of housing supply and the possible partition between the public and private sectors. Local governments perform almost identical activities as those of the LH Corporation, which operates under the direction of the central government. Even though the role of local governments is very important in the wake of the decentralization of state power under the principle of local autonomy, most local governments have not pursued their role to make housing programs for low-income residents. Two of the most important issues for local governments are the lack of funds and the lack of trained and experienced government officials for the low-income housing programs. With respect to the public housing supply, the LH Corporation has played a much more important role than local governments.

The role of the public sector in housing should be clearly different from that of the private sector. Not only from the point of view of efficiency but also of equity, it is highly desirable that the public housing for sale program be switched to rental housing. In devising future social housing programs, housing authorities have to remove ambiguities and define the target groups, taking into account local housing situations and community-wide objectives relating to maintaining "bottom-up" or popular participation in housing planning. The central government should leave more initiatives to local government in order to encourage them to meet local housing needs and promote the more active participation of the private sector.

4.3 Social mix

Contemporary urban housing policies in Korea have a common aim of balancing the "social mix" and creating communities with a blend of residents across income levels and housing tenure types, including public, private rental, and owner-occupied housing. Housing authorities believe that it is important to help create mixed and inclusive communities that offer a choice of tenure (Chun, 2004). In Korea, two major strategies are commonly adopted to achieve a more balanced social mix. First, in the case of residential development projects by the public sector, 40% of newly developed land should be used only for rental housing construction, and local planning authorities should encourage and plan to achieve the housing mix through urban redevelopment and housing environment improvement projects: between 17% and 20% of redeveloped housing should be rental housing. Second, the housing authorities purchase housing across a range of already established, privately developed neighborhoods for use as public housing. The local planning authorities have tried to encourage the development of mixed and balanced communities by avoiding the creation of large areas of housing with uniform characteristics.

In the Korean context, a balanced social mix should minimize polarization by socio-economic characteristics and tenurial polarization, and also minimize locational disadvantages. Social polarization refers to the process of differentiation among people on the basis of social class or income status. This class polarization is associated with higher-income people being clustered in homeownership and lower- income people being clustered in the public rental units. There is also socio-tenurial marginalization, which identifies a clustering and "locking in" of lower-income people in the public rental units. "Inclusive communities" and a "social mix" are based on interdependence or mutual dependence. The concept of social mix can refer to a mixing of people in a given space (region, neighborhood, housing estate) on the basis of different social classes, socio-economic statuses, tenures, and household or family types.

In terms of housing submarkets in urban Korea, the government's social mix policies in housing estates have negative side effects. A previous study in Seoul surveyed various types of tenure mixes, including homeowners, private rental tenants, and public rental tenants. About half of homeowners responded by saying that they "oppose the mixing of social housing and non-social housing within the same estate," compared to 15% among public rental tenants (Ha, 2008). This is a crucial issue: locating residents with different income levels in the same estate (neighborhood) may raise awareness of class differences and create tensions, rather than achieve the sought-after social cohesion. Some people recognize that social housing could have a negative impact on the neighborhood. Non-social housing residents speculate that, because of neighboring social housing being in line with the negative impression of poverty, there has been relative decline in the value of their housing. The result is that urban housing blocks and buildings subdivided by tenure and economic status exist within the same estate. It is apparent that residents of higher status have no intention of mixing with their poor neighbors, and they are likely to think it is desirable for similar lower-income groups to live in the same neighborhood. In order to understand the actual characteristics and nature of housing submarkets compared with housing policy in Korea, it is crucial to look at the residents' economic status and tenure types.

4.4 Housing polarization

One of the crucial housing issues in Korea is the polarization of housing conditions between the rich and the poor. Overall housing conditions have improved substantially for everyone over the past decades. However, low-income families are still suffering from comparatively poor housing conditions and a lack of affordability. The gap between the rich and the poor is ever growing. What are the sources of housing polarization in Korea?

As can be seen from the key indicators summarized in Table 7.1, the indicators of overcrowding, as well as the quality of dwellings and facilities, have improved remarkably. But the heavy subsidies generated through the price control on new housing have accrued predominantly to the middle class. Those who were lucky enough to be selected to purchase new apartments received windfall capital gains. Despite many measures to improve housing quality over the past several decades, housing has remained a persistent and divisive social issue in Korea. After the Asian Financial Crisis (1997) and the Global Financial Crisis (2008–2009), the poor have suffered more than other groups. In the wake of Korea's 1997 financial crisis and the housing crisis that have persisted for years, the roles of the public and private sectors in housing should be clearly differentiated (Ha and Lee, 2001; Ha, 2010).

The life of common individuals is unfortunate due to the unequal distribution of wealth. The vicious circle of income inequality is threatening the market economy of Korea. In particular, due to slow but important socio-economic changes, the poverty rate of people over 65 years old has reached 48.5%, which is 3.4 times higher than that of the Organisation for Economic Cooperation and Development (OECD) average. Moreover, the proportion of middle-class households has also shrunk from 75.4% in 1990 to 67.5% in 2010. Today, 55% of middle-class families are having a tough time making ends meet, as they are burdened by debt.

The problem is that income polarization and housing divides are worsening as time passes. Deterioration in wealth distribution not only makes one's life harder, but also brings about conflicts between people of different economic classes. Korea has the world's highest suicide rate and is seeing a surge in divorces and a low birthrate, all of which also contribute to the decline of the middle class as a result of unemployment and a rise in household debt.

Upward social mobility is becoming an ever-more distant dream for the poorer classes in Korea. It is becoming increasingly difficult for wage earners to rise to higher income levels. The main reasons are the dual structure of the Korean labor markets and the increasing number of contract or part-time well-educated workers, who only see marginal rises in incomes (Korea Institute for Health and Social Affairs, 2010).

Korean housing policies have been incremental, focused on short-term issues, physically oriented, and distorted during the most critical phase of Korea's rapid urbanization (Renaud, 1993). Korean housing policy has made an attempt to alleviate the problem of housing shortages and to increase homeownership through mass high-rise apartment construction. But the interventionist approach of the government to ameliorating housing shortages has not fully met the goals of efficiency and equity (Chiu, 2008). The government should recognize that housing policy should be differentiated across quasi-independent subdivisions of the overall housing market. Within these divisions, supply and demand interact to produce relatively homogeneous clusters of housing types or household characteristics, in which there is a unique set of prices (or rents).

The inequality of housing allocation and polarization is as serious as the inequality of household income. To resolve the issue of housing polarization, one of the policy options is to increase affordable housing for low-income families. While it is obvious that the supply of affordable housing will remain the first priority in the future, the revision of current public housing programs should not be neglected. The current supply of public housing driven by the national LH Corporation should be changed to a public–private partnership approach, promoting more active participation of the private sector in affordable housing programs.

5. Conclusions

The direct and indirect participation of the Korean government in housing has grown considerably over the years through an impressive number of regulations, laws, and plans for the construction of dwellings, land development, and other activities (Kim and Park, 2016; Ha, 2013). Occasionally, free markets have experienced failures and people have cried out for government control of fair housing allocation, particularly during the period of widespread real estate speculation in the 1970s and 1980s. However, the government's excessive intervention in housing markets may also result in a governmental failure. It is important to have proper involvement, transparency, and accountability.

In Korea, government intervention is still pervasive, even though the housing ceiling price system affecting the private sector has been terminated. Korea's housing market has been defined and intervened in by various governments in "a housing provision-led and anti-speculation-driven approach," with the goal of promoting homeownership and the expectation of a "trickle-down effect" on housing construction and real estate. Housing is both an asset and a consumption good. Both aspects may have to be equally considered from the policy standpoint, but the consumption aspect is much more important than the asset formation aspect when a country moves toward a welfare society. Governments may have to intervene in the market if the market cannot allocate housing resources in favor of the poor and the disadvantaged.

Good governance means coping with the conflicting goals and competing claims of different interests. The emphasis on "enabling approaches" can receive considerable support from the growing recognition that democratic and participatory government structures are not only important goals of housing development and allocation, but also important means for achieving such goals for Korea. The potential conflict in enabling shelter strategies, between the need for liberalization to give private-sector incentives and the need for regulation to correct market imperfection and curb speculation, demonstrates the wider dilemma facing Korea's economy.

Finally, it should be emphasized that the role of the government is to achieve three goals. The first is to see that growth continues, to increase incomes, and to intervene indirectly, moderating and facilitating the market once the country becomes economically mature. The second is to reduce social and spatial inequalities. The third is to minimize the fluctuations in the economic system and to ensure that the productive resources of society are used as effectively as possible.

Notes

1 ACT ON SUPPORT OF THE DISABLED, THE AGED, AND OTHER HOUSING-DISADVANTAGED PEOPLE, Enforcement Date August 12, 2016: www.law.go.kr/LSW/eng/engLsSc.do?menuId=2&section=lawNm&query=housing+act&x=23&y=39#liBgcolor3
2 In January 1988, the ceiling price was set at two levels: US$1,100 USD (KRW 1,268,800) per *pyong* (3.3 m^2) for small apartments (less than 85 m^2), and US$1,200 (KRW 1,340,000) per *pyong* for medium and large apartments (more than 85 m^2). In 1989, price ceilings for new apartments were replaced by a new type of price control, called the "construction cost linkage system," and sale prices varied in accordance with land values. The ceiling price was set at almost 90% of the market price in Seoul.
3 The collapse of Lehman Brothers, a global investment bank, in September 2008 was the immediate trigger of the world's financial system crisis. It took huge taxpayer-financed bailouts to shore up the banking industry.

4 Moon was considered the frontrunner to win Korea's 2017 presidential election, which would be the 19th term of the country's presidency, following the impeachment of Park Geun-hye.

References

Balchin, P. (2013). *Housing Policy in Europe*, London: Routledge.
Bourassa, S. C., Hoseli, M., and Peng, V. S. (2003). Do Housing Submarkets really Matter? *Journal of Housing Economics*, 12(1), 12–28.
Bourne, L. S. (1981). *The Geography of Housing*. London: Edward Arnold.
Chiu, L. H. (2008). Government Intervention on Housing: Convergence and Divergence of the Asian Dragons. *Urban Policy and Research*, 20(3), 249–269.
Chun, H. (2004). A Study on the Social Capital of Apartment Complex in Large Cities. *Korea Journal of Sociology*, 38(5), 215–247.
Downs, A. (1975). *Urban Problems and Prospects*, 2nd ed. Chicago: Rand McNally.
Ha, S.-K. and Lee, S.-W. (2001). IMF and the Crisis of the Marginalized Urban Sector in Korea. *Journal of Contemporary of Asia*, 31(2), 196–213.
Ha, S.-K. (2002). The Urban Poor, Rental Accommodations, and Housing Policy in Korea. *Cities*, 19(3), 195–203.
Ha, S.-K. (2008), Social Housing Estates and Sustainable Community Development in South Korea. *Habitat International*, 32, 349–363.
Ha, S.-K. (2010). Housing Crisis and Policy Transformation in South Korea. *International Journal of Housing Policy*, 10(3), 255–272.
Ha, S.-K. (2013). Housing Markets and Government Intervention in East Asian Countries, *International Journal of Urban Science*, 17(1): 35–42.
Ha, S.-K. (2018). Housing Policy Challenges and Social Development in Korea. In R. Chiu and S.-K. Ha (Eds.). *Housing Policy, Wellbeing and Social Development in Asia*. London: Routledge.
Harvey, D. (1977). Government Policies, Financial Institutions and Neighborhood Change in United States Cities. In M. Harloe (Ed.). *Captive Cities*. New York: John Wiley.
Jang, Y. H. (2015). Seoul Housing Policy, www.seoulsolution.kr/en/node/3448 Accessed July 23, 2018.
Kim, C.-H. and Kim, K.-H. (2000). Political Economy of Korean Government Policies on Real Estate. *Urban Studies*, 37(7), 1157–1169.
Kim, K. (1997). Government Intervention and Performance of the Housing Sector in Korea. Paper presented at the International Housing Policy Conference Seoul, Korea.
Kim, K. H. (2004). Housing and the Korean Economy. *Journal of Housing Economics*, 13, 321–241.
Kim, K. H. and Park, M. (2016). *Housing Policy in the Republic of Korea*, ADBI Working Paper 570. Tokyo: Asian Development Bank Institute. www.adb.org/publications/housing-policy-republic-korea/ Accessed July 10, 2018.
Korea Institute for Health and Social Affairs. (2010). *Supporting Vulnerable Group and Strengthen Housing Welfare Plan*, Policy Report 201-106. Seoul: KIHASA.
Korean Statistical Information Service, Housing Census, Various Years. http://kosis.kr/eng/search/search01_List.jsp Accessed May 3, 2016.
Korean Statistical Information Service. (2016). www.globalpropertyguide.com/real-estate-house-prices/S#south-korea Accessed March 29, 2016.

The Korea Herald (March 3, 2016). Savings Banks' Time Deposit Interest Dips Below 2 Percent. http://khnews.kheraldm.com/view.php?ud=20160303001025&md=20160306003333_BL) Accessed May 10, 2016.

Korea JoongAng Daily. (2008). Unsold Apartments Highest Since 1996, April 14, 2008. http://koreajoongangdaily.joins.com/news/article/article.aspx?aid=2888613.

Lee, S.-H. (2014). What are LTV and DTI? *The Korea Times*, September 9, 2014. www.koreatimes.co.kr/www/news/opinon/2015/10/162_164864.html Accessed May 10, 2016.

Meen, G. and Gibb, K. (2016). *Housing Economics: A Historical Approach*. New York: Palgrave Macmillan.

Ministry of Land, Transportation, and Maritime Affairs (MLTMA). (2012). *The 2012 Housing Condition Survey*, Seoul: MLTMA.

Oxley, M. and Smith, J. (1996). *Housing Policy and Rented Housing in Europe*. London: Routledge.

Palm, P. (1978). Spatial Segmentation of the Urban Housing Market. *Economic Geography*, 54(3), 210–221.

Renaud, B. (1993). Confronting a Distorted Housing Market: Can Korean Policy Break with the Past? In L. Krause and F.-K. Park (Eds.). *Social issues in Korea: Korean and American Perspectives*. KDI Press, Seoul.

Seoul Metropolitan Government. (2013). *Basic Seoul Urban Plan for 2020*, Seoul: The Seoul Institute.

Spence, L. H. (1993). Rethinking the Social Role of Public Housing. *Housing Policy Debate*, 4(3), 355–368.

Watkins, C. A. (2001). The Definition and Identification of Housing Submarkets. *Environment and Planning A*, 33(12), 2235–2253.

Wu, C. and Sharma, R. (2012). Housing Submarket Classification: The Role of Spatial Contiguity. *Applied Geography*, 32, 746–756.

United Nations Human Settlements Programme (UN Habitat). (2008). *The Role of Government in the Housing Market: The Experiences from Asia*. Nairobi, UN Habitat, HS/1013/08E.

Yoon, I.-S. (1994). *Housing in a Newly Industrialized Economy*. Aldershot: Avebury.

8 The land and housing delivery system in Korea

Evolution, assessment, and lessons

Man Cho
THE KDI SCHOOL OF PUBLIC POLICY & MANAGEMENT

Kyung-Hwan Kim (corresponding author)
SOGANG UNIVERSITY

Soojin Park
CONSTRUCTION AND ECONOMY RESEARCH INSTITUTE OF KOREA

Seung Dong You
SANGMYUNG UNIVERSITY

1. Introduction

Like its East Asian neighbors, Korea has experienced rapid and sustained economic growth since the 1960s. No country has industrialized without urbanization, and Korea is no exception: the share of its urban population has grown in tandem with the rapid growth of Gross Domestic Product (GDP), from 39 per cent in 1960 to 91 per cent in 2015. Again, similar to other East Asian countries, Korea had very high urban as well as rural population densities that worked as an accelerator of rural-to-urban migration during the high-growth period (Renaud, Kim, and Cho 2016). In particular, between 1960 and 2015, the population of the Seoul Capital Region (SCR, including Seoul, Incheon, and Gyeonggi Province, which surrounds Seoul) grew from 5.1 million to 25.4 million. The SCR's population share also increased steeply from 20.8 percent to 49.4 percent.

Rapid urbanization and sharp increases in population in large cities during the period of rapid growth have created a large demand for affordable housing and urban services. In response, the government has instituted a series of policy measures to increase the supply of affordable housing in urban areas, while trying to tame occasional rampant price hikes of urban land and housing through demand-side interventions.

A key feature of the Korean experience in addressing the housing challenge since the 1960s is that the government took an interventionist approach, and public-sector institutions played a crucial role. In the process of establishing the system for the delivery of developable land and housing, the national government provided strategic guidance in charting the course of development of the housing sector in a comprehensive way. But it also relied on the market mechanism in allocating scarce resources, and the role of private-sector institutions expanded over time in both the real sector and the financial sector. In overall terms, housing conditions have improved remarkably in both quantity and quality, although housing remains unaffordable for some segments of the population.

The objective of this chapter is twofold: first, to document the evolution of the Korean land and housing delivery system over time in the context of changes in the market environment and government policies; and second, to draw implications for other countries faced with the challenge of providing adequate and affordable housing. Instead of providing a very detailed history of the development of Korea's housing sector, we focus on the key building blocks and salient features of the system, and examine how the system was established and evolved over time. Then we offer a summary of the findings that might be relevant for other countries. We also suggest some areas for further research.

This chapter is organized in five sections. Following the introduction, Section 2 documents the evolution of the land and housing delivery system in Korea. Section 3 assesses the performance of the delivery system using a set of housing indicators. Section 4 presents the current system and its notable attributes. Section 5 summarizes major findings and draws policy implications.

2. Evolution of the land and housing delivery system

Korea was a low-income, agrarian society in the early 1960s, with a share of urban population below 40 percent. During the half-century that followed, however, the country has completely transformed itself into a fully industrialized, high-income, urban economy. During the early phase of economic growth, the government channeled scarce financial resources to support the development of industries, while assigning a low priority to the housing sector. The sharp increase in demand for housing combined with inadequate supply response led to serious housing shortages in Seoul and other major cities. However, the government took an interventionist approach to address the housing challenge early on. A series of public-sector housing institutions was established, and new financing mechanisms were instituted.[1] A more market-based and mature land and housing delivery system emerged in later years.

To consider the evolution of the system of land and housing supply over the past 50 years or so, we divide the whole period into four phases: the period of institution building, which started in the 1960s and lasted until the late 1980s (Phase 1); the period of massive supply of new housing from the late 1980s until the outbreak of the Asian Financial Crisis (AFC) (Phase 2); the period from the post-AFC recovery until the outbreak of the Global Financial Crisis (GFC) of 2008–09 (Phase 3); and the post-GFC period (Phase 4). The Appendix provides more details on changes in macroeconomic and housing market conditions, together with the key housing policy adjustments for each of these four phases.

2.1 Phase 1 – Institution building: from the early 1960s till the mid-1980s

Several important institutions for physical housing development and its financing were established during the early stage of rapid economic growth and urbanization. The notable ones are:

- The Korea National Housing Corporation (KNHC) (1962), to implement housing development projects nationwide as a partially government-funded agency;
- The Korea Housing Bank (KHB) (1969), to provide both consumer financing (generally for moderate-income households) and corporate financing for homebuilders. The KHB was privatized in 1997 as the Housing and Commercial Bank, which was merged later with the Kookmin Bank (KB) in 2000;
- The Korea Land Development Corporation (KLDC) (1979),[2] to expropriate land for government-initiated residential development via eminent domain. The KLDC merged with the KNHC and was renamed the Land and Housing Corporation (LH) in 2009; and
- The National Housing Fund (NHF) (1981), to finance the construction and purchase of affordable housing, with funds raised from the contract savings by prospective homebuyers and the issuance of housing bonds.

Various laws related to housing development were also enacted, such as the Land Readjustment Act of 1966; the Housing Construction Promotion Act of 1972; the Urban Redevelopment Act of 1976; and the Public Land Development Act of 1982.

The land readjustment scheme was the dominant channel for supplying developable land throughout the 1960s and 1970s. Under the land readjustment scheme, the local government (or public-sector developers) assembled parcels of land in private ownership, made an area-wide land-use plan, provided infrastructure, and adjusted the boundaries to accommodate roads and other on-site infrastructure. Newly demarcated parcels of somewhat smaller sizes, but in more regular shapes and of more value, were returned to the original landowners. The government retained selected parcels and auctioned them or sold them at market prices to recover the costs of infrastructure investment. The scheme has had a long history and made a substantial contribution to urban land development through the 1960s and 1970s, but it proved to be time consuming and unfit for large-scale development projects. Another criticism of the scheme was that it failed to capture windfall gains accruing to the landowners due to land value appreciation. Starting in the early 1980s, the government discouraged the application of the land readjustment scheme and opted for a new scheme called the public land development method.

Under the public land development scheme, the government (the provincial government in the case of smaller projects) designates a large area located outside the built-up urban area for development. Then the public sector developers (the KLDC, the KNHC, and local government) purchase non-urban land in the designated area from the landowners through negotiation or expropriation with compensation. They service the land with infrastructure according to the land-use plan and sell the rezoned and subdivided plots to construction companies to build houses and apartments. The KNHC also built apartments for sale to eligible homebuyers and public rental housing for target households. The public land development scheme overtook the land readjustment scheme in terms of the quantity of supply of developable land in 1982 and has become the dominant mechanism for supplying developable land ever since. The new scheme revealed some problems as well. Questions were raised about the fairness in land-taking and compensation in some projects. The new scheme favored green land development outside built-up areas, where it was easy to find a large land area, and high-density development in the farther-out locations distorted the spatial structure of the metropolitan area.

On the demand side, an array of policy measures has been implemented since the 1960s in order to contain speculation on housing and land. It includes the anti-real estate speculation tax in 1967 (predecessor to the capital gains tax), the price ceiling on newly constructed housing units in 1977, the land transaction approval system, and the special taxes on real estate holdings by corporations unrelated to their core businesses in 1978.

Housing can be provided through new development of land, but also through redevelopment of built-up areas. A challenge to policy makers during the early stage of urbanization was the redevelopment of the informal settlements in Seoul and other major cities. Given the severe housing shortage in the 1960s and 1970s, many low-income households resorted to makeshift dwelling units built anywhere they could find vacant land. About 29 percent of all houses in Seoul as of 1970 were such illegally built properties (Kim, 1980).

The initial government approach to dealing with squatter settlements was to bulldoze them without compensation, or to provide alternative housing, which was met by fierce resistance from the residents.[3] The next approach, as implemented by the Seoul Metropolitan Government (SMG) in the mid-1960s, was to provide informal settlers a 25-square-meter plot of land in public ownership per family to resettle in outer areas of the city. In most cases, relocated residents were given an implicit leasehold of the land and full ownership of the structure that they built on it. However, even basic infrastructure such as piped water and access roads was not provided, and many squatter families gave up these new locations and moved back to better-located squatter areas (Chang, 1989). Given the scarcity of vacant public land to accommodate informal settlers, the city government purchased 10 square kilometers of land beyond the city boundary in 1968 and began to build the satellite city of Seongnam to house 100,000 families evicted from Seoul. This time, again, basic services were not provided and living conditions were extremely poor. In addition, due to the lack of transportation services, most residents could hardly commute to their place of work in Seoul. A violent riot broke out in 1971, and the project could not be completed. Eventually the SMG settled on a different approach, whereby those whose informal houses were torn down were given cash compensation or offered the right to purchase apartments built by the city at subsidized prices, a profitable option for the resettler, as he or she could resell the right with a handsome gain. At the same time, the city tried to ensure that no new informal houses were built through close on-site monitoring and the punishment of local public officials if new informal houses were built in their jurisdictions.

2.2 Phase 2 – Massive supply: from the late 1980s till the outbreak of the Asian Financial Crisis

The population of the SCR kept growing during the 1970s and 1980s as economic growth continued, which necessitated the development of new and large-scale urban clusters outside its urban core to accommodate the increased demand for housing and urban space. The first such development took place in the Gangnam area (south of the Han River), within Seoul's city boundary, in the early 1970s. The second spatial expansion occurred in the late 1980s, when the government implemented a drive to build two million new dwellings over a five-year period, 1988–1992. Two million units were equivalent to about 30 percent of the total housing stock as of 1989. Included in the government drive was the development of five new towns beyond the outer edge of Seoul's Green Belts. The locations of these first-phase new towns were selected based on several criteria: commuting time of one hour or less to Seoul (20–25 kilometers to Seoul city center); good linkage to the existing urban infrastructure; a pleasant living environment; and availability of developable land at reasonable costs.

The driving force behind the Two Million Housing Drive (TMHD hereafter) was a sharp run-up in housing prices and the consequent deterioration of housing affordability, largely due to the booming macroeconomy in the mid- to late 1980s and the accumulated shortage of quality housing. Inadequate access to affordable housing for low- and middle-income families became an overarching political issue. During the 1987 presidential election campaign, the then ruling-party candidate Mr. Noh Tae-Woo, who later won the election, announced a plan to build four million housing units during his six-year term. The plan was modified to a more realistic scale when the government announced the plan in 1988 to build two million housing units during the five-year period of 1988–92. Although some concerns were expressed about the feasibility of the plan within the government, the Ministry of Construction (MOC) and the Office of the President held on to the position of fulfilling the election campaign pledge (Cho, 2014).

The TMHD became a milestone in housing policy in Korea. The target itself was ambitious as it called for more than doubling the annual volume of new construction right away. More remarkably, it successfully delivered the target quantity of new housing ahead of schedule and contributed to stabilizing housing prices in Seoul and other major cities throughout the 1990s. In terms of dwelling type, the high-rise apartment has become the dominant kind since the late 1980s, representing 59 percent of the total housing stock in 2010, compared to only 7 percent in 1980. The large share of apartment buildings contributed to the efficient delivery of the large quantity of housing units within the planned time horizon.

Nonetheless, the massive residential development during such a short time period produced some side effects, including the poor quality of some buildings, a lack of diversity in building designs and floor plans, and inflation in prices of construction materials and wages. By the time the drive was over, an excess capacity was created in the home-building industry. (Cho and Kim, 2011; Kim and Cho, 2010). The project is also believed to have led to a shift of an inelastic housing supply to the right, rather than making supply more elastic, which raised the risk of another price hike in response to a new demand shock (Green, Malpezzi, and Vandell, 1994).

Several factors were instrumental in implementing the TMHD in particular. The first was the application of the public land development scheme, which proved to be effective for large-scale land development projects. Secondly, the government incentivized the home-builders by modifying the price control on new apartments to reflect the production costs and a mark-up. It also provided financing to both homebuyers and home-builders through the NHF. Finally, the implementation of the TMHD mobilized all relevant central ministries, local governments, and state enterprises with coordination by the Office of the President.

One important institution for consumer protection was established in 1993. The government entrusted the Korea Homebuilders' Cooperative to offer a guarantee program to protect those who purchased apartments on a pre-sale contract against the risk that the contractor failed to deliver the completed dwelling unit as promised. However, the cooperative also provided guarantees on loans its members took from banks, and the loss on the loan guarantee operation pushed the cooperative into financial hardships in the mid-1990s, which were further aggravated by the outbreak of the AFC. The government injected new capital and transformed the cooperative to the Korea Housing Guarantee Corporation (KHGC) in 1999.

2.3 Phase 3 – The recovery from the AFC till the outbreak of the Global Financial Crisis

The AFC of 1997–98 landed a very severe shock to the Korean economy. Unlike in some other Asian countries, where the real estate sector was a cause of the crisis, the real estate sector in Korea suffered greatly from it. The real housing prices in Seoul and other cities fell sharply, by 30 percent within two years of the outbreak of the crisis. Many construction companies that relied heavily on debt financing were severely hit by the credit crunch and the sudden rise in the market interest. Consequently, 14 percent of housing construction companies went bankrupt in 1998.[4]

As the housing industry had to go through organizational restructuring to reduce indebtedness, a notable change was made to the mode of operation so that functions of development and construction were separated from each other. Prior to the outbreak of the AFC, construction companies served the dual functions of developers and contractors. After the crisis, the home-building business became a distinct two-tier process. In the first phase, a developer secures land, applies for building permits, arranges initial financing, and conducts marketing under the pre-sale system. Then a contractor takes over the project in the second phase and carries out the construction work. Although the division of labor can enhance efficiency through specialization, the new business practice is claimed to have exacerbated volatility in housing markets. One reason is that most developers have little of their own capital and depend heavily on debt financing to secure land, and the lenders require a guarantee by the contractor, a much larger business entity. This practice exposed contractors to the risk of non-repayment of the loan on behalf of the developer in case it failed to meet its obligation.

Fortunately, the AFC ended quickly, and a strong and sustained recovery followed. The economic recovery and two other circumstances – one related to housing markets and the other to institutional changes – led to a housing boom in the early 2000s. First, the very low interest rates and the consequent reduction in financing costs boosted housing demand and home-building activities. Second, financial sector liberalization measures instituted after the AFC – mainly the liberalization of lending and deposit rates and the deregulation of real estate lending – helped expand residential mortgage lending and construction financing.[5] At the turn of the new millennium, Korea's residential mortgage system changed from a small, government-subsidized one to a much larger, market-oriented one. As a result, home purchases became more affordable for moderate-income and other underserved households. For example, the typical loan-to-value (LTV) ratio, which was in the 20–30 percent range in the 1990s, rose to the 50 percent range, with a maximum set at 70 percent for loans securitized by the Korea Housing Finance Corporation (KHFC), and at 60 percent for those originated by the commercial banks and other private financial institutions. The volume of residential mortgage lending increased dramatically after the AFC, both in absolute amounts and in terms of the mortgage debt outstanding relative to GDP ratio.

Propelled by these two favorable circumstances and several years of sluggish supply of new housing due to the AFC, housing prices began to rise in 2002. This led to a sharp rise in apartment prices in the hot submarkets of Seoul. The government made all-out efforts to contain the housing price hike, mobilizing various policy instruments. A punitive capital gains tax was levied on the owners of two or more houses and on short-term transactions. The comprehensive real estate tax, a national tax levied on the aggregate value of real estate holdings by an individual, was introduced in 2006. The LTV ceiling was tightened, and the debt service-to-income ratio (DTI) regulation was added. But the housing boom lasted until 2007.

A charge was introduced on the unrealized capital gains of the high-density redevelopment of low- and mid-rise apartments. As these apartments become obsolete and unsafe for residence after passing the age of 30 years from construction, they can be rebuilt in higher densities. Owners of these properties form an association that commissions home-builders to undertake the redevelopment project. These projects generate large profits as they are in good locations, and the higher density allows many more units to be built for sale to a wider clientele. The government and the SMG have been regulating the process and the terms of redevelopment, such as the minimum age of the buildings, the maximum height and floor-area ratio, the contribution to expand infrastructure to absorb added pressure on the existing infrastructure, and the prices of new apartments offered for sale to non-members of the association. The association is also required to provide a specified number of rental units. This regulation and the development charge have been controversial and have changed from time to time to reflect market conditions as well as the policy stance of the government in power.

On the supply side, the second-phase new towns were developed about 30 to 50 kilometers from downtown Seoul, starting in 2003. However, these projects were not as successful as their first-phase cohort due to their poorer access to Seoul and weak demand for housing in these farther-out locations. Consequently, a large stock of finished housing units remained unsold, which was a reason for the prolonged recession in the housing market in the SCR following the outbreak of the GFC in 2008.

During this period, housing welfare emerged as a new area of focus in housing policy. Although the problem of absolute housing shortage was resolved by the early 2000s thanks to the accumulated increases in new housing supply, housing affordability was not achieved for some segments of the population. The government made a 10-year plan to provide one million public rental units to address the housing needs of low-income households. This initiative continued in subsequent administrations with some modifications.

2.4 Phase 4 – Post-GFC market reshaping

The impact of the GFC on the overall economy was much milder than that of the AFC, but the impact on the construction sector was substantial. The incidence of bankruptcies among construction companies was minimal, but the number of financially distressed companies – or "zombies"[6] – in the construction sector surged after the GFC, both in numbers and as a share of all listed construction firms. The latter figure rose from 8 percent in 2007 to almost 16 percent in 2012, which was even higher than the level experienced during the AFC (Renaud et al., 2016).

A root cause of the problem is widely believed to be the so-called project finance (PF) loans, the construction financing scheme introduced by the commercial banks and other financial institutions after the AFC. Under this scheme, the developers took excessive leverage, often with a minimal equity investment of 2–3 percent of the land acquisition cost. With so little at stake, developers engaged in land development in an aggressive manner during housing booms, and the subsequent oversupply in later years depressed the market, leading to a highly pro-cyclical leverage cycle.

The demand for housing shrank in the aftermath of the GFC and due to the pessimistic market sentiments. Potential buyers took the softening of housing prices as a signal of a forthcoming long-term depression in the housing market. This belief was based on the recognition that declining fertility, population aging, and the mass retirement of the baby-boomers would create an excess supply of housing and possibly cause housing prices to collapse. The government introduced various incentives to encourage home purchases, such as favorable loan terms, tax breaks, and relaxation of regulations. In fact, most of the measures introduced to contain housing price hikes by the previous administration in the early to mid-2000s were lifted or amended. The housing market began to recover around 2014.

As the demand for owner-occupied housing stagnated, the demand for rental housing increased. The structure of the rental market changed so that the traditional *chonsei* lost some ground to monthly rental leases. Under a *chonsei* lease contract, the tenant gives a large sum of deposit, usually 40–60 percent of the purchase price, to the landlord at the start of a two-year lease contract, pays no monthly rents during the lease period, and gets the entire deposit back at the end of the contract. *Chonsei* is also an informal housing finance mechanism whereby the landlord borrows from the tenant, and the tenant is given the right to occupy the house instead of receiving interest on the loan.

Chonsei dominated the Korean rental market against the background of rising housing prices, housing shortages, and high interest rates. Rising housing prices allowed the landlord to take a fraction of the asset price of housing as a deposit for the occupation of the house in anticipation of a substantial capital gain. Housing shortages gave an upper hand to the landlord through the *chonsei* system, which entails a very low risk of rent delinquencies. The high interest rate enabled the landlord to invest the deposit for a handsome return. As all these favorable factors disappeared, landlords began to demand a smaller deposit and monthly rental payments instead of a large deposit and no monthly rents. On the other hand, the tenants preferred a *chonsei* lease over a monthly rental lease because the interest rate used to convert the deposit to monthly rents was much higher than the bank lending rate.

A tight rental market resulted in a high rate of appreciation of deposit or rents. The government responded by expanding the supply of rental housing in both the public sector and the private sector. A new program called New Stay was introduced in 2015 to engage the private-sector developers and investors in the provision of high-quality rental housing combined with various services to the middle class. Incentives were provided to the developers on the condition that the units would be rented out for a minimum of eight years and the rent increase would be kept within 5 percent a year. Real estate investment trusts (REITs) were mobilized to facilitate the financing of rental housing. Several types of public rental housing were provided, including one targeted at young renters. On the demand side, a housing benefit scheme targeted at low-income renters and homeowners was launched in 2015.

The institutions described in the earlier subsections were instrumental in producing affordable housing units and allocating them to low- and moderate-income households. Nonetheless, the social missions of some of these institutions weakened as the private home-building sector developed in later years, and organizational changes followed to realign their missions accordingly. For example, the KNHC, which merged with the KLDC in 2009, has redirected its operations towards the development and allocation of rental housing units and away from developing owner-occupied units. The context was the need to increase the stock of public rental housing, while letting the private sector supply owner-occupied housing. And the NHF was reorganized as the Housing and Urban Fund in 2015, with an additional mandate of support for financing urban regeneration as regeneration became a new focus area of government policy. The fund now provides equity investments and guaranties, in addition to issuing conventional loans. The KHGC was transformed to the Korean Housing and Urban Guarantee Corporation (HUG) as a state enterprise to support both housing and urban development.

3. Performance of the supply system and housing outcomes

In this section, we present various data to assess the performance of the system of land and housing supply in improving the housing conditions in Korea.

3.1 Housing conditions

Housing conditions in Korea have improved dramatically over the past 50 years, and especially since 1980, in terms of both quantity and quality, as shown in Table 8.1. The housing supply ratio, defined as the ratio between the number of dwellings and the number of households, has improved steadily and surpassed the 100 percent mark. This means that there are at least as many housing units as there are households for the nation as a whole.[7] A more widely used indicator of overall quantitative measure is the number of dwellings per 1,000 inhabitants. The figure for this indicator has increased from 142 to 383 since 1980. Popular measures of crowding (such as the number of persons per room or per household), as well as those of housing consumption (such as floor space per person or per household), have all improved significantly. The two measures of the quality of dwellings in the table – the share of dwellings equipped with hot-water baths or showers and modern toilets – exhibit a remarkable improvement. Although it is not shown in the table, the share of dwellings that do not meet minimum standards fell substantially since the standards were established in 2000 and made stricter in 2011 (Kim and Park, 2016).

3.2 The supply of new housing

The remarkable improvements in overall housing conditions are attributable to the cumulative increase in the supply of new housing of good quality. We will look into several measures of the new supply of housing. Figure 8.1 illustrates the annual construction of new housing since 1962, broken down into the public

Table 8.1 Improvement in housing conditions in Korea.

	1980	1990	2000	2010	2015
Housing supply ratio (%)	71.2	72.4	96.2	100.5	102.3
Dwellings per 1,000 inhabitants	142	170	249	302	383
Number of rooms per family	2.2	2.5	3.4	3.7	3.6
Number of persons per room	2.1	1.5	0.9	0.7	1.3
Floor space per person (m²)	10.1	14.3	20.2	25.0	26.9
Floor space per household (m²)	45.8	51.0	63.1	67.4	68.4
Dwellings with hot baths/showers (%)	9.9	34.1	87.4	96.9	99.8
Dwellings with flush toilets (%)	18.4	51.3	86.9	97.0	97.3

Source: MOLIT (a) (various years).

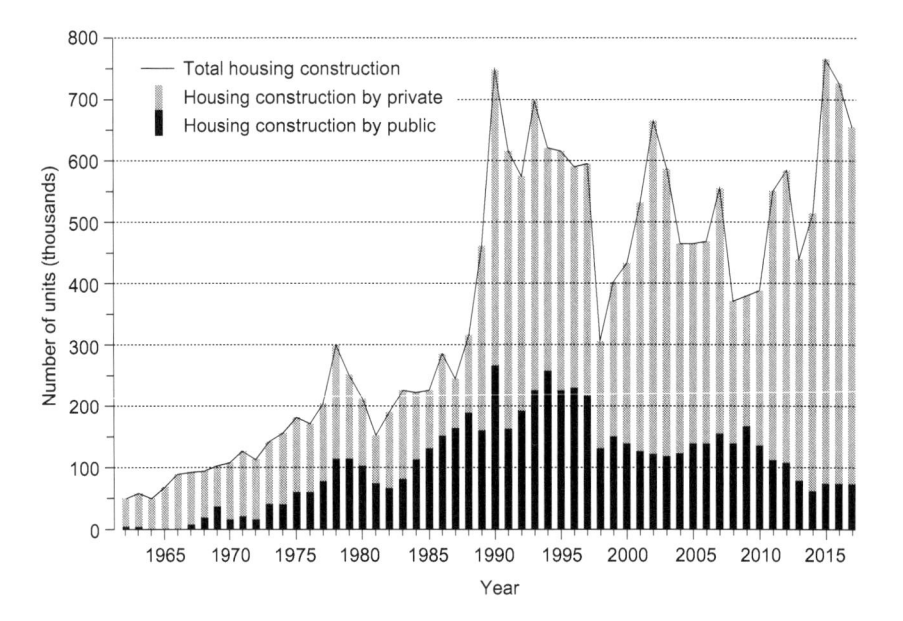

Figure 8.1 New housing supply, 1962–2017.
Source: Ministry of Land, Infrastructure and Transport (MOLIT).

sector and the private sector. The graph is based on building permits issued, because the data on completions is available only from 2011. It should also be noted that the units supplied by the public sector are defined as those financed by the NHF, regardless of who actually built the units.[8]

The first thing to notice from the graph is that the number of new housing units constructed in the 1960s and 1970s was very modest. Second, the annual supply of new housing took a quantum leap around 1989, the first year of the TMHD, when the figure more than doubled. Third, the new elevated level of housing construction was sustained thereafter, excepting the aftermath of the

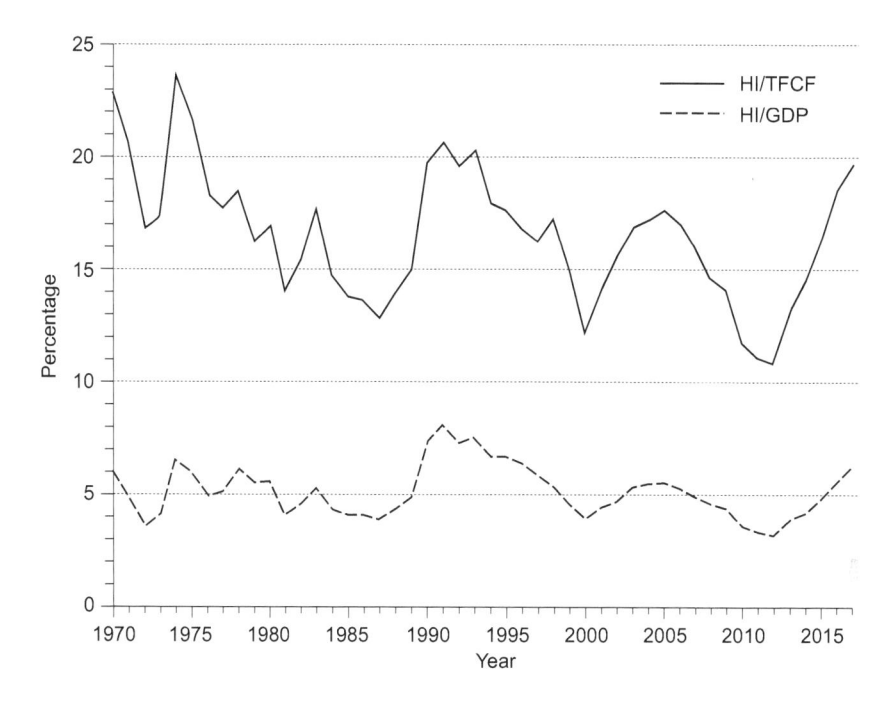

Figure 8.2 Housing investment as a share of GDP and of total investment (TFCF).
Source: The Bank of Korea, ecos.bok.or.kr.

AFC and the GFC. As for the division of the new supply between the public sector and the private sector, the latter dominates the former, and this dominance has become more apparent in recent years.[9] The split between the public sector and the private sector for the entire period 1962–2017 was 30 percent and 70 percent, respectively. The dominant contribution of the LH (including the KNHC) is noteworthy, although it cannot be seen from Figure 8.1.[10]

Another measure of new housing supply is housing investment or residential construction in the national income account defined as the value of residential structures added through new construction and improvement on existing units.[11] Figure 8.2 exhibits the share of housing investment as a percentage of GDP and its share as a percentage of total investment (total fixed capital formation, TFCF). The graph shows that housing investment has averaged 5.2 percent of GDP over the period 1970–2017[12] and increased significantly in the late 1980s as a result of the TMHD of 1988–1992. The share of housing investment in total investment exhibits a similar pattern.

New dwellings are provided through new residential land development or through redevelopment. Two types of redevelopment project have been implemented. The oldest method involves the clearing and rebuilding of old residential areas, through which just over one million dwellings were added

Table 8.2 The supply of public rental housing.

	Private	KNHC	Local gov't	Sum	Total housing	Public rental share (%)
1987–1991	138,370	193,272	75,904	407,546	2,386,821	17.1
1992–1996	240,917	108,322	23,949	373,188	3,104,854	12.0
1997–2001	372,744	131,875	5,810	510,429	2,270,523	22.5
2002–2006	49,219	388,401	48,283	485,903	2,648,867	18.3
2007–2011	42,812	400,812	86,795	530,419	2,245,000	23.6
2012–2016	75,614	267,043	54,203	402,400	3,033,627	13.3
1987–2016	919,676	1,489,725	294,944	2,704,345	15,689,692	17.2

Source: MOLIT (a) (various years).
Note: The figures for public rental housing include the units to be sold to the tenant at the end of a 5 or 10 year lease period. Rental housing is defined to be any housing built on the plots provided by the public sector and/or financed by public funds.

between 1973 and 2016. The second type is the redevelopment of low-rise and mid-rise apartments into higher densities. About 260,000 units were built by redeveloping 160,000 units since 1983; many more projects are under way or in the planning stage.

The supply of new housing consists of owner-occupied units and rental units.[13] The rental housing includes rent-to-own units, whereby the renter is given the right to purchase the unit after renting it for five or 10 years. Most of the rental housing is provided by the owners of two or more houses on a two-year lease. As of the end of 2016, the total housing stock was 19.88 million, and the rental stock was 7.79 million, about 39 percent of the total housing stock.

The government has been trying to build a sizeable stock of long-term public rental housing to meet the housing needs of low-income households. Table 8.2 presents the data on the supply of public rental housing since 1987. The share of public rental units in the total supply of new housing has varied widely and averaged about 17 percent over the past 30 years. As of the end of 2016, the long-term public rental housing stock was 1.25 million, which represented 6.3 percent of the total housing stock (MOLIT (a), Various years, p. 465).

3.3 The supply of developable land

The massive supply of new housing has been possible because of the supply of developable land. Prior to the adoption of the public land development scheme in the early 1980s, the land readjustment scheme was the primary source of the supply of developable land. The land readjustment scheme was time consuming, and the scale was not large enough to meet the needs. The average duration of the land readjustment projects in Seoul that were completed by the end of 1989 was five years, with 19 percent lasting longer than 10 years (Seoul Metropolitan Government, 2017). As of the 2000 year-end, a total of about 477 square

Table 8.3 Supply of developable land by the public sector (unit: 1,000 square meters).

	KLDC	KNHC	Local Gov't	Sum
1962–1966	-	1,122	16,804	17,926
1967–1971	-	6,042	80,265	86,307
1972–1976	-	2,735	59,436	62,171
1977–1981	7,601	10,203	56,077	73,881
1982–1986	34,955	13,291	38,652	86,898
1987–1991	55,789	21,735	35,338	112,862
1992–1996	68,352	20,308	76,666	165,326
1997–2001	51,703	18,626	54,464	124,793
2002–2006	73,529	63,502	45,050	182,081
2007–2011	167,202	40,060	39,150	246,412
2012–2016	37,424	-	22,897	60,321
1962–2016	496,555	197,624	524,799	1,218,978

Source: MOLIT (b) (various years); Cho (1991).

Table 8.4 Supply of developable land by region (unit: 1,000 square meters).

	Seoul	ICN+GG	SCR	5 MC	Others	Korea
1981–1991	44,318	72,606	116,924	35,709	74,450	227,083
1992–1996	3,594	73,900	77,494	34,503	53,329	165,326
1997–2001	1,864	53,979	55,843	21,653	47,297	124,793
2002–2006	7,691	97,288	104,979	28,732	48,370	182,081
2007–2011	14,935	147,418	162,353	16,516	67,543	246,412
2012–2016	1,974	38,019	39,993	3,984	16,344	60,321
1981–2016	74,376	483,210	557,586	141,097	307,333	1,006,016

Source: MOLIT (various years).
Note: ICN=Incheon, GG=Gyeonggi Province, SCR=Seoul Capital Region=Seoul+ICN+GG, MC=provincial metropolitan cities.

kilometers of land was provided through 542 projects, implying an average size of 0.88 square kilometers.

Table 8.3 shows the volume of developable land provided by the KLDC, the KNHC, and local government since 1962. One can see that land supply has increased remarkably since 1982, as the public land development method overtook the land readjustment method. According to the data on land use, the stock of buildable land increased by 1,219 square kilometers between 1962 and 2016. The total supply of land over the period was 1,052 square kilometers and was equivalent to 82 percent of the increase in land zoned residential during the same period. This outcome demonstrates the dominant role played by the public-sector land developers in residential land development.

Table 8.4 presents the breakdown of the supply of developable land since 1981 by region. The figures show that about 55 percent of total land supply was allocated to the SCR, where the housing demand pressure was strongest. This is

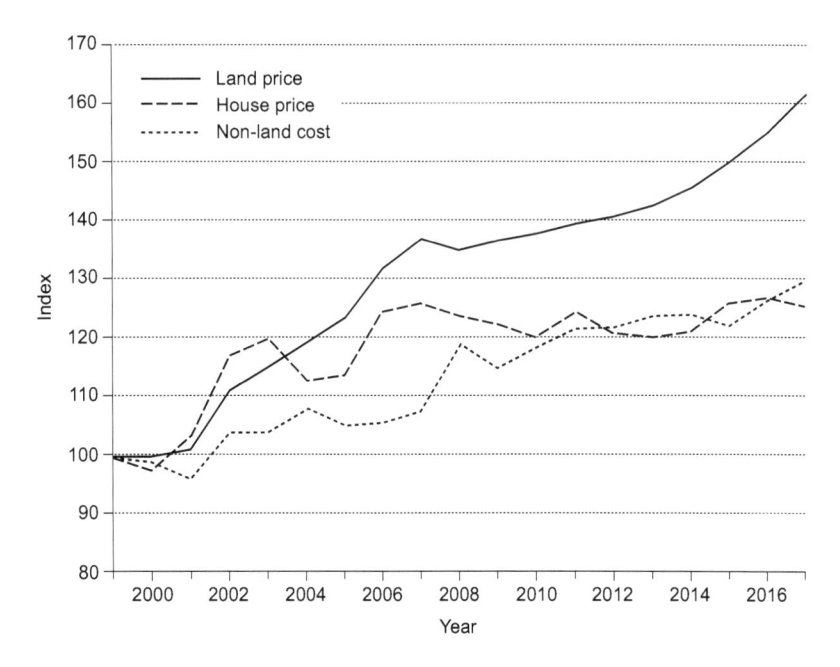

Figure 8.3 Housing price, land price, and construction cost in real terms.
Source: Based on the data provided by the Construction Associaion of Korea.

somewhat larger than the share of the region's population in national population, which stands at 49 percent.

A policy goal of increasing developable land is to expand housing supply, thereby stabilizing the housing price. The price of housing is determined by the interplay between demand and supply in the housing market and also by macroeconomic factors, the interest rate in particular. On the supply side, the price of new housing is affected by the availability and prices of inputs, especially of land. Figure 8.3 illustrates the trends in real housing price against real land price and real construction cost (non-land cost inclusive of materials and labor)[14] since 1999. Figure 8.3 shows that land price increased much faster than housing price, while the pace of increase in non-land construction cost was slower than that of housing price, although the gap has been closed in recent years. Since land cost takes about one-third of the total cost of new housing on average, the fact that real housing price has stayed almost constant during the past 10 years despite the sharp increase in land price implies that home-builders somehow absorbed the incremental cost.[15]

3.4 Housing cycles

Housing construction activity goes through cycles reflecting demand and supply factors as well as macroeconomic conditions. Figure 8.4 illustrates the fluctuations

Land and housing delivery in Korea 165

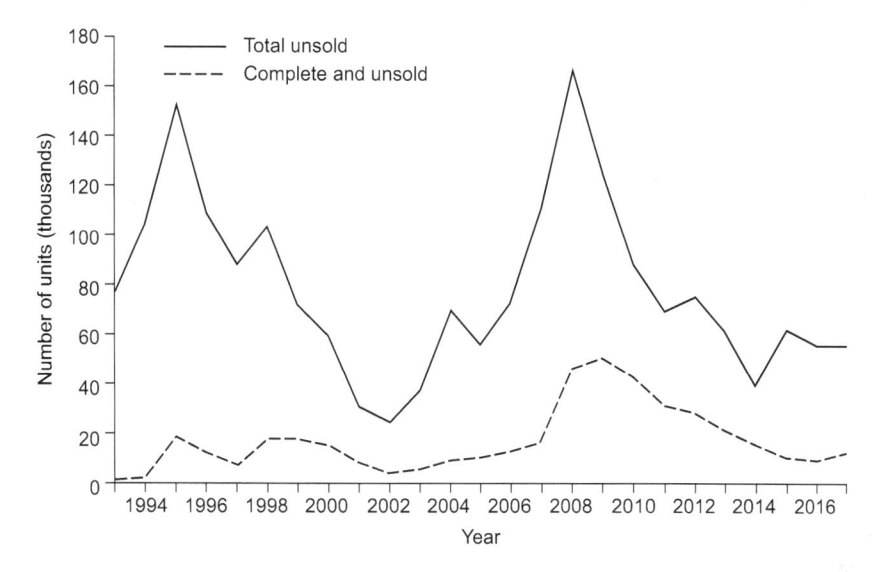

Figure 8.4 Number of unsold dwellings.
Source: MOLIT.

in the number of unsold dwellings and the number of completed but unsold units since 1993. The total number of unsold units had two peaks, one in 1995 as a result of massive increase in supply and softening of housing prices, and the other in 2008 due to the GFC. The number of completed but unsold new dwellings increased significantly in 2008 and 2009.

Housing cycles can be seen in terms of fluctuations in housing price. Figure 8.5 exhibits real house price indices for Korea, Seoul, Gyeonggi Province, and the six provincial cities. A few patterns are noticeable. First, after the completion of the TMHD, housing prices in both groups fell in real terms throughout the 1990s, before they took a sharp drop in the wake of the AFC. Housing prices in the two regions reveal a fairly strong co-movement until the outbreak of the AFC, with the provincial cities registering a slightly higher appreciation rate. Second, after a sharp but short-lived drop in housing prices in 1998, both markets experienced a strong price recovery starting from the early 2000s, with Seoul and Gyeonggi Province in the SCR showing a much higher growth rate than the provincial cities. Third, housing prices in Korea as a whole have registered a moderate fluctuation in real terms since 2007. Finally, a pattern of price decoupling emerged after the crisis: while the SCR experienced a price decline in real terms between 2009 and 2014, the provincial cities registered a fairly strong upward price trend. Housing price appreciation in the former group has exceeded that of the latter group since then.

Figure 8.5 Real housing price cycles.
Source: KB Kookmin Bank.

4. The current system of housing delivery and its notable attributes

4.1 An overview of the system of residential development

The supply of new housing comes from either new residential land development or the redevelopment of built-up areas. Although the latter is responsible for most of the new supply in the city core, the vast majority of new units are produced through the former, as the data presented in the previous section show. Therefore, our discussion in this section will focus on the former.

A typical residential development project in Korea consists of three phases, and each phase is associated with a distinct risk-return profile, as illustrated in Figure 8.6: (1) the land acquisition and development phase (time t_0 to t_1 in the figure); (2) the pre-sale and construction phase (time t_1 to t_3); and (3) the occupancy and operation phase (after t_3). The vast majority of apartments are pre-sold in Korea. Under the pre-sale system, a homebuyer enters into a contract with a home-builder at the beginning of the construction phase (t_1) and pays an initial deposit, usually 10–20 percent of the price of the unit. Subsequently, the buyer makes a series of interim payments totaling about 60 percent of the price (t_2), and the final payment is made at the time of occupancy (t_3). Various institutions, both public and private, play a role in each phase. We will elaborate on the notable attributes of the current system of residential development in Korea.

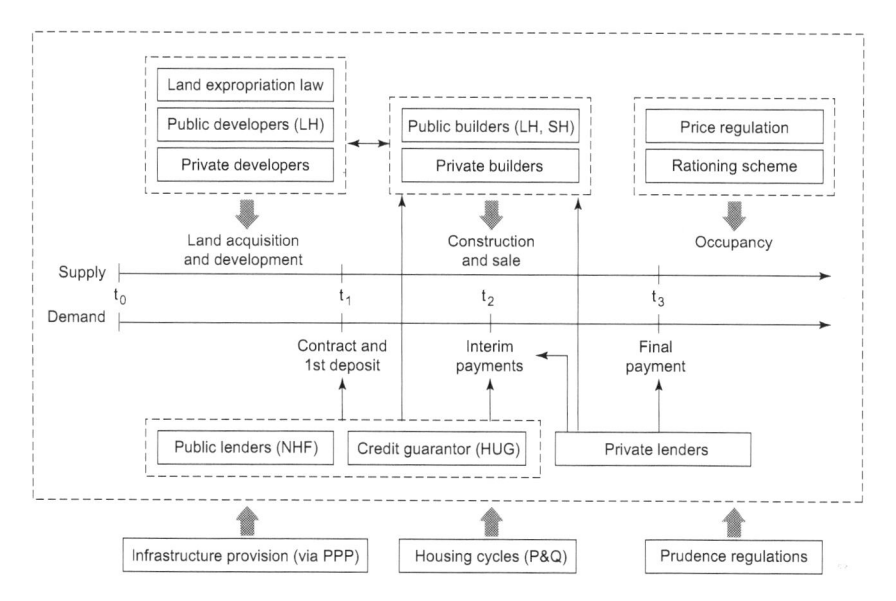

Figure 8.6 The process of land and housing development in Korea – a schematic view.
Note: P and Q refer to price and quality, respectively.

4.2 Land development

The residential development process begins with the acquisition of properly zoned land. For large-scale housing development projects, LH purchases raw land designated for development through negotiation or expropriates paying compensation to landowners, services the land with onsite infrastructure, subdivides the land into blocks of plots, and sells the subdivided plots to private home-builders. In some cases, LH builds apartments for renting or owner-occupation. As mentioned earlier, land expropriation by public-sector developers is an important building block in delivering housing in large quantities and at more affordable prices than otherwise. The public-sector monopoly on large-scale land development has been justified on the grounds that the windfall gains accruing to landowners due to the rezoning of land should be recouped for the benefit of society. Private-sector developers are allowed to participate in redevelopment projects and small projects.

4.3 Provision of infrastructure

Provision of transportation and on-site infrastructure is a prerequisite for successful land and housing development projects. On-site infrastructure includes water and sewage, electricity, communication networks, co-generation plants, garbage incinerating plants, parks, and man-made lakes.

In large-scale land development projects, including the new towns in the SCR, efforts were made to coordinate land and housing development with the provision of infrastructure. The first-phase new towns were equipped with urban parks, internal roads, and parking lots of higher quality than other cities in the SCR.

In general, the on-site infrastructure was financed by developers based on the beneficiary-pays principle. For example, the Korea Land Corporation, the land developer of Bundang New Town project, had a total budget of 4.16 trillion KRW. Approximately 1.33 trillion KRW (32 percent) was allocated for on-site improvement and infrastructures, 1.58 trillion KRW (38 percent) for roads, and the remaining 1.25 trillion KRW (30 percent) for land purchase. The cost of water, sewage, and other on-site infrastructure has been factored into the sale price of buildable land and subsequently passed onto the homebuyers, who are the beneficiaries of the infrastructure.

Although the public–private partnership (PPP) method has been applied in various infrastructure projects since 1994 to alleviate the fiscal burden on the government and to accommodate innovative ideas, applications to infrastructure related to land and housing development are limited. In the case of the first- and second-phase new towns, the school board of Gyeonngi Province applied a build-transfer-lease (BTL) scheme to close the revenue shortfall from public financing. The cost of constructing school buildings was paid by the school board. A total of 885 billion KRW was invested in 69 schools through BTL.

4.4 Financing

During the construction phase, a prospective homebuyer enters into a pre-sale contract with a home-builder. The down payment and interim payments during the construction phase are an informal financing mechanism for development finance offered by the homebuyers to the contractor. The contractor is mandated to buy insurance from the HUG for the delivery of the completed dwellings. In the event the contractor goes bankrupt, the project will be taken over and completed by a new contractor, or homebuyers can get a refund of the pre-paid amount if two-thirds of the homebuyers in the project agree to it. The consumers tend to take the first option when the housing market prospect is good, and the second option otherwise.

The HUG insurance also makes it possible for home-builders to acquire bank loans during the construction phase. Another public sector financier, the National Housing and Urban Fund (NHUF), provides affordable mortgage loans to moderate-income homebuyers and construction loans to small and medium-sized construction firms. The KHFC also provides a guarantee for bank loans to finance interim payments, as well as one for construction loans for home-builders.[16] The loan guarantees provided by the HUG and the KHFC are similar in risk exposure to the mortgage insurance products in the U.S., in that the guarantee protects the lenders from the credit risk embedded in the lending. Bank loans for interim payments are made to the homebuyers as a group without assessing the credit risk of individual buyers. For that reason, there is a concern about credit risk, as the volume of these loans has increased in tandem with an increase in new housing supply in recent years.

Another important vehicle for construction financing is the PF provided by private lending institutions during the phase of land acquisition and permit issuance. The PF scheme emerged during the period of economic recovery from the AFC. The developers who take PF loans are often small and highly leveraged. The PF loans are securitized into asset-backed securities (ABS) by creating a bankruptcy-remote special purpose company (SPC).

Along with the rising share of the private home-builders' contribution to total new housing construction over time (see Figure 8.1), PF loans increased rapidly in the early to mid-2000s, thanks partly to the record-low interest rate. The increase led to the volatile construction boom-bust for both residential and non-residential developments before and after the GFC, which generated a debate on how to enhance the soundness of construction financing. Areas for improvement include more solid valuation of the cash flows involved (by the lenders in particular), restriction on maximum leverage by developers, and more prudent loan underwriting (Son, 2012).

4.5 Allocation of new housing

In the third phase of housing development in Figure 8.6, newly built homes are allocated to target consumer groups based on an elaborate rationing scheme (in terms of duration of contract savings, size and composition of household, and so on), and the prices of these units have been regulated via the price ceilings set by the government. At the time of the final payment by consumers, conventional mortgage loans are issued by commercial banks and other financial institutions, at which point the lending risk is shifted from the public sector (i.e., the HUG and the KHFC) to the private sector (i.e., commercial banks and other financial institutions).

The long-held objective of housing policies in Korea has been to deliver new housing units at "affordable" prices to "non-speculators." To achieve the goal, the government has relied heavily on direct intervention through various regulations. The price regulation on new apartments, first introduced in 1977 in the form of a uniform ceiling applicable to all apartments, is a good example. The price ceiling was raised each year, until it was lifted on units with a net floor space of 85 square meters and above in July 1981. Then, in December 1982, price control was reintroduced and a different ceiling was set for each of two size categories: units with net floor space of 85 square meters and above, and units with 85 square meters or below. The price ceiling on the smaller units was adjusted upward slightly, but that on the larger units remained unchanged until 1988. Recognizing that the erosion of home-builders' profit constrains the expansion of new housing supply to fulfill the target of the TMHD, the government devised a new pricing scheme in November 1989, based on the cost of production.

The price control has spawned several subsidiary regulations. Since the price ceiling on new apartments creates a windfall gain for the buyers, and the amount of the windfall gain rises with the size of the dwelling unit, the government imposes a set of criteria on the eligibility of buyers, as well as controlling the size distribution of the dwelling units produced. The rules governing the allocation of the dwellings provided at controlled prices have become very complex, even for the government officers in charge of the process. The latest twist to the price regulation is the requirement to release itemized costs associated with the sales price proposed by the suppliers since 2007. Initially, 61 cost items were released on apartments built on land provided by the public sector, and seven items on those supplied by the private sector. Since 2014, only the former has needed to be revealed on 12 items. The proposed price of the apartments subjected to the price ceiling must be submitted to the local government for review by a committee consisting of local government officials, representatives of non-governmental organizations (NGOs), and consumer advocates.[17]

5. Concluding remarks and implications for other countries

This chapter has surveyed the evolution of the land and housing delivery system over the past 50 years or so. Korea has experienced rapid economic growth and urbanization since the mid-1960s. During the early stage of economic development, the housing sector was given low priority in the allocation of financial resources, and an insufficient amount of developable land was provided for housing. Consequently, a chronic shortage of developable land and affordable housing developed in Seoul and other major cities. The government has increased its efforts to address the housing challenge since the late 1980s. The drive to build two million new dwellings between 1989 and 1992 was a milestone event in the provision of housing in Korea. The sustained supply of new housing, disturbed only temporarily due to the AFC, helped to resolve the problem of absolute housing shortage by the early 2000s. Housing conditions have improved remarkably since 1980s in both quantity and quality, although affordability of housing for some segments of the population remains a problem.

We have identified four features of the Korean system that other countries faced with housing shortages might find relevant: (1) the early creation of public-sector institutions to deal with land acquisition, construction, and finance; (2) the reliance on conventional as well as non-conventional financing methods; (3) the development of elaborate rationing mechanisms for allocating new housing units; and (4) the integration between land development and infrastructure provision.

First, the institutional framework of the housing delivery system was laid out as early as the 1960s and became the cornerstone for efficient production and allocation of affordable housing units over the years.[18] Two public-sector institutions – the KNHC and the KLDC – have been instrumental in securing land for large-scale residential development projects, the actual production of affordable housing units, and also in the allocation (or rationing) of those units to the target consumers. Over time, the role of the private-sector institutions (e.g., construction companies and developers) has expanded, while the missions of those public-sector institutions have changed as a more mature and market-oriented housing sector has emerged. In order to better cope with the changing environment, the KNHC and the KLDC merged in 2009 to create a new organization, the LH, whose main mandate is to develop and operate public rental housing. Establishing the right balance between markets and government intervention is a difficult task.

Second, the financing mechanism has been a critical component of the Korean housing delivery system. The two key housing finance institutions were the KHB, established in 1968, and the NHF, created in 1981. Both institutions have injected liquidity into both suppliers (construction companies) and potential buyers (target consumers) via low-cost, subsidized lending. In addition, the risks associated with pre-sale contracts, the dominant method for the provision of new housing in Korea, have been hedged by another public-sector financial institution – the HUG – and its predecessors, the KHGC and the Homebuilders' Cooperative, established in 1993. The performance insurance product offered by the KHGC served as an important market-maker, protecting consumers from these risks, as well as enabling home-builders to obtain construction loans from banks. In recent years, both the NHF and the HUG were assigned a mandate of facilitating urban redevelopment projects in addition to residential development. REITs are being used more actively in the financing of rental housing and urban regeneration.

Third, completed housing units were allocated to targeted buyers based on elaborate eligibility rules and regulations. Eligibility criteria included a minimum period of contract saving for home purchase, length of residence, prior history of homeownership, and number of children. In addition, the government has regulated the sales price of new apartments since 1977 to make new housing more affordable for eligible homebuyers. The buyers of the new apartments were forbidden to flip them within a specified time period. However, the validity and effectiveness of the price regulation and its subsidiary regulations have been disputed for many years.

Fourth, large-scale residential development was integrated with land-use planning and the provision of urban infrastructure. Investment in infrastructure was made within the context of a comprehensive land-use plan, as best illustrated by the case of the five first-phase new towns around Seoul. The cost of on-site infrastructure was financed by the profits from land development, while that of the extension of trunk infrastructure was shared among beneficiaries. Although Korea has utilized various forms of PPP arrangements to alleviate the fiscal burden caused by the rising infrastructure spending since 1994, PPP has so far played a marginal role in financing infrastructure related to housing development. PPP does have potential, and the right modalities need to be further explored.

Any particular country at a given point in time has its own housing challenges. The Korean experience summarized above might be useful to some countries, provided that it is interpreted with reference to the specific context.

In closing, it is worth mentioning several topics that deserve more research and policy attention regarding land and housing supply. First, given the high priority given to housing price stability, it is important to improve our understanding of the patterns of housing prices and quantity cycles at the city level and their determinants. Our preliminary analysis suggests that the supply elasticity is smaller in Seoul and the SCR than in other provincial cities. More thorough investigation is needed of the city-level supply elasticity and its determinants, such as the natural constraints and regulatory barriers.

Second, as the overall housing shortage is no longer the main problem and the demand for large-scale developments decreases, the government is trying to promote urban regeneration and rehabilitation of older neighborhoods instead of greenfield development outside built-up areas. The success of the new approach hinges on the governance in the planning and implementation of projects, the community engagement, and the financing mechanism.

Third, PF needs to be strengthened to enhance the feasibility of residential development, as well as to mitigate market cycles. PF is more relevant as the pressure to limit the pre-sale scheme rises on the grounds that the financing scheme is prone to flipping and compromises the quality of the product. If apartments must be sold at completion or close to it, the developer/contractor must secure construction loans, which might be challenging to smaller players.

Notes

1 Efforts to address the issue of housing shortage started right after the Korean War ended. In September 1953 the government announced a presidential proclamation to build one million housing units within five years, but it was not realized due to a lack of appropriate implementation means for land expropriation, financing, and home-building.
2 The Land Bank, its predecessor, was established in 1975.
3 The discussion of the informal sector redevelopment in this section is based on Son (2015).
4 Lim, 2005.
5 See Cho and Kim (2011) for details of those post-AFC financial deregulations.
6 A zombie company is defined as one whose debt service coverage ratio is less than one (i.e., net operation profit is less than the amount of debt service) for three consecutive fiscal years. See Renaud et al., 2016, pp. 272–273.
7 The definition of the housing supply ratio has been modified since 2005. As a result, the figure became smaller. See Kim and Park, 2016, p. 94.
8 Over the period of 1981–2015, the NHF has funded 100 trillion KRW for 4.6 million dwellings, including 2.4 million rental units and 1.9 million owner-occupied units.
9 A similar pattern emerges from the data on new housing construction per 1,000 inhabitants, although the data are not reported here.
10 Between 1962 and 2014, LH constructed 2.43 million units, or 13.5 percent of the total (LH, 2014). Among the private-sector players, the combined share of the 10 largest contributors between 1998 and 2016 was about 22 percent.
11 The value of land is not included in housing investment, as land is not newly produced.
12 The average between 1995 and 2017 was 4.9 percent. According to the data from Eurostat, the average in 28 European countries during the same period was 4.4 percent, with Germany and France at 5.9 percent, and the UK and Sweden at 3.3 percent.
13 The homeownership rate was 59 percent, and the owner-occupancy rate was 57 percent in 2016.
14 The data on non-land cost is from the Construction Association of Korea.
15 Ideally, the comparison should be made against the sales price of new housing, as opposed to the market price of existing housing, which the housing price index used here measures. The share of land cost in the total cost of new housing production varies across location as well, as do the cost and the time of purchase of land. The share of land cost is believed to be around 30 percent on average.

16 In the absence of bank loans to finance interim payments, homebuyers lend money to finance construction activity. In this sense, the arrangement is an example of informal lending to construction companies.
17 The description of the price ceiling system and its assessment builds on Cho and Kim (2011).
18 The institutions are defined here as the combination of the housing-related legal entities established by the government, and the rules and regulations that impact the behavior of economic agents on the demand and supply sides of housing markets. Alternatively, institutions can be more broadly defined by adding governance, human capital, and informal rules that affect labor and/or capital productivity. See Rodrik (2013).

Appendix

Milestone events and housing policy interventions in Korea

Time Period	Macroeconomic social changes	Housing market conditions and policy involvements
• Phase 1 – Institution building (from the early 1960s till the mid-1980s)	• The first and second five-year Economic Development Plans (1962–66, and 1967–71); objectives – to prepare for industrialization, expand infrastructure, and achieve food self-sufficiency; planned annual growth rate: 7% • Construction of the Seoul–Busan expressway (length: 428 km, width: 22.4 meters) with partial government financing (23.6% from the national budget) (1968–70) • Development of the export industrial parks: the Guro Industrial Park on the outskirts of Seoul (focusing on light industrial products) and the Ulsan Industrial Park in the southeastern region (focusing on petro-chemical industries) • Establishing the first Comprehensive National Territorial Plan (CNTP) for 1972–81, to achieve "efficient utilization of territory for efficient economic growth"; subsequent revisions of the CNTP to accommodate changes in development needs	• Large-scale demolition of the housing stock during the Korean War (18% of 3.3m housing units destroyed during the war) • Increase in substandard housing units (the share of total housing stock in Seoul in the 1960s was about 20%, and 32% in the 1970s) • Establishing the Korea National Housing Corporation (KNHC) in 1962 to conduct housing construction and supply projects with partial government financing • Establishing Korea Housing Bank (KHB) in 1969, to finance private citizens and housing suppliers; privatized in 1997 as the Housing and Commercial Bank • The National Housing Fund (NHF) accounts installed in 1981 to supply funds for affordable housing

Time Period	Macroeconomic social changes	Housing market conditions and policy involvements
	• The Capital Region Population Rearrangement Plan in 1978, to control building and extending of factories within the Seoul Capital Region, and to accommodate residential developments to the south of Han River • Development of the Gumi Industrial Park and the Changwon Industrial Park to foster heavy and chemical industries	• A series of anti-speculation measures adopted: the anti-real estate speculation tax (1967), the price limits on newly constructed housing units (1977), the land transaction approval system, the anti-speculation taxes, and the special taxes on non-essential real estate holdings by corporations (1978)
• Phase II – Massive supply (from the late 1980s till the outbreak of the AFC)	• Population in the Capital Region grew from 5.2m (20%) in 1960 to 13.2m (34%) in 1980 to 18.6m (43%) in 1990 • Volatile macroeconomic conditions: high growth propelled by the construction boom in the Middle East in the 1970s and 1980s; cooling down in exports due to the recessions experienced by trading partners in the early to mid-1980s; and a boom following the 1988 Seoul Olympics • Privatization and deregulation in the early to mid-1990s: the sequential abolishment of the price ceiling on new housing from 1995, and privatization of the KHB in 1997 • Establishing the "real name" registration for real estate asset ownership in 1995, contributing to enhancing transparency in the real estate markets • The Asian Financial Crisis (AFC) as a large-scale macroeconomic shock, causing a broad-based financial liberalization, deregulation, and industrial restructuring	• Sluggish housing supply was due to the anti-speculation policies and the weakening macroeconomic conditions; the policy package to stimulate the real estate sector in the early 1980s • The housing supply ratio (total number of housing units/total number of households) dropped from 73.0% in 1980 to 70.9% in 1989 • Unprecedented and nationwide real estate price hikes in the late 1980s; the policy shift toward managing the real estate speculation • The Two Million Housing Drive (1988–92), including 900k units in the Capital Region; development of the five new towns around Seoul • Increase in urban land supply via deregulation in the "semi-agricultural" land area (26.5% of total land area) • Vast improvement in the housing supply ratio from 72.4% in 1990 to 93.3% in 1999 • Housing prices stabilized in Seoul and other major cities, and plummeted during and right after the AFC

Time Period	Macroeconomic social changes	Housing market conditions and policy involvements
• Phase III – The recovery from the AFC till the outbreak of the GFC	• Expansion in consumer lending, both credit card (non-collateralized) loans and mortgages backed by (real estate) loans; establishment of the credit data repository (Korea Credit Bureau) • Development of markets for advanced financial products, including ABS, MBS, and REITs • The Global Financial Crisis (GFC) became another milestone in rethinking housing policies: stabilization of real estate prices through macroprudential regulations, and the promotion of housing welfare (via public rental housing and housing vouchers)	• A set of lending restrictions (LTV/DTI limits, designation of speculative zones, changing risk weights for short-term adjustable-rate mortgages) to dampen house price increases • Twin housing price cycles in the early to mid-2000s, especially in the Gangnam submarket in Seoul, due to sluggish supply of housing • Announcement of construction of one million national rental housing units in 2003–12 (subsequently revised by various affordable housing initiatives in the 2000s) • Announcement of a plan to build 1.5m affordable housing units (0.8m for owning and 0.7m for renting) in 2008 • The housing supply ratio surpassed 100% in 2002
• Phase IV – Post-GFC market reshaping	• Acceleration of population aging and retirement of baby-boomers • Expansionary monetary policy (a low and stable interest-rate environment after the GFC) • The fourth Industrial Revolution and the rising impact of ICT innovations (e.g., Smart City, FinTech)	• Aftermath of project finance lending crisis and an increase in the non-performing ("zombie") construction companies • Demand for owner-occupied housing stagnated, while demand for rental housing increased, resulting in rising *chonsei* deposits and rents. • Policy responses: increased rental housing stock; both public rental housing and corporatized high-quality, long-term private rental housing

Sources: Cho, 2010. "The real estate stabilization policies in Korea," presented at the CASS Conference in Beijing, September 2010; Korea Development Institute (2012); Cho and Kim, 2011.

References

Chang, S. H. (1989). Urbanization, State, and Urban Poor. In H. K. Kim (Ed.). *Informal Housing Area and Redevelopment*. Seoul, Korea: Nanam Publishing (in Korean).

Cho, J. H. (1991). Policies for Developable Land for Housing. In *Solutions for housing problems*, Seoul, Korea: Hyundai Research Institute.

Cho, M. (2010). The Real Estate Stabilization Policies in Korea, Presented at the CASS Conference in Beijing, September 2010.

Cho, M. (2014). Housing Price and Mortgage Credit Cycles. In Susan Wachter, Man Cho, and Moon Joong Cha (Eds.). *The Global Financial Crisis and Housing: A New Policy Paradigm*, Edward Elgar, 82–111.

Cho, M. and Kim, K-H. (2011). Housing Sector Reform in Korea. *Revised draft of presentation at the OECD Conference Making Reform Happen*. Paris, November 2010.

Cho, M. and Kim, K-H. (2015). Real Estate Cycles: International Episodes. In P. Chinloy and K. Baker (Eds.). *International Real Estate Investment*. New York: Oxford University Press.

Green, R., Malpezzi, S., and Vandell, K. (1994). Urban Regulations and the Price of Land and Housing in Korea. *Journal of Housing Economics*, 3, 330–356.

Kim, J. W. (1980). *A Study on Informal Housing Policies: Case of Seoul*. Master's thesis, Graduate School of Public Policy. Seoul, Korea: Konkuk University (in Korean).

Kim, K.-H. and Cho, M. (2010). Structural Changes, Housing Price Dynamics and Housing Affordability in Korea. *Housing Studies*, 25(6), 839–856.

Kim, K.-H. and Park, M. S. (2016). Housing Policy of the Republic of Korea. In N. Yoshino and M. Helble (Eds.). *The Housing Challenge in Emerging Asia: Options and Solutions*. Tokyo: Asian Development Bank Institute, 92–125.

Korea Development Institute (2012). *Policy for the Construction and Supply of Affordable Housing in Korea*. Seoul, Korea: KSP Modularization Project.

Land and Housing Corporation (LH). (2014). *Land and Housing Statistics*, Seoul, Korea.

Lim, S.H. (2005). *Housing Policies over the past Half Century*, Seoul, Korea: Gi-Moon-Dang.

MOLIT (a) (various years). Compiled data from several documents published by the Ministry of Land Infrastructure and Transport. www.index.go.kr/potal/main/EachDtlPageDetail.do?idx_cd=1230 (in Korean); www.index.go.kr/potal/stts/idxMain/selectPoSttsIdxSearch.do?idx_cd=1229

MOLIT (b) (various years). http://kosis.kr/statHtml/statHtml.do?orgId=116&tblId=DT_MLTM_5546&conn_path=I3

Renaud, B. R., Kim, K.-H., and Cho, M. (2016). *Dynamics of Housing in East Asia*. West Sussex, UK: Wiley Blackwell.

Rodrik, D. (2013). *The Past, Present, and Future of Economic Growth*. Working Paper 1, June 2013, Global Citizen Foundation.

Seoul Metropolitan Government. (2017). *White Paper on Land Readjustment*. Seoul, Korea.

Son, J.-Y. (2012). Korea's Development Finance at the Crossroad. In S. Wachter, M. Cho, and M. Tcha (Eds.). *The Global Financial Crisis and Housing: A New Policy Paradigm*. Cheltenham, UK: Edward-Elgar-KDI, 208–228.

9 Housing price, housing mobility, and housing policy in Taiwan

Chien-Wen Peng

NATIONAL TAIPEI UNIVERSITY

Bor-Ming Hsieh

CHANG JUNG CHRISTIAN UNIVERSITY

Chin-Oh Chang

NATIONAL CHENGCHI UNIVERSITY

1. Introduction

The factors influencing the housing market are various and complicated. In addition to demand and supply conditions in the market, the macro-economic and financial environments have important effects on the housing market. The disequilibrium between demand and supply in the housing market has not only caused severe fluctuations in housing prices and transaction volumes, but has also had serious impacts on economic growth and on the stability of the financial market; it has even led to social inequity. As a result, in making decisions on housing policy, the government should consider the diversity and complexity of the housing market.

Taiwan's real estate market has experienced four dramatic cycles over the past four decades.[1] The first and second cycles occurred in 1973–74 and in 1979–80, respectively. These two cycles resulted from the inflation caused by rising oil prices in the 1970s. Many households eagerly purchased real estate to protect themselves against inflation. The third real estate cycle, which occurred in 1987–89, resulted from a rapid growth of economic and money supply, excess saving, appreciation of currency, and deregulation of the financial market; the cycle thus drove up real estate prices, and the stock price index reached a historic high.[2] This economic chaos and asset price bubble triggered a substantial housing protest in Taiwan on August 26, 1989. Thousands of protesters camped out overnight on Zhongxiao East Road, Section 4 in Taipei City, near Sogo Department Store, and became known as the "Shell-less Snail Movement." To restrain a rapid rise in housing prices, the central bank launched a selective credit control policy on February 28, 1989;[3] this policy gradually came into effect after 1990.

The fourth and most recent cycle started during the second quarter of 2003, just after the SARS outbreak, and lasted to the end of 2014. During this period, housing prices increased around 100% to 200% across the whole island, which is a much higher increase than economic growth rates and the consumer price

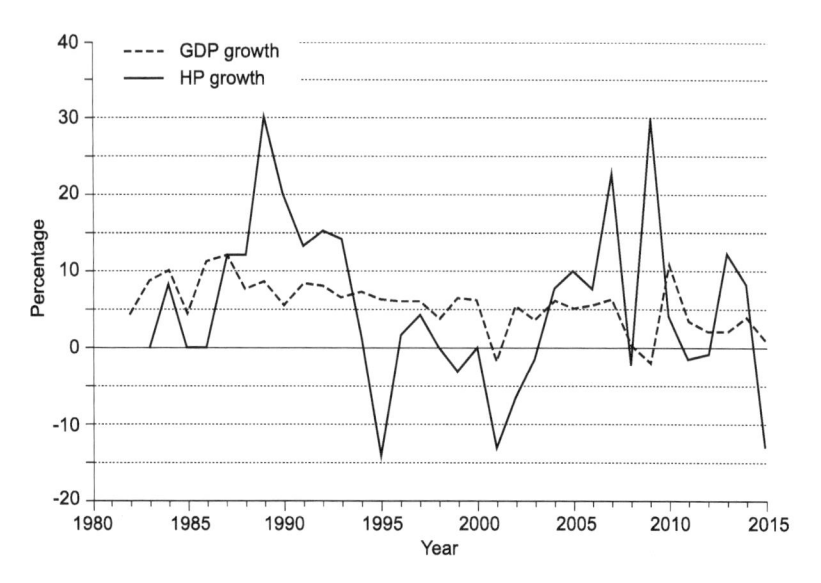

Figure 9.1 Annual house price (HP) growth and annual Gross Domestic Product (GDP) growth.
Source: 1.GDP data is from National Statistics, Directorate-General of Budge, Accounting and Statistics, Executive Yuan, R.O.C. (Taiwan).

index (Figure 9.1, Figure 9.2). Furthermore, the household income growth rate during this period was less than the inflation rate, effectively making the income growth negative. The fourth cycle has been totally different from the previous three cycles because the rising housing prices were triggered by low interest rates and excess money supply in the financial market, rather than being driven by economic growth.

The rapid increase in housing prices has caused a serious affordability problem, which affects not only many young and renter households but also many homeowners who cannot find houses suitable to their needs, thus reducing housing mobility and depriving them of the chance to improve their living conditions. In response, the government has proposed policies to address this severe affordability problem, such as a series of selective credit controls, increasing transaction costs for non-occupant households, and a new registration regime based on the true market prices to provide transparent information on the housing market (the details of these policies will be discussed in Section 4).

These policies have caused a significant decline in transaction volumes in the housing market. In 2011, a total of 406.7 thousand units were sold, but by 2016 the number of units sold was down to 245.4 thousand units, which is the historical low in Taiwan. (Figure 9.3). Yet housing prices steadily rose during this significant reduction in the number of housing units transacted.

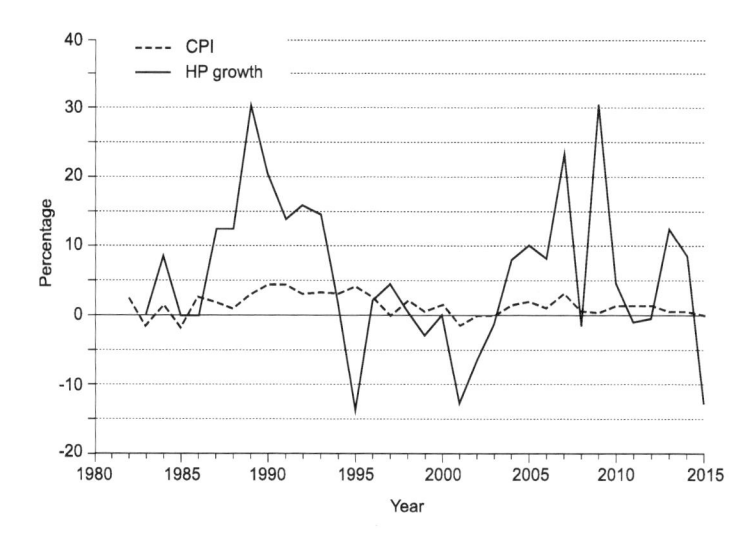

Figure 9.2 Annual house price (HP) growth and annual change of Consumer Price Index (CPI).

Source: 1. Annual percent change of consumer price index is from National Statistics, Directorate-General of Budge, Accounting and Statistics, Executive Yuan, R.O.C. (Taiwan).

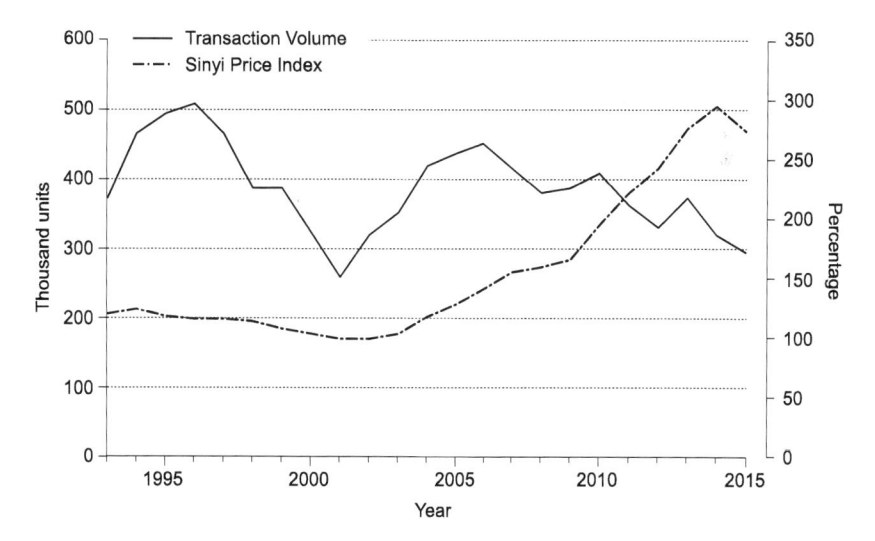

Figure 9.3 House price index and transaction volume.

Source: 1. Transaction volume data is from the Real Estate Information Platform, Ministry of the Interior, R.O.C. (Taiwan).

In the past four decades, the goal of housing policy in Taiwan was to "help every household to own a home." Therefore, the government spent most resources to build public housing units for sale and to provide mortgage subsidies to homebuyers. These owner-oriented subsidies, however, have raised inequality in the housing market. Most low-income households cannot afford to own their homes and therefore cannot benefit from the homeowners' subsidies. As a result, Taiwan's government launched the Overall Housing Policy in 2005, and the goal of housing policy has changed to "help every household to live in a decent home." The focus of housing subsidy programs has shifted to rental subsidies. Given the long imbalance in the allocation of housing subsidies, the number of public rental housing units accounts for only 0.08% of the total housing stock.[4]

To protest unreasonably high housing prices, thousands of citizens from all over the island camped out overnight on Taipei's Renai Road in front of The Palace, a luxury residential complex, on October 4, 2014. During that night, the organizer, the Housing Movement, raised five requests for housing market reforms as follows: (1) ensure housing rights in the Constitution;(2) reform housing taxation to inhibit speculation; (3) build more social housing units to reach 5% of the total housing stock; (4) stop building affordable housing units for sale; and (5) reform the rental housing market. This Housing Movement also commemorated the Shell-less Snail Movement held 25 years earlier. Previous housing policies and subsidy programs that focused on the owner-occupied sector have caused an unbalanced development of Taiwan's housing market. A series of strategies and suggestions are offered in this study to enhance a healthy development of Taiwan's housing market.

The analysis is structured as follows. The next section reviews the problems in Taiwan's housing market and their causes. The third section analyzes the change in housing prices and housing mobility. The fourth section discusses the current housing policy reforms. The fifth section analyzes the challenges and future development of housing policy. The final section provides our conclusions.

2. Housing problems and causes

There are three specific, unusual phenomena in Taiwan's housing market: high price-to-income ratios across markets and exceptionally high homeownership rates for a market economy co-exist with very high housing vacancy rates. As discussed earlier, the fourth real estate cycle started during the second quarter of 2003 and lasted until the end of 2014. During that cycle the average housing price increased around 100% to 200% across the whole island. In particular, the price-to-income ratio (PIR) in Taipei City reached a historic high of 16.2, and the mortgage payment-to-income ratio (MIR) also rose to 68.6% in the first quarter of 2015. Taipei City has become one of the world's worst cities in the difficulty of achieving owner-occupation.[5] In its neighbor city, New Taipei City, the PIR and MIR also reached 13.0 and 55.4%, respectively. Meanwhile the average PIR and MIR were 8.5 and 35.9% for the island. Those figures show

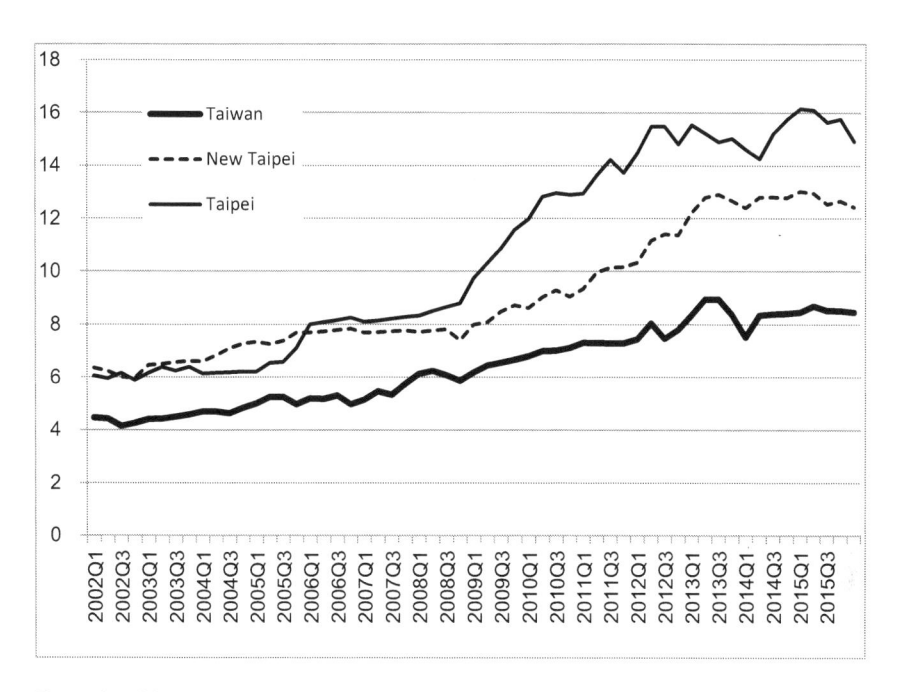

Figure 9.4 House price-to-income ratio.
Source: House price-to-income ratio data is from the Real Estate Infromation Platform, Ministry of the Interior, R.O.C. Taiwan.

that housing prices are especially high and unaffordable in northern Taiwan (Figure 9.4, Figure 9.5).[6]

The rapid increases in housing prices are affected by several factors, namely quantitative easing (QE) in the monetary policy of major central banks, active economic interactions and active trade between Taiwan and mainland China, low interest rates, and especially low rates on estate and gift taxes, which have attracted overseas hot money to keep flowing back into Taiwan.[7] Furthermore, the tax base of real estate taxes in Taiwan is based on the assessed value, which is significantly lower than the market value. As a result, the effective rate for real estate taxes accounts for only about 0.09% to 0.13% compared to the market value, and this is much lower than the weighted nominal rate 0.6% (Peng et al., 2007). The tax basis of the Land Value Tax is based on public assessed value, which accounts for merely 15–20% of the market value. The tax basis for the Land Value Increment Tax is based on assessed present land values. which account for merely 50–60% of the market value.[8] In view of these long-standing conditions, the Taiwanese have seen real estate as the most attractive investment due to its high return and low risk.

Even though skyrocketing housing prices have caused a severe affordability problem for most people, the homeownership rate in Taiwan is still rising. From

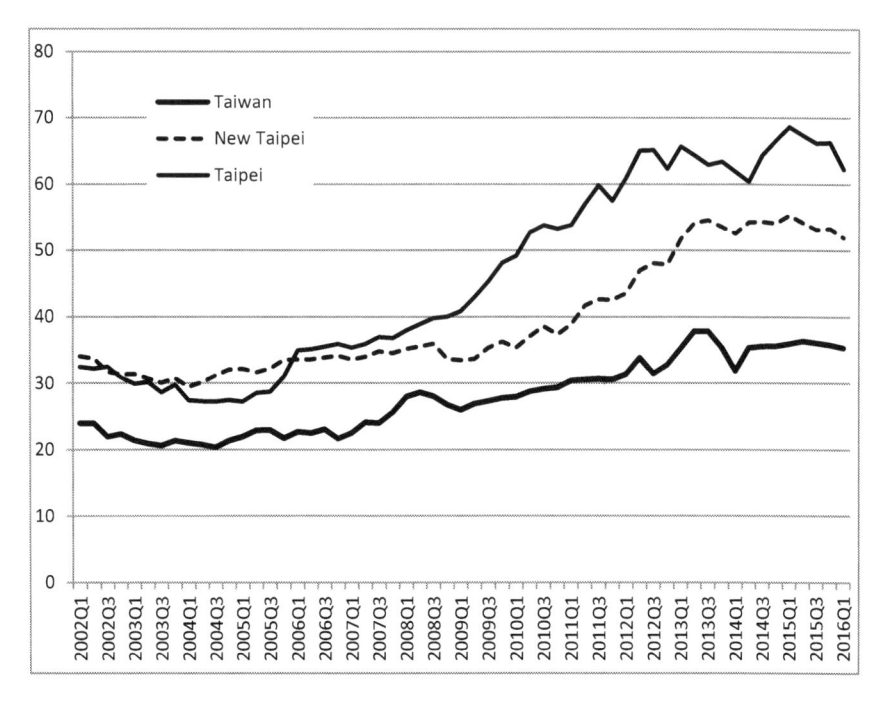

Figure 9.5 Mortgage payment-to-income ratio.
Source: Mortgage payment-to-income ratio data is from the Real Estate Infromation Platform, Ministry of the Interior, R.O.C. Taiwan.

1980 to 2010, according to the statistics of various Population and Housing Census Surveys, the overall homeownership rate rose from 79.1% to 84% island-wide. For major metropolises, the homeownership rate has increased from 59.2% to 81.5% in Taipei City, from 69.7% to 84% in New Taipei City, and from 77.04% to 86.5% in Taoyuan City (Figure 9.6).

Why is the homeownership rate so high in Taiwan? As Bourassa and Peng (2011) have shown, the rapid rise in housing prices has comparatively reduced the transaction cost for homeowners, making owning is much more attractive than renting. Even though the homeownership rate is as high as 85%, the severe problem of housing affordability still remains. The very high homeownership rate for a market economy implies that households do not face a neutral housing tenure choice. As mentioned earlier, the government's subsidy program has favored homebuyers for a long time and has resulted in a small and vulnerable rental sector of rather low quality. These prevailing conditions have forced households to buy their homes rather than renting, in order to achieve living conditions of higher quality.[9] Furthermore, in the face of rising housing prices over the long

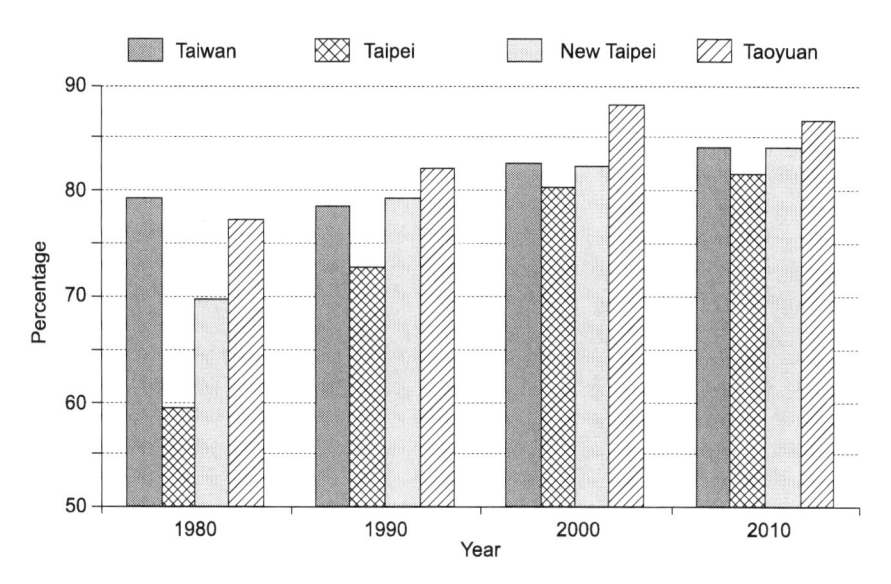

Figure 9.6 Homeownership rates.
Source: Home ownership rate data is from Housing Census of 2010, Directorate-General of Budge, Accounting and Statistics, Executive Yuan, R.O.C. Taiwan.

term, many households would rather buy their homes at the present time than buy later, which also means that they must pay very high mortgage payments and sacrifice some important family life cycle decisions, such as having babies. The results from Peng and Tsai (2012) show that many young households are buying homes but are forced to decide not to have children. Housing market conditions are the major reason for the low fertility rate in Taiwan, which is currently among the three lowest in the world according to the respected Population Reference Bureau.

The steady increase in housing prices has led households to pay more attention to the high returns from sales than to leasing, because the rental yield is incredibly low.[10] In addition, landlord–tenant disputes have raised transaction costs in renting. Under these conditions, many homeowners would rather leave their investment homes vacant than rent to tenants. According to the Population and Housing Census Surveys, the overall vacancy rate increased from 13.1% to 19.3% between 1980 and 2010 island-wide. For major metropolises, the housing vacancy rate rose from 11.3% to 13.8% in Taipei City, from 16.7% to 22.1% in New Taipei City, and from 16.7% to 21% in Taoyuan City (Figure 9.7). This study claims that the high vacancy rates not only seriously distort the allocation of national housing resources, but also worsen housing justice: many Taiwanese households still cannot afford to buy their own homes in the presence of such a high vacancy rate in the housing market.

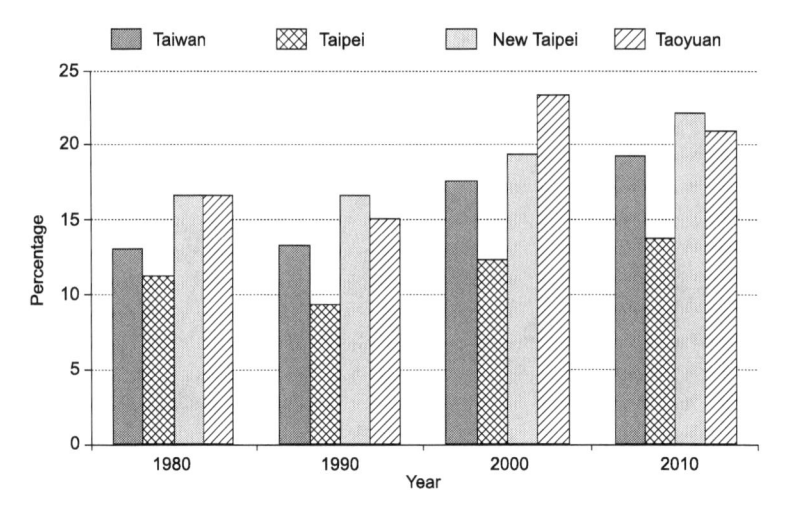

Figure 9.7 Housing vacancy rates.
Source: Housing vacancy rates data is from Housing Census of 2010, Directorate-General of Budge, Accounting and Statistics, Executive Yuan, R.O.C. Taiwan.

3. Housing prices and housing mobility

Moving to a different house causes many tangible and intangible costs, especially for homeowners. Theoretically, households will measure the costs and benefits of moving carefully before they act. Any voluntary household move can be seen as an improvement of housing quality, but a long-distance move might occur due to job-related rather than housing-related reasons.

According to previous studies, demographic factors, especially migration and intra-city mobility, influence the demand for housing and then the transaction price and transaction volume of the housing market. From a national perspective, an increase of immigrants is one important trigger of housing rent or price appreciation. From a regional perspective, an increase of net migration into a region will also increase housing demand and push housing prices up, and vice versa. However, housing demand contains both consumption and investment demands. The influence of migration on housing price depends on the relative change of these two demands (Saiz, 2007; Akbari and Aydede, 2012).

Conversely, a change of housing price will also influence mobility. Sterk (2015) finds that a decline in housing price reduces geographic mobility because homeowners face declines in their home equity levels, which make it more difficult to provide the down payment required for a new mortgage loan. However, Nenov (2015) reveals that regional reallocation has a limited importance for the aggregate labor market, and the effects of housing markets on relocation are small.

Peng et al. (2009) find that the intra-regional mobility rate is negatively associated with the homeownership rate, but positively associated with the marriage rate, floor area of occupancy permit, housing vacancy rate, and ratio of price to income. Both homeownership rates and marriage rates are among the key determinants. From the empirical results follow some implications. First, owning a house will increase a household's cost of moving, so the possibility of a household's moving will be lower. Second, married households have higher likelihoods to go through a change in their family life cycle, so the possibility of a housing demand adjustment will also be higher and cause a higher residential mobility rate. Third, the lower level of housing affordability in a city means that households cannot gain access to their ideal house in one step. They need to adjust in several steps in order to reach their ideal house, so the mobility rate will be higher. Finally, households have more choices to move if the housing supply in a city is elastic, either by new construction or vacant houses, so the housing mobility rate will be higher.

We can summarize that an increase of housing price in a region will have four different partial effects on residential mobility:

1. An increase in housing prices will increase the home equity of homeowners. To improve housing quality or realize capital gain, the households in this region are more capable of moving to locations with lower housing prices in the same or other regions. This capability implies that an increase of housing price has a positive effect on inter-region or intra-region residential mobility.
2. An increase in housing prices will worsen the housing affordability of renters in a region, which may make them give up on moving or force them to move to locations with lower housing prices in the same or other regions. The net effect of rising housing prices on renters' mobility is ambiguous.
3. If some households expect that house prices will rise continually in the future, they will choose to stay in place, instead of moving. There is a negative effect of rising housing prices on households' relocation decisions, which causes mobility to decrease.
4. If the housing price appreciation rate in one region is lower than in other regions, then the households will be less capable of moving to other regions, causing lower residential mobility. Conversely, if the housing price appreciation rate in one region is higher than in other regions, it will block the households in other regions from moving in, and cause a lower in-migration rate.

Since the relative housing price levels and the quality of public facilities and services are quite different between and within regions, the factors influencing the impact of rising housing prices on residential mobility will be very diverse and complicated (Cebula and Vedder, 1973; Cebula and Belton, 1994; Grassmueck, 2011; Cebula an Nair-Reichert, 2012). We use three cities in Taipei metropolises—Taipei, New Taipei, and Taoyuan—to examine the changes in residential mobility during booms and busts in the housing market. Taipei is the core city of the Taipei Metropolitan Region, which is surrounded by New Taipei and Taoyuan, located in the south of New Taipei (Figure 9.8). The housing price

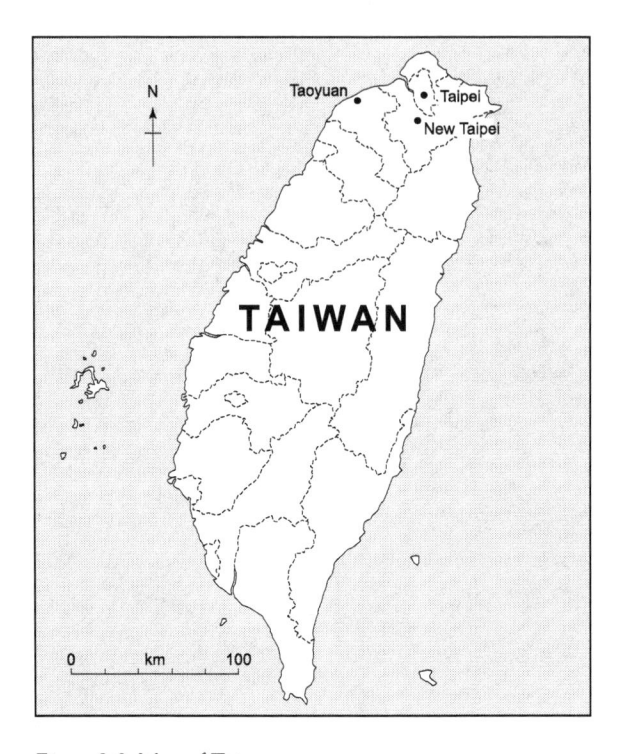

Figure 9.8 Map of Taiwan.

level of Taipei is the highest, around double that of New Taipei and quadruple that of Taoyuan.

According to the real estate cycles, we define the bust period as being between 1994 and 2002, and the boom period as being between 2003 and 2014. The housing prices started to fall in the beginning of 2015. We measure the housing mobility by using four different indicators: net migration rate (NMR), migration rate (MR), intra-city mobility rate (ICMR), and intra-district mobility rate (IDMR) (Figures 9.9, 9.10, 9.11, and 9.12, respectively).[11] The NMR is the difference between in-migrants and out-migrants divided by the midterm population of a city. The MR is the sum of in-migrants and out-migrants plus intra-city movers divided by the mid-term population of a city. The ICMR is the sum of interdistrict and intra-district movers divided by the mid-term population of a city. The IDMR is the intra-district movers divided by the mid-term population of the city.

3.1 Taipei

The average NMR between 1994 and 2015 is -0.4%. The average NMRs during the bust period and boom period are -0.78% and -0.11%, respectively. The NMR reveals that an increase of housing prices reduces the out-migrants, especially

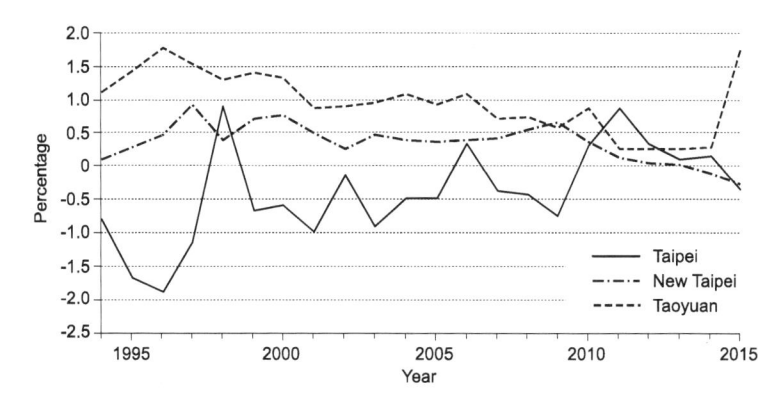

Figure 9.9 Net migration rate.
Source: Net migration rate data is calculated from National Satistics, Directorate-General of Budge, Accounting and Statistics, Executive Yuan, R.O.C. Taiwan.

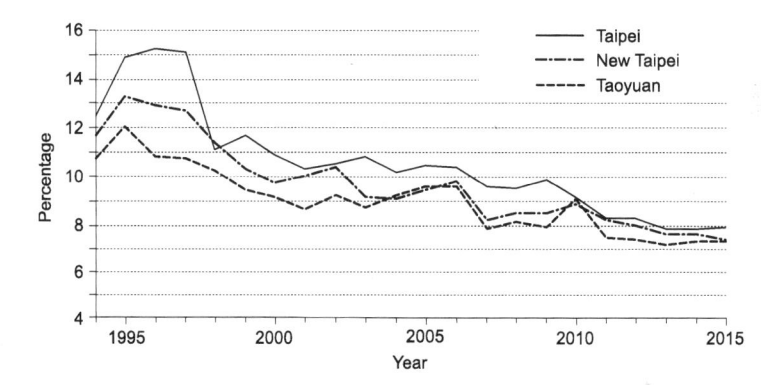

Figure 9.10 Migration rate.
Source: Migration rate data is calculated from National Satistics, Directorate-General of Budge, Accounting and Statistics, Executive Yuan, R.O.C. Taiwan.

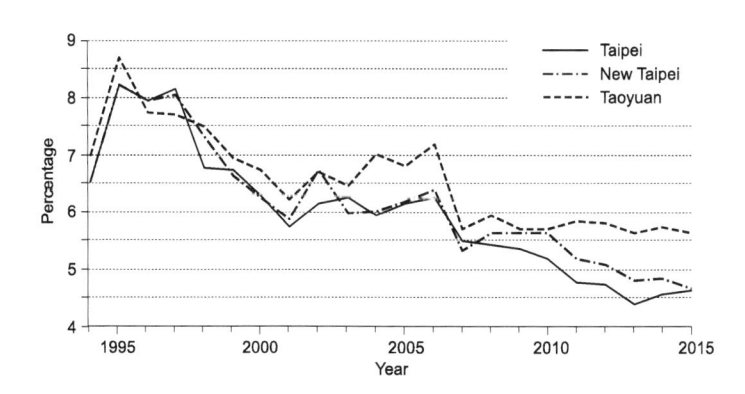

Figure 9.11 Intra-city mobility rate.
Source: Intra-city mobility rate data is calculated from National Satistics, Directorate-General of Budge, Accounting and Statistics, Executive Yuan, R.O.C. Taiwan.

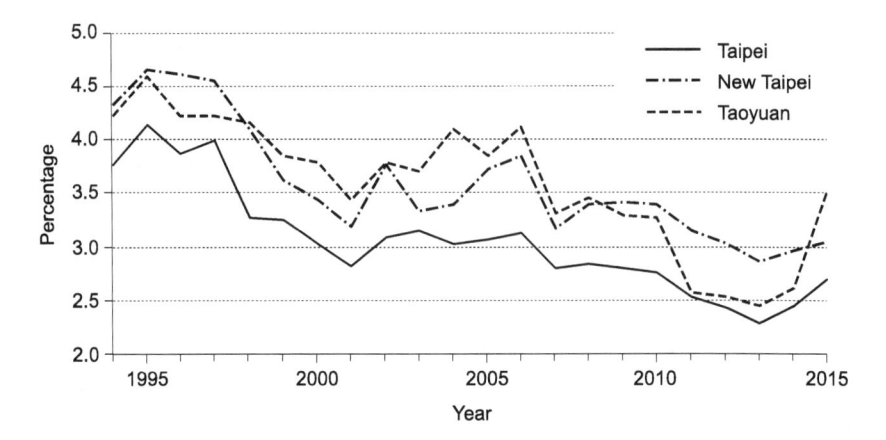

Figure 9.12 Intra-district mobility rate.
Source: Intra-district mobility rate data is calculated from National Satistics, Directorate-General of Budge, Accounting and Statistics, Executive Yuan, R.O.C. Taiwan.

since the NMR in Taipei has turned into a positive value since 2010. The average MR between 1994 and 2015 is 10.6%. The average MRs during the bust period and boom period are 12.47% and 9.36%, respectively. The MRs also reveal that increased housing prices reduce population migration.

The average ICMR between 1994 and 2015 is 6.0%. The average ICMRs during the bust period and boom period are 6.95% and 5.38%, respectively. The ICMR reveals that an increase of housing prices reduces residential mobility in a city. The average IDMR between 1994 and 2015 is 3.06%. The average IDMRs during the bust period and boom period are 3.47% and 2.77%, respectively. The IDMR also reveals that an increase of housing prices reduces intra-district residential mobility in a city.

3.2 New Taipei

The average NMR between 1994 and 2015 is 0.36%. The average NMRs during the bust period and boom period were 0.49% and 0.31%, respectively. The NMR reveals that an increase of housing prices reduces the out-migrants of Taipei and also the in-migrants of New Taipei. The average MR between 1994 and 2015 is 9.71%. The average MRs during the bust period and boom period are 11.40% and 8.60%, respectively. The MR also reveals that increased housing prices reduce population migration.

The average ICMR between 1994 and 2015 is 6.14%. The average ICMRs during the bust period and boom period are 7.06% and 5.56%, respectively. The ICMR reveals that an increase in housing prices reduces residential mobility in a city. The average IDMR between 1994 and 2015 is 3.60%. The average IDMRs during the bust period and boom period were 4.03% and 3.31%, respectively. The

IDMR also reveals that an increase in housing prices reduces intra-district residential mobility in a city.

3.3 Taoyuan

The average NMR between 1994 and 2015 is 0.97%. The average NMRs during the bust period and boom period were 1.29% and 0.67%, respectively. The NMR reveals that an increase in housing price reduces the out-migrants of Taipei and New Taipei and also the in-migrants of Taoyuan. The average MR between 1994 and 2015 was 9.03%. The average MRs during the bust period and boom period were 10.13% and 8.31%, respectively. The MR also reveals that an increase in housing prices reduces population migration.

The average ICMR between 1994 and 2015 was 6.58%. The average ICMRs during the bust period and boom period were 7.24% and 6.13%, respectively. These ICMRS reveal that an increase in housing price reduces residential mobility in a city. The average IDMR between 1994 and 2015 was 3.60%. The average IDMRs during the bust period and boom period were 4.03% and 3.02%, respectively. It also reveals that an increase in housing price reduces intra-district residential mobility in a city.

3.4 Comparison of the three cities

Taipei is an out-migration city, and New Taipei and Taoyuan are in-migration in the Taipei Metropolises. The NMR in Taoyuan is double that of New Taipei, which might be due to the difference in housing price levels in these cities. However, an increase in housing prices in Taipei reduces the out-migration of Taipei and also the in-migration of New Taipei and Taoyuan. The MR of Taipei is the highest, followed by New Taipei, and Taoyuan has the lowest MR. The MRs reveal that the higher the housing price, the more housing demand adjustments are needed, causing higher residential mobility.

As for the ICMR, the average value for Taoyuan is the highest, followed by New Taipei, and Taipei is the lowest. These values might be due to the fact that Taoyuan and New Taipei have larger areas of land. The differences in the quality of public facilities and services in different districts are also larger, which motivates households to move. However, there is only a slight difference in average ICMR in the three cities. Finally, the IDMR is lowest in Taipei, and New Taipei and Taoyuan have almost the same values. The average IDMR is about 3.5% in the three cities, with only slight differences.

4. Housing policy and affordable housing strategies

4.1 Changes in housing policy

For a long time, the goal of housing policy in Taiwan was "to help every household own a home." According to the Public Housing Construction Loan Act passed in 1957, the government started to build public housing, provided mortgage subsidies for households to build their homes, and encouraged real estate developers to build public housing and to redevelop military dependents' villages.[12] In January 1999, due to a rapid increase in vacant units and recession in the housing market, the government announced that it would stop building public housing. The government then abolished the encouragement of private investment in public housing and instead moved to implement mortgage interest subsidies to homebuyers in 2001. Since then, the policy focus has shifted from a supply-side subsidy to a demand-side subsidy.[13] A series of subsidies to homebuyers of various types has further been enacted, including mortgage interest subsidies to military officers, government employees, teachers, laborers, aboriginal people, and so on.

It was found that repeated subsidies have targeted the same households, and thus they have caused inequity. As a result, the government proposed the Overall Housing Policy in 2005 and amended the goal of housing policy to state "to help every household live in a decent home." Several objectives developed under the new policy goal include promoting a healthy housing market, developing a reasonable subsidy program with equity and efficiency, and promoting a better living environment. In 2007 the government further implemented a comprehensive housing subsidy program targeting low-income and vulnerable households. The housing subsidy takes three forms: mortgage interest subsidies to homebuyers, mortgage subsidies to existing owners for home improvement, and rent subsidies to tenants.

As stated earlier, the rapid boom in housing prices has caused a serious affordability problem since 2003. In fact, the problem was not eased during the Global Financial Crisis of 2008–09. As a result, high housing prices have become the greatest cause of popular discontent in the metropolitan areas, according to the government's polls on the Internet in 2010. In response to public opinion, in 2011 the government passed the Housing Act, which has been the law for current housing policy and social housing provision since then. However, the number of social housing units, as discussed in a previous section, accounts for only 0.08% of the total housing stock, which is far below any reasonable figure. Due to financial difficulties, the number of social housing units built by the government is very small.

With respect to housing subsidy programs, the number of applications for rent subsidies is steadily increasing and is more than the planned number. In contrast, the number of applications for mortgage interest subsidies for first-time homebuyers and for existing homeowners to do home improvements is declining annually. As a result, the government has gradually adjusted the allocation of housing resources across the three types of housing subsidies during the period 2007 to 2014. Recently, the housing subsidy program has been further adjusted across urban areas. In 2015, the total number of mortgages in below-market rates to homebuyers in Taipei City increased from NT$200 million to NT$250 million. It increased to NT$230 million in New Taipei City; and in other local authorities it increased to NT$210 million. The loan period was extended up to 20 years, with the first five years being interest-only repayments.

The amount of rent subsidy is now also adjusted by area. The amount of rent subsidy in Taipei City is NT$5,000 per month per applicant. The amount in other cities such as New Taipei City, Taoyuan City, Taichung City, Hsinchu City, and Hsinchu County is NT$4,000 per month per applicant. The amount in Tainan City and Kaohsiung City is NT$3,200 per month per applicant, while it is NT$3,000 in other local authorities. The central government has spent a total of NT$3.3 billion to assist local authorities in issuing rent subsidies. The percentage of rent subsidies provided from the central government to local authorities ranges from 60–90% of the rent subsidies to recipients.

In addition to the regular housing subsidy, the government has also provided a series of policy-oriented mortgage subsidies to various categories of homebuyers in order to improve their ability to afford homes. For example, a total of NT$120 billion was provided to young homebuyers applying for mortgages at below-market rates in August 2000. A total of NT$1,680 billion in mortgage subsidies was provided to first-time homebuyers from 2000 to 2009. However, for other homebuyers, the amount saved from mortgage subsidies is far less than the amount they needed to pay for the rising housing prices during the price boom. As a result, the goal of mortgage subsidies to improve homebuyers' ability to afford homes is unmet. In addition, it seems that real estate developers may gain more benefits than homebuyers from the mortgage subsidies.

The central government amended the Overall Housing Policy in September 2015. The four strategic objectives of housing policy are now to strengthen housing market mechanisms, to uphold social justice, to encourage private participation in the social rented sector, and to secure housing rights. The main channels to achieve those objectives include strengthening the rental housing market, providing diversified housing assistance, and promoting better living environments. Based on these goals, every level of government, from central to local, needs to set up housing plans and financial plans. The central government acts as an enabler, while local authorities are executors of housing policy and strategies.

4.2 Strategies for strengthening the housing market

To cope with the steady rise in housing prices, the central government's Executive Yuan called for a seminar on the strategies for strengthening the housing market on January 12, 2009. A series of strategies was proposed in the seminar, including securing the homebuyer's rights, providing adequate assistance to homebuyers and rent subsidies to tenants, providing open and transparent information on real estate transactions, and establishing a reasonable housing price adjustment mechanism.

Selective credit controls

The central bank has also implemented a series of policy instrument strategies to cope with skyrocketing housing prices over the past several years. In June

2010 the central bank launched selective credit controls in 10 specific districts of Taipei City and New Taipei City, where the household's mortgage-to-price ratio (MPR) was limited to a maximum of 70% when buying second and more homes. The areas where selective credit controls are applied were further enlarged to include Linkou, Sanxia, and Tamsui in New Taipei City on December 30, 2010. The maximum MPR was further reduced to 60%. The central bank also limited the maximum land loan ratio to 65%, and 10% of the loan was provided after the construction.

In June 2012 the central bank further limited the maximum mortgage ratio to 60% for residential dwellings whose transaction prices were higher than NT$80 million in both Taipei City and New Taipei City, and when they were higher than NT$50 million in other cities. Moreover, a series of tighter credit controls was implemented in June 2014 to other areas. The maximum mortgage ratio was also reduced to 50% for those who bought third and more homes nationwide. The minimum house price qualifying for credit controls (50% of maximum mortgage ratio) was set to NT$70 million in Taipei City and to NT$40 million in other cities; in New Taipei City the minimum house value was set at NT$60 million. The maximum mortgage ratio was also restricted to 50% for business owners nationwide.

Recently, given the decline in the volume of real estate transactions in the market, the central government has adjusted some of its credit controls. On August 13, 2015 the areas where credit controls apply were reduced. Six areas, such as Bali and Yingge in New Taipei City and Taoyuan, Luzhu, Zhongli, and Guishan, were excluded from credit controls. The maximum mortgage ratio in these areas increased from 50% to 60%. Since the credit controls have effective impacts on the housing market, the central bank ended credit controls for the whole nation on March 23, 2016; however, controls remained for high-priced residential dwellings.

Reforms of real estate taxation

In addition to selective credit controls, a series of reforms of real estate taxation was implemented to secure more justice in terms of living conditions for all households over the past several years. On June 1, 2011 the government enacted the Specifically Selected Goods and Services Tax (the so-called Luxury Tax), according to which any property held for less than two years is taxed. A property may be taxed at 15% of sale price when held for less than one year, and at 10% of sale price when the property is held for less than two years without any consideration of gains or losses in the transaction. The tax basis of the Luxury Tax is the actual transaction price of the property rather than the assessed present value used for the Land Value Increment Tax, which is considerably lower. The high tax burden of the Luxury Tax has pushed many existing homeowners to hold their properties for more than two years. As a result, the transaction volume has declined significantly in the housing market. However, the housing price level has steadily risen as the number of dwellings for sale on the market was clearly reduced.

Taipei City government has also launched a series of reforms of real estate taxation. Starting in July 2011, additional house taxes called the Luxury House Tax have been levied on high-quality and luxury dwellings. Assessed standard values have increased up to 100% since July 1, 2014 for the Luxury House Tax. A tax for hoarding houses has been levied since May 2015 for owners having more than three homes. The tax rates are 1.2%, 2.4%, and 3.6% for three homes, four to five homes, and six or more homes, respectively.

Furthermore, the central government launched the Immovable Property Gains Tax on January 1, 2016, which aims to correct the drawbacks of the Land Value Increment Tax regarding the assessed present value that is announced once a year and is significantly below the market value as stated before. The tax, known as the Integrated Housing and Land Tax, refers to the levy applied to the combined profits from housing and land sales. These profits are calculated by subtracting the actual cost of acquiring the property from the sale revenue. The rate on housing and land transaction profits ranges from 15% to 45%. An owner-occupied residence will receive preferential taxation with a 10% rate. A property worth up to NT$4 million is tax exempt. The tax revenue will be used for housing policy and social welfare policy.

Transparency in real estate transaction

To promote transparency in real estate transactions, the Legislative Yuan passed amendments to the Real Estate Brokerage Management Act, the Land Administration Agent Act, and the Equalization of Land Rights Act in December 2011. The actual price registration regime was also implemented on August 1, 2012, with the regulation that all property transactions should be registered within one month after closing the deal. Any violation is fined from NT$30,000 up to NT$150,000.

5. Challenges and perspectives

5.1 Challenges

Even though some new and reasonable concepts have been introduced in the newly updated overall housing policy, some policies contradict each other. On the one hand, some policies intend to suppress the rise in housing prices. On the other hand, other policies aim to promote transactions in the housing market. Furthermore, some specific interest groups have actively lobbied lawmakers and policy makers and have hindered the reform of current housing policies. As a result, the effectiveness of policy reform has been very limited over the past three decades.

The government's responses to the five issues raised by the Housing Movement are seen as passive and have not satisfied most people; this weak governmental response has been a major factor for the loss of the ruling party, Kuomintang, in the presidential election. Why did the government respond to people's call for housing justice in such a passive way? There are some misleading concepts or myths behind the government's decision-making. It might be right that housing construction is an important investment in aggregate fixed asset investment. Construction can create many job opportunities, and it has also stimulated the development of other housing-related industries in the past, especially during the periods of insufficient housing supply. However, the Taiwanese economic structure is changing; the drivers of economic growth have now shifted from manufacturing to information, high-tech, and service industries. We can say that the housing industry is still an important industry, but it should not be seen as the only trigger of economic growth, especially when we now face an oversupply in the housing market, and a misallocation of resources in housing as shown by the very high vacancy rates.

Residential mortgages loans are also an important asset class for the financial and insurance industries. If housing prices fall dramatically, the decreasing value of collaterals might cause a potential financial crisis. Further, since most households are homeowners, the decline in housing wealth will lower the aggregate demand and affect public willingness to support the ruling party. Real estate businessmen are also very important financiers of the main political parties, including the ruling party. Falling housing prices will also affect their profits, so they will try their best to persuade the government to soften the impact of those housing market reforms. Their lobbying behavior can be well explained by capture theory and rent-seeking, which is a form of government failure.[14]

Why do the people protest the unreasonably high housing prices on the street instead of following the normal legal procedures? This is because those activists believe that the government and Legislative Yuan are controlled by interest groups, especially big enterprises.[15] The announced goal of the new housing policies is to solve the problem of unaffordable housing, but the actual result turns out to be an even worse situation. The policies to build a sound real estate market actually benefit real estate-related businessmen instead of the general public. The people who support the housing movement have totally lost their faith and trust in the government to achieve housing justice because many very high-ranking housing-related officers have been involved in corruption cases.[16]

Housing prices have started to fall by about 5–10% since the beginning of 2015, and the house price-to-income ratios in Taipei and New Taipei are 14.9 and 12.43, respectively. However, housing is still unaffordable for most households. Furthermore, Taiwan is now facing a dramatic structural change in population. The trends of having fewer children and an aging population dominate the demography of the island. The percentage of elderly people (above 65 years old) in Taiwan reached a historic high of 12.5% in 2015, changing Taiwan from an "aging" to an "aged" society, according to the United Nations' demographic definitions. The total fertility rate (TFR) and general fertility rate (GFR) are the two most important indicators for measuring the trends in childbirth. Taiwan's TFR and GFR were 1.175 children and 35‰ in 2015, respectively, with 35‰ being one of the lowest fertility levels in the world. At the same time, there are insufficient facilities to meet the needs of the aging population. In view of these challenges, this study suggests that Taiwan's government should put more efforts into housing market reforms.

5.2 Perspectives

We want to emphasize that the goals of housing policy should not just be housing affordability but also good housing quality. To achieve these two important goals, we propose six components of a new housing strategy and we make four suggestions for future policies as follows:

Speed up urban regeneration and redevelopment

Urban regeneration is an important tool to improve the quality of the housing environment, especially in old and crowded areas. However, urban regeneration is extremely slow in Taiwan, because the negotiations between the builders and the households, and also the agreed and the disagreed households, are very difficult, particularly for those highly multi-owned properties. The most important thing for the government is to build a transparent mechanism for profit-sharing to increase the mutual trust between builders and households. Further, central and local governments should try to exclude notorious builders from joining urban regeneration projects, and also have a critical solution for those so-called "nail households," the minority disagreed households.

Increase social housing units

The current stock of social housing units is extremely low in Taiwan. The government has promised to build 200,000 units in the coming eight years (by 2026). We encourage the government to try its best to keep this promise and also to consider how to manage these social housing units equally and efficiently. Social housing is a part of housing subsidy programs, and it is not a profitable sector. We would suggest that the government build effective mechanisms to encourage more non-governmental and non-profit organizations to join in. In addition, the government needs to increase the reserved ratio of social housing units for economically and socially disadvantaged groups from the current 10% to at least 50%.

Enlarge the rental housing market

The current very high homeownership rate in Taiwan is not a good phenomenon, and it can be called a "distortion" of the housing market. We suggest that government makes the housing tenure choice more neutral through the improvement of the rental housing market and other tax incentives for owner-occupied housing and rental housing. The most important task is to establish a specific rental housing law to make the legal positions for landlords and renters more equal. Furthermore, the government needs to provide more incentives to encourage more non-governmental and non-profit organizations to join in this important industry. We expect the rental housing industry to serve the housing demand of at least one-third of the total households.

Reform real estate taxation

To reduce speculation activities in the real estate market, real estate tax reform is the most important but difficult task because of the current structure of the legal environment. The strategy of the government's real estate tax reform is to "have" in the first step and then to "be good" in the next step. But even though the government has adopted such a strategy, it is still not easy to realize real estate tax reform. The current best result of real estate tax reform is to pass an acceptable instead of an ideal project, so its effects on real estate markets usually differ from its goals. We argue that the design of real estate taxes should fully consider the relationships among all individual taxes and take them as a package. Otherwise, the real estate taxes cannot be shared equally to reduce the uncertainty of real estate investment.

We would suggest that the tax basis of the real estate tax combine the value of land and building to avoid conflicting valuation problems. Furthermore, the tax basis should be the market value instead of the assessed value. To reduce the tax burden of the real estate owners, especially the owners who own a housing unit only for their own use, the local governments can have a discount strategy according to their fiscal situation. The tax burden adjustment process can be gradual and flexible to reduce potential opposition from the real estate owners.

Strengthen real estate industry management

Many conflicts exist between the households and the real estate-related enterprises, especially the builders and brokers, during the transaction process. The main conflicts are the deficits in construction quality and cheating in advertisements and contracts, which have great influence on the safety and asset value of housing. Taiwan's real estate market has too many builders and brokers because the industry's management is too loose. Some builders are even one-project companies and take no responsibility after their sale of the project. To increase the quality and safety of housing and build a sound housing market, the most important job for the government is to strengthen the management of real estate-related industries.

Promote a sound housing finance system

It is the government's responsibility to build a sound housing finance system to help households while they are buying a house. In particular, there should be a serious underwriting review process that distinguishes between homebuyers and investors. If there are too many investors who can borrow money easily from the financial institutions, then they will crowd out the real consumers and become assistant contributors to housing injustice.

In addition to the above six strategic components, we also have the following four executive suggestions. First, we suggest that the Executive Yuan call for a national land and housing conference to discuss controversial land- and housing-related issues widely, and to try to reach some consensus for problem-solving.

Second, land and housing affairs are currently assigned to different departments and organizations in Taiwan. However, the planning of land- and housing-related policies should have a very comprehensive consideration and multi-disciplinary integration. Otherwise, the benefits of those policies will be very limited and sometimes have undesirable side effects. We suggest that the Executive Yuan establish a mechanism for cross-sector communication and coordination.

Third, the current application of housing subsidies and social housing is based on the household registers. Since there are many inter-region migrations, it is better to view and solve the housing problem across metropolises instead of individually. We suggest that the government establishes a mechanism for cooperation between the central government and local authorities, particularly in metropolises, to allocate the limited housing resources more equally and efficiently.

Finally, some housing-related policies cannot reach their goals because the government overemphasizes the planning and regulation phases but neglect the management phase that is the next step. There is a huge gap between the announced housing policy and its actual implementation. We suggest that the government strengthens housing policy management, narrow the gap, and realize the goals of housing policies.

6. Conclusions

The rapid rise in housing prices is the most serious problem in Taiwan's housing market, and it has been exacerbating the housing affordability problem. The skyrocketing housing prices have not only worsened living conditions, but also reduced housing mobility and thus lowered households' chances to upgrade their living quality through adjustments in their demand for housing. In addition, scarce resources have been overwhelmingly concentrated on the housing market and have damaged overall economic development. Over the past several decades, we have witnessed financial institutions financing a great number of mortgages in the real estate market. The high leverage of the mortgage market has implicitly raised the risks of a financial crisis and has also worsened income and social inequality between rich and poor people. In this situation, many young people are falling into poverty and cannot see a future, thus generating more severe economic and social problems in the nation.

Over the past several decades, we have seen that improper housing policy, imperfect taxation, and inadequate financial programs have generated an unhealthy housing market in Taiwan, where homeowners can generate surplus profits from transactions and thus attract numerous investment demands in the housing market. This has been the major driving force for rising housing prices.

Furthermore, Taiwan is now facing a dramatic structural change in population. The trends of having fewer children and an aging population are very apparent all over the island. At the same time, current structures are too old and lack sufficient facilities to satisfy these trends. In view of these challenges, this study suggests that Taiwan's government should put more efforts into housing market reform. This study proposes six strategic components to promote a healthy housing market: to speed up urban regeneration and redevelopment, to increase the number of social housing units, to enlarge the rental housing market, to reform real estate taxation, to strengthen management of real estate industries, and to promote a sound housing finance regime.

We also urge the government to hold a national land and housing seminar in order to generate a consensus on the components of reform strategies for the housing market, nationally and locally. Because land and housing issues are complicated and widely related to other economic affairs, we suggest that the Executive Yuan needs to establish a mechanism for cross-sector communication and coordination, and to clarify the allocation of duties and responsibilities between the central government and local authorities, particularly within metropolises.

In summary, during recent years we have seen many policies that failed due, in part, to a lack of effective management. Therefore, we suggest that governments at all levels pay more attention to this management issue. In view of the complexity of the political environment and past government failures, we believe that citizens should develop alternative ways to pressure the government to implement housing market reforms.

Notes

1 For a detailed discussion of the real estate cycles and policies against bubbles in Taiwan, please refer to Chang and Chen (2012).
2 The rapid rise in real estate prices was triggered in February 1987, when Cathy Life Insurance bid on a parcel of state-owned land in Nanjing E. Road, Taipei City for an amazing price. The stock price index reached a historic high of 12,682 points on February 12, 1990.
3 The selective credit control policy consists of stopping uncollateralized loans for land, restricting the amount of construction loans, and keeping the LTV (loan-to-value ratio) on residential loans below 50%.
4 For a detailed discussion of the social housing problems in Taiwan, please refer to Chen (2011).
5 According to the Demographia 12th Annual International Housing Affordability Survey 2016, the ratio is 19.0 for Hong Kong, 12.2 for Sydney, and 10.8 for Vancouver, three of the most unaffordable cities in the world; Taipei is not included in this survey.
6 The Demographia International Housing Affordability Survey uses the median multiple (similar to the PIR) to measure housing affordability. If the median multiple is 5.1 and over, then it is considered severely unaffordable. When the value is between 4.1 to 5.0, housing is seriously unaffordable; between 3.1 and 4.0 is moderately unaffordable. Housing is affordable when the value is 3.0 and under.

7 In January 2016, the marginal tax rate of estate and gift taxes was reduced from 50% to 10%. The reason the government lowered the inheritance tax was to boost domestic investment and offset the decline in foreign investment. This idea followed the cancelation of inheritance tax in Hong Kong and Singapore.
8 The Land Value Increment Tax rate was reduced to half of the normal rate on January 30, 2002. Then the progressive tax rates were reduced to 20%, 30%, and 40%, compared to previous rates of 40%, 50%, and 60%. According to statistics published by the Department of Land Administration, Ministry of Interior, the assessed present land value is announced to reach 90% of market value; in practice, it accounts for only 50–60% of the market value.
9 A more detailed discussion of Taiwan's taxation and homeownership subsidy programs appears in Bourassa and Peng (2011).
10 According to the Global Property Guide, the overall rental yield in Taiwan is merely 1.57%, which means that the landlord needs to accumulate about 64 years of returns to make a turnover. However, the average economic life for a house is only 50 years; as a result, many landlords do not maintain their properties well and thus worsen the quality of rental properties.
11 There are 12, 29, and 13 districts in Taipei, New Taipei, and Taoyuan, respectively.
12 Public housing physically constructed by the government has the following characteristics: (1) very low numbers of dwelling units for disadvantaged groups; (2) units for sale, not for rent; and (3) loosely defined criteria for the selection of households. For a detailed discussion of the history of public housing policy in Taiwan, please refer to Chang and Yuan (2013).
13 Taiwan's government has played a passive role both as a supplier and regulator of housing in Taiwan, especially when it ended new construction of public housing in the 1990s. Chiu (2008) noted that Taiwan has the least regulated housing system in East Asia.
14 Rent-seeking is an attempt to obtain economic rent by manipulating the social or political environment in which economic activities occur, rather than by creating new wealth. In many market-driven economies, much of the competition for rents is legal, regardless of the harm it may do to an economy. However, some rent-seeking competition, such as bribery or corruption, is illegal.
15 Chang and Yuan (2013) argue that Taiwan's public housing policy was captured because it had no control over interest groups' lobbying and political donations. A series of acts related to the Sunshine Law were enforced one by one after 2000, such as the Act on Recusal of Public Servants due to Conflicts of Interests, the Political Donation Act, the Act on Property-Declaration by Public Servants, and the Lobbying Act. However, the effectiveness of these laws remains doubtful because many people still doubt the impartiality of the judge.
16 There were two very famous corruption cases in recent years. The first case involved the Taoyuan Deputy Mayor and former head of the Construction and Planning Agency, Yeh Shih-wen. Yeh was charged by prosecutors for taking bribes from Farglory Construction Company relating to two public development projects in his two separate official positions. The second case is the former New Taipei City Deputy Mayor Hsu Chih-chien, also the former head of Department of Urban Development Bureau in Taipei City government. Hsu was charged with corruption over alleged bribes to facilitate urban renewal projects on bids by construction contractors.

References

Akbari, A. H. and Aydede, Y. (2012). Effects of Immigration on House Prices in Canada. *Applied Economics*, 44(13–15), 1645–1658.

Bourassa, S. C. and Peng, C.-W. (2011). Why is Taiwan's Homeownership Rates So High? *Urban Studies*, 48(13), 2887–2904.

Cebula, R. J. and Vedder, R. K. (1973). A Note on Migration, Economic Opportunity, and the Quality of Life. *Journal of Regional Science*, 13(2), 205–211.

Cebula, R. J. and Belton, W. J. (1994). Voting with One's Feet: An Empirical Analysis of Public Welfare and Migration of the American Indian, 1985–1990. *American Journal of Economics and Sociology*, 53(3), 273–280.

Cebula, R. J. and Nair-Reichert, U. (2012). Migration and Public Policies: A Further Empirical Analysis. *Journal of Economics and Finance*, 36(1), 238–248.

Chang, C.-C. and Yuan, S.-M. (2013). Public Housing Policy in Taiwan. In J. Chen, M. Stephens, and Y. Man (Eds.).*The Future of Public Housing: Ongoing Trends in the East and the West*. Berlin Heidelberg: Springer-Verlag. DOI: 10.1007/978-3-642-416224-4_6.

Chang, C.-C. and Chen, M.-C. (2012). Taiwan: Housing Bubbles and Affordability. In A. Bardhan, R. H. M. Edelstein, and C. A. Kroll (Eds.). *Global Housing Markets: Crises, Policies, and Institutions, Kolb Series in Finance: Essential Perspectives*. Hoboken, New Jersey: John Wiley & Sons, Inc.

Chen, Y.-L. (2011). New Prospects for Social Rental Housing in Taiwan: The Role of Housing Affordability Crises and the Housing Movement. *International Journal of Housing Policy*, 11(3), 305–318.

Chiu, R. L. H. (2008). Government Intervention in Housing: Convergence and the Asian Dragons. *Journal of Urban Policy and Research*, 26(3), 249–269.

Demographia. (2016). Rating Middle-Income Housing Affordability. *12th Annual Demographia International Housing Affordability Survey: 2016*. Belleville, IL: Demographia.

Grassmueck, G. (2011). What Drives intra-country Migration: The Impact of Local Fiscal Factors on Tiebout Sorting. *The Review of Regional Studies*, 41(2, 3), 119–138.

Nenov, P. T. (2015). Regional Reallocation and Housing Markets in a Model of Frictional Migration. *Review of Economic Dynamics*, 18(4), 863–880.

Peng, C.-W., Wu, S.-T., and Wu, S.-H. (2007). The Influences of Effective Property Tax Rates on Housing Values: Evidence from Ta-tung and Nei-hu Districts in Taipei City. *Journal of Taiwan Land Research*, 10(2), 57–66 (in Chinese with English abstract).

Peng, C.-W., Wu, W.-C., and Kung, S.-Y. (2009). An Analysis of Determinants of Residential Migration. *Journal of Population Studies*, 39, 85–118 (in Chinese with English abstract).

Peng, C.-W. and Tsai, I.-C. (2012). Long- and Short-term Influences of Homeownership Rates on Fertility Rates: An Application of the Panel Cointegration Model. *Journal of Population Studies*, 44(1), 57–86 (in Chinese with English abstract)

Saiz, A. (2007). Immigration and Housing Rents in American Cities. *Journal of Urban Economics*, 61(2), 345–371.

Sterk, V. (2015). Home Equity, Mobility, and Macroeconomic Fluctuations. *Journal of Monetary Economics*, 74(3), 16–32.

10 Housing challenges in Hong Kong's dualistic housing system

Implications for Chinese cities

Rebecca L. H. Chiu

DEPARTMENT OF URBAN PLANNING AND DESIGN
THE UNIVERSITY OF HONG KONG

1. Introduction

Hong Kong has a dualistic housing system, composed of a public and a private sector accommodating 46% and 54% of the total population in 2016, respectively (Census and Statistics Department, 2016). The challenges confronting the public housing sector, which is 67% rental and 33% owner-occupied, are to meet the rapidly growing needs and demands from lower-income families, engendered by the skyrocketing private housing prices since 2010, and to lift income eligibility following the promulgation of the minimum wage policy. The acute housing affordability problems of families ineligible for subsidized housing, epitomized by a price-to-income ratio of 17, pose an unprecedented challenge to the government. Convinced that the suspension of new land formation in the early 2000s, and the reduced new land supply from then until 2010, were the major causes of the land and housing supply shortage today, the current government has tried on all fronts to increase land supply in the short, medium, and long terms. But is land shortage the only major cause of the enormous price hikes? In an economy where land is state owned and land-use zoning is statutory,[1] why is the land shortage for housing development so serious? Apart from the short supply of land and housing, are other housing challenges facing Hong Kong today? Is a housing demand projection model driven by demographic trends, as adopted in the 2014 Long-term Housing Strategy, sufficient to prevent future market shocks? Given the similarity in land ownership and the statutory status of land-use zoning, what are the implications of Hong Kong's housing challenges for mainland Chinese cities?

This chapter attempts to answer the above questions by first looking into the overt challenges of Hong Kong's housing sector, and then considering the covert challenges. Finally, based on the major similarities and differences between the housing systems of Hong Kong and urban China, and taking Shenzhen's housing subsidy policy as an example, this chapter attempts to draw the implications of Hong Kong's housing challenges for Chinese cities facing continued housing price hikes.

2. Overt challenges

2.1 Challenge 1: unaffordable and unavailable homeownership

Despite its reputation as one of the freest economies in the world, 31% of Hong Kong's housing stock is made up of public rental housing (765,080 units) with another 15% (370,200 units) of subsidized owner-occupier housing directly produced and provided by the government, whereas only 54% (1,332,720 units) was private housing (Hong Kong Housing Authority, 2016a). As such, its homeownership rate is expectedly modest, standing at 51% in 2016 (Census and Statistics Department, 2016). Homeownership, in fact, started to rise only in the late 1970s, when the major economic activity switched from manufacturing to the service industries. It reached the first peak in the late 1980s, and another in 2004, then plateaued from the mid-2000s onward, with a slight dip since 2011 (Figure 10.1). The homeownership growth in the private housing sector was found to correlate positively with economic growth and household income, and negatively with housing price. Expectedly, no such correlations were found in the subsidized homeownership sector, as it has been driven by housing policy, subsidy, eligibility, and supply (Chiu, 1998; Yip et al., 2007).

Affordability in purchasing private homes has been one of the biggest housing challenges in Hong Kong. As shown in Figure 10.2, since the future of Hong Kong was settled by the Chinese and the British governments in 1984, housing prices had been rising continuously, averaging an annual increase rate of 19%, until the Asian Financial Crisis began to impact the housing market in 1998. To assess affordability for homeownership, we refer to the most common housing unit sizes in Hong Kong and to the median household income. As shown in Table 10.1,

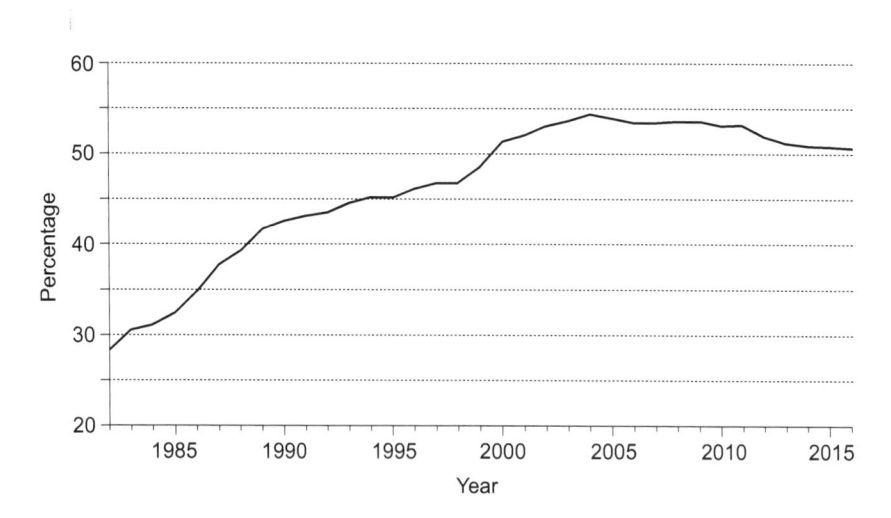

Figure 10.1 Homeownership trend, 1982–2016.
Source: Census and Statistics Department, 2016a, 2016b.

Table 10.1 Changing composition of housing by unit size over time (%), selected years.

Year	A	B	C	D	E
	Less than 40 m²	40–69.9 m²	70–99.9 m²	100–159.9 m²	160 m² or above
1985	41.7	40.9	9.4	5.2	2.8
1991	38.9	43.8	9.6	5.1	2.6
1997	37.0	45.6	9.9	5.1	2.4
2003	34.1	47.5	11.2	5.1	2.2
2015	31.1	48.9	12.2	5.5	2.2

Source: Rating and Valuation Department, 2016.

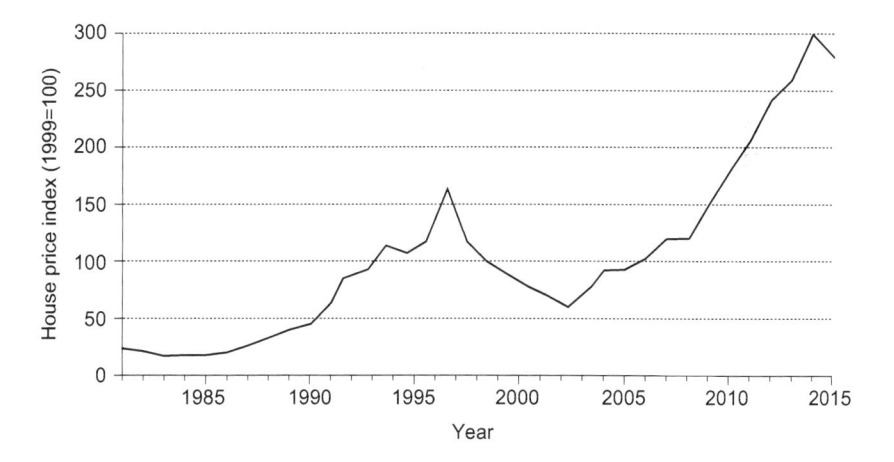

Figure 10.2 The nominal price trend of private housing, 1981–2015.
Note: The price index was calculated based on nominal prices.
Source: Rating and Valuation Department, 2016a.

the most common housing types are flats of 40–69.9 square meters (49%) and less than 40 square meters (31%). Thus, the reference price is that of a housing unit of 500 square feet in a medium-price area. As shown in Figure 10.3, affordability before 1998 declined sharply, with the price-to-income ratio reaching 9.5 and the affordability ratio (monthly mortgage repayment/household income) rising to almost 90% in 1997, marking a historic peak at the time. The affordability problems subsequently eased off when the housing price dropped by 60% in the post-crisis economic depression. Nonetheless, transactions also fell by 40–70% (Figure 10.4) due to the poor economic outlook, job insecurity, and salary cuts.

As the economy picked up again in the mid-2000s, housing prices resurged (Figure 10.2). The negative impact of the Global Financial Tsunami was short-lived nonetheless. Housing prices eventually surpassed the 1997 historical peak by

204 Rebecca L. H. Chiu

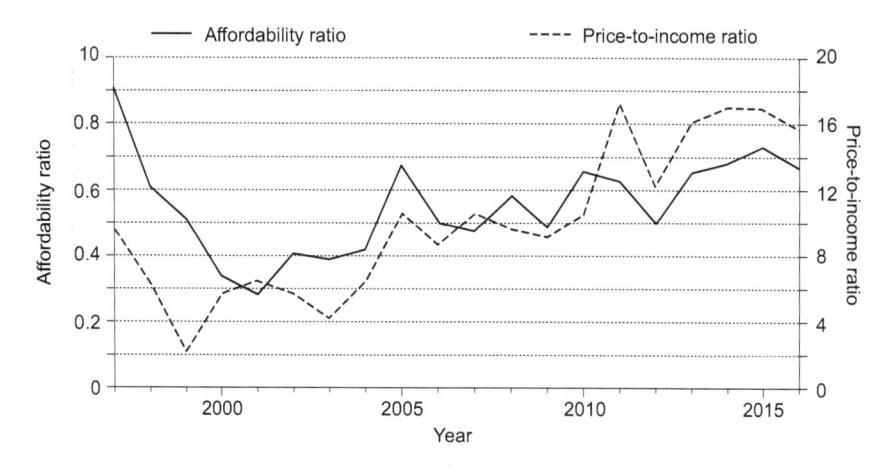

Figure 10.3 Housing affordability trends, 1997–2016.
Note: Housing price-income ratio = (price per sq ft) X 500 / annual household income. Affordability ratio = Monthly repayment/Monthly household income. Interest rates in 2016 was 2.15%.
Source: Bank of China, as of September 2016.

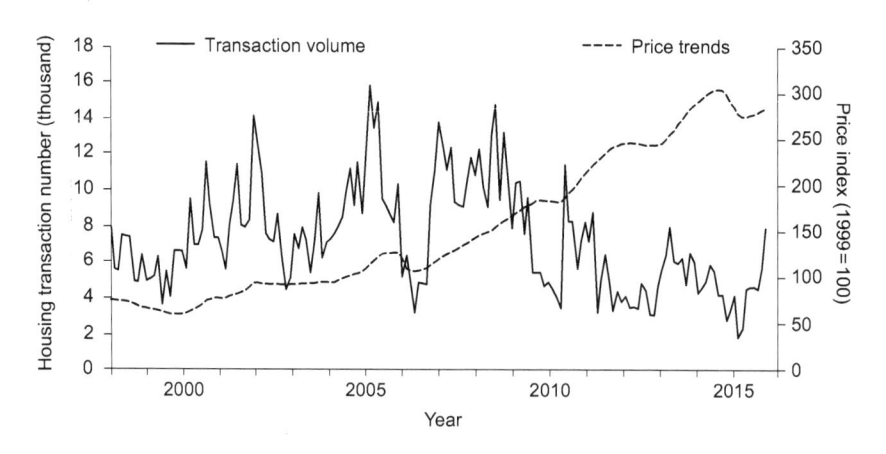

Figure 10.4 Trends of housing price and housing transaction volume, 2002–2016.
Source: Rating and Valuation Department, 2016a, 2016b.

31% (Financial Secretary's Office, HKSAR Government, 2013), imposing even greater housing affordability problems on middle-income families and below. The price-to-income ratio reached an incredible level of 17 in 2011, dropped in 2012, but resurged in subsequent years. The affordability ratio nonetheless remained around 60% due to the historical low level of mortgage interest rates, bordering around 2.75% (Figure 10.3). A subsidized housing scheme directly provided by

Housing challenges in Hong Kong 205

the government, which had been effective in enabling lower-income families to acquire homeownership, was suspended in 2002 as private housing had deflated to affordable levels. It was reintroduced in 2010, though on a much smaller scale owing to land supply constraints.

Overall, Hong Kong's homeownership growth was fostered by economic restructuring and development in the 1970s and 1980s. However, affordability deteriorated as the economy thrived and as housing demand escalated. Government subsidy, basically direct provision of housing at below-market rates, assists lower-income families to buy their own homes, but for those who are outside the subsidy eligibility net, to enter the private market has been formidable, especially under the current market conditions. As such, the transitional zones to homeownership are two-fold in Hong Kong's dualistic system, bordering the public and private housing sectors.

2.2 Challenge 2: shortage of affordable rental housing

The rental trend in the private sector shows a similar trajectory to that of housing price, though the upsurge was of a lesser scale: 60% increase between 2009 and 2015 compared to a 150% rise in housing price (Figures 10.1 and 10.5). As expected, rents, which reflect consumption, are less volatile than housing prices, reflecting investment value. Likewise, the rent-to-income ratio peaked in 1997 and 2015 at 38%, whereas the price-to-income ratio of the same years was at 9.5 and 16.6, respectively. However, with overall rent-to-income ratios in the private market close to or exceeding 30% since 2008, and moving towards 40% since 2013, rental costs have been increasingly burdensome. Tracking down the age of homeowner household heads, Tang, Zhang, and Chiu (2015) found that

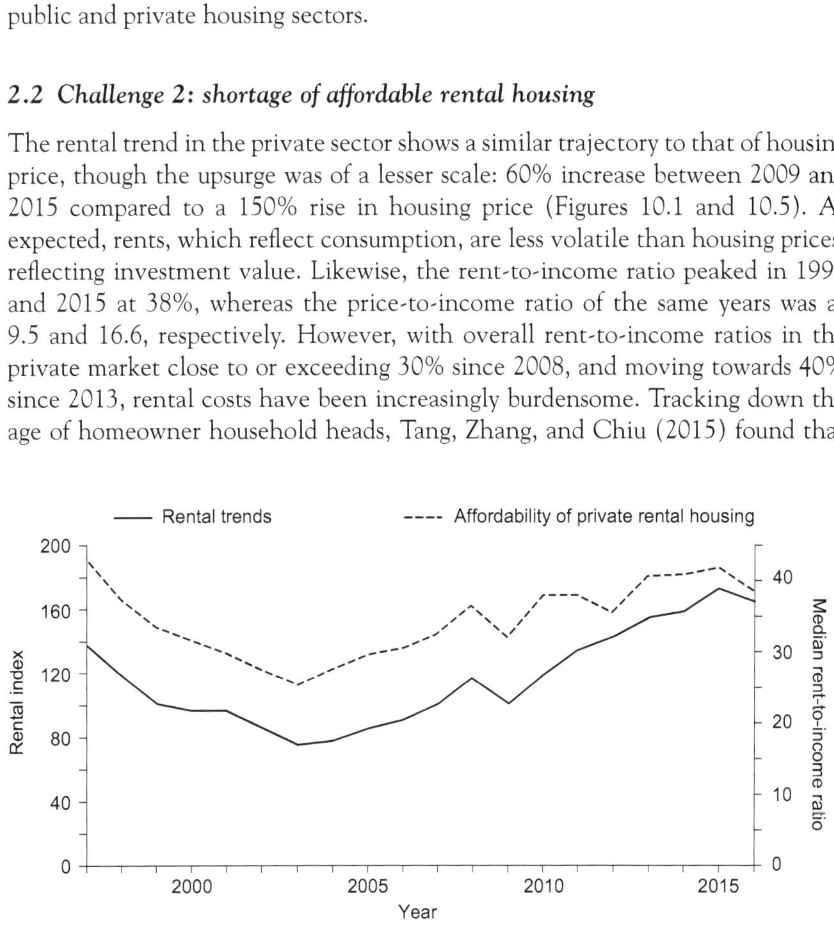

Figure 10.5 Rental trends and rental affordability, 1997–2016.

Source: Compiled by author based on private housing rents of properties of medium size (i.e. 40 m²–69.9 m²) from Rating and Valuation Department, 2016b, 2016c.

206 Rebecca L. H. Chiu

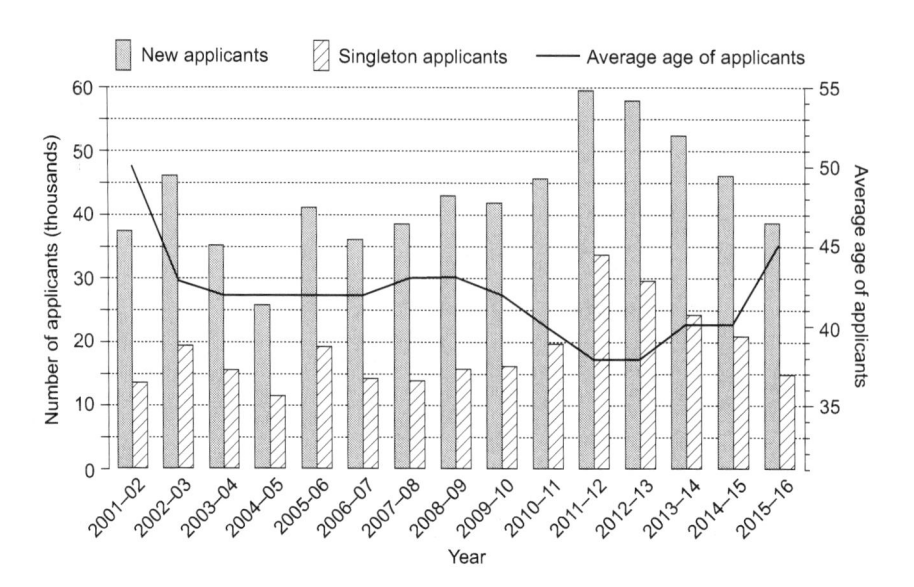

Figure 10.6 The trend of public rental housing applications, 2001–2016.
Source: Hong Kong Housing Authority, 2016c.

the percentage of homeowning household heads in the cohort below age 30 has dropped since 1991, from 46% to 39% in 2013. Thus, rental demand has been on the rise, especially among low-income families.

Figure 10.6 shows that the number of new applicants for public rental housing, expectedly low-income families, has been on the rise since 2005, when the economy picked up again after the Asian Financial Crisis and when the rent-to-income ratio resurged, especially among singleton (one-person households) applicants. The sharp increase in 2011 was partially due to the increase of the income eligibility criterion following the implementation of the Minimum Wage Policy in order to enable low-income employees such as security guards and cleaners to remain eligible for public housing. This increase also explained the lower increase in applications in the following year. The increase in young singleton applicants and the decline in the applicants' average age are outcomes of both deteriorating rental affordability and the growing social trends of young people leaving parental homes and seeking independent lives. Indeed, singleton families with ages below 30 accounted for 55% of total singleton applicants in March 2015 (Hong Kong Housing Authority, 2016c). Coupled with a supply bottleneck, as discussed later, the average waiting time of nuclear families for admission to public rental housing has been rising from 2.1 years in 2010 to 4.1 years in September 2016 (Hong Kong Housing Authority, 2016c). Consequently, the supply of subdivided housing units – that is, rooms with or without toilets carved out from original dwelling units in high-rises – has been increasing to meet the market demand.

The number of subdivided units has increased rapidly by 31%: from 66,900 in 2013 to 87,600 in 2016 (Census and Statistics Department, 2016; Transport and Housing Bureau, 2014).

Thus, a big challenge for the government is not only to curb the increase in subdivided units, but actually to reduce the waiting time for public rental housing by increasing supply. The government also needs to cool down rental costs in the private housing market. The recent trends in both, unfortunately, speak to the contrary. The average waiting time for admission to public rental housing was 4.1 years for nuclear families in June 2016. The average waiting time for elderly singleton households also increased to 2.4 years from 1.9 years in 2015 (*Sing Tao Daily*, August 12, 2016b).

2.3 Challenge 3: low housing space standards

Hong Kong's average housing space consumption of 16.7 square meters per capita and 48.4 square meters per household in 2014 was the lowest among major Asian cities, including the prime cities in mainland China (Table 10.2). These low space standards are attributable to two factors: (1) a large public housing sector, which provides only rental housing meeting the basic housing standard, including space

Table 10.2 Living space by type of dwelling in Hong Kong, 2013/2014.

Housing type	Provider	Total housing area (m²)	Stock	Residents	Per capita living space (m²)
Public rental housing	HKHA HKHS	26,712,440 1,027,470	748,605 33,055	2,055,380 87,679	12.9 (8.6)
Subsidized owner-occupied housing	HKHA HKHS	18,926,270 974,697	452,614 19,280	1,312,581 55,912	14.5 (14.3)
Private housing	HKHS Developers	184,799 65,418,230	3,473 1,123,633	10,071 3,258,536	20.1 (16.4)
Total		113,243,906	2,380,660	6,780,159	16.7 (13.3)

Note: *Figures in brackets are per capita living space of 1998/99; **HKHA – Hong Kong Housing Authority; HKHS – Hong Kong Housing Society (an NGO providing subsidized housing with government land). Its housing space standards are similar to that of HKHA, but more flexible, and it also offers a rental scheme for lower-middle income groups; ***Public rental housing and interim housing are internal floor areas. Others are saleable areas.

Source: Based on Centadata Company Limited, 2014; Chiu, R. L. H., 2005; Hong Kong Housing Authority, 2014a; Hong Kong Housing Authority, 2014b; Hong Kong Housing Society, 2014a; Hong Kong Housing Society, 2014b; Rating and Valuation Department, 2014; information supplied by Hong Kong Housing Authority; information supplied by Hong Kong Housing Society.

208 *Rebecca L. H. Chiu*

per capita, and publicly supplied owner-occupier housing of standards not higher than that of the lower-end private housing; and (2) poor affordability for private housing. As discussed, affordability of private housing has always been tight except during the long spell of economic depression after the Asian Financial Crisis of 1997. Given the exceedingly high housing price relative to income since 2011, there is a tendency for developers to build even smaller housing units to suit affordability. A case in point is a recent project supplying flats of 15.3 square meters to 18.1 square meters, which is 66% lower than the average flat size of 45 square meters in 1999 and 48.4 square meters in 2014 (*Sing Tao Daily*, June 10, 2016a). The outlook on improving the housing space standards of Hong Kong is, in fact, dismal: the share of public housing in the total housing stock is very unlikely to reduce, and the affordability of private housing eases off only during market decline, which is also associated with economic depression and job insecurity.

An analysis of per capita housing space improvements between 1998/99 and 2014 among the major housing types found that the biggest jump was in public rental housing, which increased from 8.6 square meters to 13, whereas that for subsidized owner-occupier housing increased from 14.3 square meters to 14.5, and for private housing from 16.4 square meters to 20.1, with an overall increase from 13.3 square meters to 16.7 (Table 10.2). The significant rise in the space standard of public housing was unintentional: it is due to the transfer of subsidized housing projects under construction to public rental housing in 2002, following the termination of the subsidized housing scheme after the Asian Financial Crisis; and to the reduction in household size in recent years. Intentional significant boosting of the housing space standard in public housing is hard to justify when non-subsidized households struggle to pay for homes of basic standards.

The only solution to ameliorate the low housing space standards and their problems, and to improve livability, is perhaps by way of improving the provision and design standards of public or communal spaces and commercial spaces (e.g., restaurants and cafés) in neighborhoods, especially within housing estates, to enable some of the household activities (such as recreation and social activities) to be carried out outside of the housing units. Indeed, public spaces and restaurants have been generally well used in subsidized and private housing estates.

2.4 Challenge 4: housing mobility bottlenecks in a dualistic system

Based on the discussion above, this section attempts to identify the challenges to the government in enabling housing users to climb up the housing ladder (Figure 10.7), which has been set as a policy objective. The housing ladder concept was first emphasized by the new government formed after Hong Kong's return to China in 1997, to symbolize its intention to promote upward housing mobility through an expanded housing subsidy policy. The rungs of the ladder could be defined in terms of housing tenure, or housing types, or subsidy eligibility. More relevant to the discussion of this chapter is subsidy eligibility. According to the

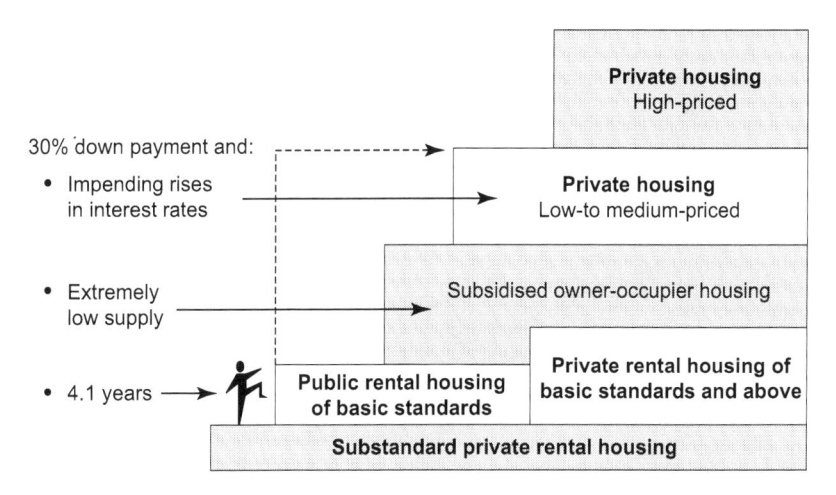

Figure 10.7 Housing ladder and housing tiers – by subsidy eligibility only.
Source: Author.

latest census data, the public housing sector accommodated nearly half of the households (46%) in Hong Kong in 2015: 31% in public rental housing and 15% in subsidized owner-occupier housing. The remaining 54% lived in private housing: 21% as renters and 33% in owner-occupier housing (Census and Statistics Department, 2016). Income levels and residency requirements form the dividing lines, as income limits and a minimum stay of seven years by not less than half of household members in Hong Kong define eligibility for housing subsidies. The income eligibility was set according to the affordability of renting or purchasing private homes commensurate with the lifestyles of the respective socio-economic groups in the city.

The eligibility criteria for housing subsidy, excluding the meager income tax credit for mortgage interest, cover six housing tiers within the dualistic system, as detailed in Table 10.3. Owing to the leniency in evicting public tenants whose housing affordability has improved over time, there are mixed income groups in the public sector. The income heterogeneity, producing diversified lifestyles and life chances, and the absence of the ethnicity factor in the eligibility criteria and in the actual housing allocation process render the housing class concept of Rex and Moore (1971) inadequate to conceptualize the housing tiers in Hong Kong.

As discussed above, the lengthening waiting time for admission into public rental housing poses the first challenge to private renters of substandard housing taking the first step on the housing ladder (Figure 10.7). Apart from the short supply, the residency requirement defers the admission of new migrant families to public housing, but the deterrence is transient as a seven-year stay qualifies a family for permanent residency. The second challenge is moving from renting to homeownership. Between 1980 and 2016, two major socio-economic groups have

Table 10.3 Housing tiers defined by housing subsidy eligibility.

Housing tier	Housing subsidy	Eligible households	Household number in 2011 (% of total household)
Tier 1	Public rental housing to be made available to tenants of substandard private housing	Private tenant households with monthly income < HK$20,000*, subject to their household size and residency status	117,681 (4.8%)
Tier 2	Public rental housing	Public housing tenants with income below the eligibility limit at the time of admission	720,892 (30.3%)
Tier 3	Subsidized owner-occupier housing to be made available to eligible private and public tenants	Private tenant households with monthly income HK$20,000–$40,000**	92,223 (3.7%)
		Public tenants with monthly income higher than HK$20,000***	188,962 (7.7%)
Tier 4	Subsidized owner-occupier housing, with mortgagors eligible for meager income tax exemption for interest payment	Households with income below the eligibility limit at the time of admission	377,615 (15.3%)
Tier 5	Private rental housing meeting basic standards and above, without subsidy	Private tenant households with monthly income > HK$40,000	119,821 (4.9%)
Tier 6	Private owner-occupier housing, with mortgagors eligible for meager income tax exemption for interest payment	Owners of private housing (all income)	845,272 (34.3%)

Note: *The income eligibility for public rental housing for four-person households was HK$25,168 per month in 2014/15. HK$20,000 is used here as the average eligibility criterion across all household sizes, and it is also used because income bracket breakdowns in census data use round-off figures; **Assuming that those who earn less than HK$20,000/month are not interested in buying under the Home Ownership Scheme due to unaffordability and their eligibility to apply for public rental housing; ***It is assumed that public tenant households earning less than HK$20,000 a month cannot afford homeownership.

Source: Based on Census and Statistics Department, 2012.

been latent homeowners waiting for their opportunities to move to the homeowner rung: lower-middle- to middle-income private and public tenants eligible for subsidized owner-occupier housing (Tier 3 in Table 10.3), contingent upon the income eligibility criteria and the availability of subsidized housing supply; and the middle- to upper-middle-income private tenants (Tier 5), whose chances to own homes hinge upon their affordability among the volatile private housing prices, as discussed above.

Unlike the English-speaking economies, which devise mortgage innovations to facilitate homeownership at times of price rises, and indirectly make housing wealth more fungible (Smith, 2010), the Hong Kong government facilitates ownership by tightening up mortgage loans to suppress demand, or expands the subsidized housing scheme, at least rhetorically. However, as Hong Kong's housing market has gone through three price cycles, as discussed above, and is going through the fourth one, there have been "churners" who have made use of the property cycles to accumulate wealth by going in and out of the homeownership sector. These churners are mostly owner-occupiers of private housing in Tier 6. Owner-occupiers of subsidized housing in Tier 4 were much less interested in churning, as the price trend in this sector was much more stable. Within the three property cycles, there were two major price crashes, between 1982 and 1984, and 1997 and 2003. Many more drop-outs in Tier 6 were engendered in the second bust due to the more significant price plummet and the longer-lasting downturn, as well as the government's ambitious homeownership promotion strategy prior to the market crash assisting marginal homebuyers, as discussed later. These two classes of homeowners, the churners and the drop-outs, in the terminology of Wood et al. (2013), could be called the winners and the losers, respectively, in the housing ladder. But the most frustrated are the latent homeowners, who pose the greatest challenges to the Hong Kong government's efforts to institute a functioning housing ladder.

The latent homeowners refer to exasperated private and public tenants (Tier 3 in Table 10.3) who have been aspiring and saving hard to buy their own homes but still stay in the transitional zone, either because of their deteriorating ability to afford homes due to the continuous or intermittent acute price rises, or due to not being picked in the ballots in the sale exercises of the subsidized homeownership schemes. Although these people might have stayed in the rental sector for a long time, it is not their intention to remain there, and hence they are regarded as having initially fallen into and then been trapped in the transitional zone before homeownership. Between 1990 and 2014, income increased 2.5 times but fell much behind the five-fold increase in housing prices (Census and Statistics Department, 2015; Rating and Valuation Department, 2015). However, results of a face-to-face survey of a sample of 5,000 respondents, drawn from different districts in Hong Kong and conducted in 2014 by the author's research team, indicated that 69% of private and public tenants earning household income above HK$20,000 aspired to buy their own homes if they could afford to do so (Chiu et al, 2015), indicating the significant number of latent homeowners trapped in the transitional zone.

The private tenants in the lower-middle-income bracket (earning incomes between approximately HK$20,000 and HK$30,000)[2] in this transitional zone are the most frustrated latent homeowners, as subsidized owner-occupier housing supplied under the major scheme, the Home Ownership Scheme, is the least available to them, and private housing was only marginally affordable when prices dropped to low levels around 1999–2003 (Figure 10.2). Before and after the crash of the housing market in 1998, only 20% of the newly produced subsidized owner-occupier housing was allocated to this group of tenants. The number of applications was found to be 12 times that of the housing units available to this tenant group. The bulk of these subsidized flats were allocated to public tenants because of the prospect of restoring their rental units for further reallocation to the wait-listed applicants for public housing. Thus, as higher priority was given to enabling low-income households to acquire standard housing as soon as possible, and with a lenient policy about evicting public tenants whose income had risen to levels needed to afford private housing, the private tenants' homeownership aspirations received much less attention. Indeed, it was only during the economic depression in 1998–2003, when the take-up of subsidized housing by public tenants was inadequate and when an overall over-supply existed, that applicants who were private tenants were all offered chances to buy. Yet the gloomy economic and employment outlook deterred many from committing to homeownership.

Without the subsidized scheme, the lower-middle-income private tenants could only resort to private housing. The private housing price trends of low- to medium-priced housing initially rose much more slowly than that of the high-priced tier (Figure 10.8), but they have shown double-digit growth since 2008, rendering these properties unaffordable even to middle-income families. Since 2011, the overall transaction numbers have reverted to the same as the post-Asian Financial Crisis period. A closer look at the affordability problem reveals that the affordability ratio could remain at around 40% (a norm used by the government and generally accepted by the community) only due to the historically low interest rates. However, 13 years of saving were required for the down payment on a housing unit commensurate with the housing consumption behavior of the lower-middle- and middle-income families, even if households saved 25% of their monthly income to purchase housing, and if housing price increases did not surpass income increases significantly. Eventually, in 2010, the government had to acknowledge the affordability problem and introduced a pilot scheme aiming to help middle-income families save for down payments (the Government of the HKSAR, 2010). This scheme was subsequently amalgamated with the original subsidized owner-occupier scheme when it was revived by the new Chief Executive in 2011.

A major change in the reintroduced schemes was to raise the income limit to include middle-income families in order to quench their soaring resentment against the rapidly rising housing prices. This rhetorical measure nonetheless further diluted the chances of lower-middle-income families to acquire subsidized homes. What further aggravated the situation was the severe shortage of housing sites, which is a corollary of the suspension of identifying and forming new land

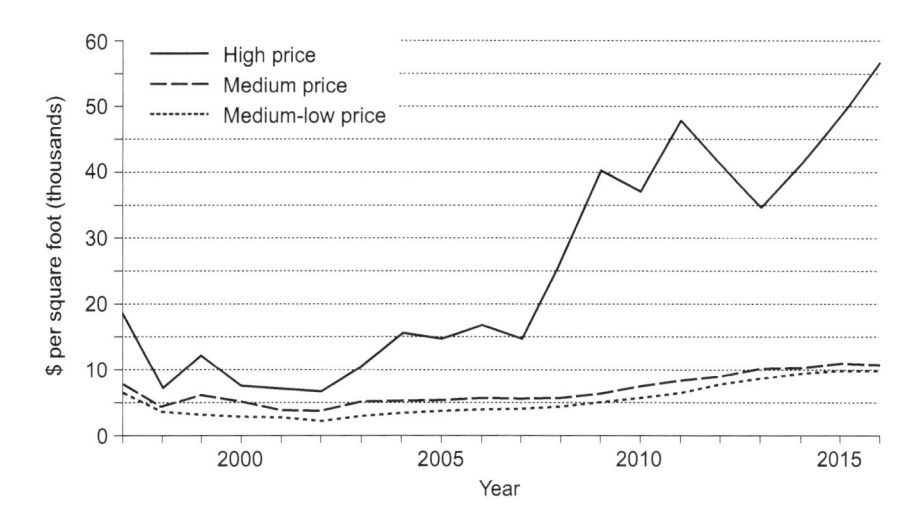

Figure 10.8 Price changes of different housing market niches, 1997–2016.

Note: The reference prices of high-priced housing, medium-priced housing, and medium-low-priced housing were drawn from Repulse Bay, Whampoa Garden, and City One Shatin respectively.

Source: Compiled by Author based on transaction prices from Centaline Property (http://hk.centanet.com/home/Index.aspx).

Table 10.4 Projected supply of subsidized owner-occupied housing and unmet demand, 2013.

Demand for subsidized owner-occupier housing (households with monthly income $20,000–< $40,000)		Projected supply in 2013*	
Private tenants	92,223 50% = 46,112	Average anticipated annual new supply of subsidized owner-occupier housing**	4,570***
Public tenants	165,660 50% = 82,830	Secondary subsidized owner-occupier housing available to private tenants in 2013	5,000

Note: *Excludes the open secondary HOS market, which sells units to public tenants retaining original subsidy, and to private tenants without original subsidy. Total HOS stock is 377,615; **Available to both private and public tenants; ***17,000 units in three to four years.

Source: Census and Statistics Department, 2012, and The Government of the Hong Kong SAR, 2013.

for housing development since the early 2000s, as detailed later. As indicated in Table 10.4, the supply was meager compared with the demand, even if only half of the eligible households with relatively sufficient income were interested in joining the scheme. In the sale exercise conducted in March 2016, the over-subscription rates for private tenants were 26.4 times, and 2.2 times for public housing tenants (*HKSAR Government Press Releases*, May 10, 2016).

Thus, the availability of affordable housing to lower-middle-income private tenants has been limited; when the supply was forthcoming, the economic and market outlook was so dismal that homeownership was risky. The government's supply of new subsidized homes to lower-middle-income private tenants, as deliberated above, is always smaller than that of subsidized rental housing to low-income families living in substandard rental housing, and smaller than that of subsidized owner-occupier housing to public tenants of all income levels. In contrast to private tenants, public tenants have been entitled to join all subsidized homeownership schemes regardless of their income, because of the urgent need to recover public rental units from tenants who could afford to move up the housing ladder.

The public–private duality of the Hong Kong housing system has given rise to a rigidly stratified housing system and more conflicts among the housing tiers in Hong Kong. It is also owing to duality that a large transitional zone to homeownership has been created, featuring mainly latent homeowners. Among these aspiring but frustrated homeowners, the lower-middle-income private tenants are the most disadvantaged. On the one hand, they are accorded the lowest priority in the subsidized homeownership schemes as the government uses these schemes to mitigate the over-subsidization problems in the public rental sector; on the other hand, their chances to purchase private homes are often shattered by market volatility. It may be argued that they lose out because of the government's lack of political commitment and competence to strictly enforce the target-subsidy policy, and because of the government's liberal policy in the private housing market, which is made possible by the safety net provided by the public housing sector, which accommodates half of the population. This situation is currently worsening as the income limit eligible for subsidized homeownership is relaxed as a stopgap measure to quench soaring social discontent over the rapidly deteriorating affordability situation. With the expected low supply of new subsidized owner-occupier housing units in the next few years, the lower-middle-income tenants' dreams of homeownership have become more dismal for the near future.[3]

Thus, the dilemma of a dualistic housing system plays out in that those who are at the margins of the public housing sector – that is, households who are qualified but not prioritized – are trapped in the transitional zones to both subsidized and private homeownership. Yet, for those who have secured subsidized rental housing, moving up the housing ladder is just a matter of having saved and earned enough to pay for subsidized homeownership, with plenty of choices in the secondary subsidized housing market, though more limited at times in the primary segments when new supply is limited. Thus, the blatant discontinuity and structural inequality in achieving homeownership between those who are in and those who are just out of the subsidy net marks the main feature of the transitional zone of homeownership in a dualistic housing system. This discontinuity worsens during periods of short housing supply.

In a dualistic housing system, the direct provision of subsidized housing may cause sharply different housing outcomes in terms of homeownership, hinging upon whether one can succeed in getting onto the first rung of the public housing ladder, and whether subsidized homeownership is used as a means to redress anomalies in the public housing sector. Hence, the biggest challenges in Hong Kong's dualistic housing system are to enable the low-income families living in substandard housing units to move to public rental housing promptly, and also to permit the lower-middle-income and middle-income households to acquire owner-occupier housing with or without subsidies. Both challenges are currently looming large, especially the latter. Low housing space standards are corollaries of these limitations, even if livability problems are mitigated by improving public space provision and design in housing estates. Underlying these overt challenges are the covert challenges pertaining to topographical, structural, and institutional problems or weaknesses.

3. Covert challenges

3.1 Challenge 1: constrained land supply

Hong Kong has a land area of 1,106 square kilometers, and it had a population of 7.35 million in 2016. Its average population density is 6,642 per square kilometer, but that of developed areas is much higher given Hong Kong's composite urban form comprising a main urban area and nine discrete new towns (e.g., reaching 130,000 per square kilometer in the busy old district of Mong Kok). Developable land in Hong Kong is constrained by its hilly terrain, which comprises 78% of the total land area. Even with the cutting back of hills and reclamation of land from the sea, its developed area only accounts for 24% of total land area. To combat the natural constraints, reclaiming land from the harbor has been a major source of large-scale land provision to support economic development and meet the housing needs of the growing population, accounting for 26% (6,824 hectares) of total developed land in Hong Kong (Panel on Development, Legislative Committee). However, as detailed in Gurran, Gallent, and Chiu (2016), since the promulgation of the Harbour Protection Ordinance in 1998, and with the community's growing concern for environmental protection and preserving Victoria Harbor as a natural heritage site, this source of new land supply has dissipated, as shown in Figure 10.9.

Currently, based on the reasons that severe land shortage is regarded as the root cause of the housing affordability problem and that reclamation is the most effective option in providing large land areas for development, the government has proposed creating an artificial island between Hong Kong Island and Lantau Island as the major source of new land supply beyond 2030, expected to house 600,000 people (Development Bureau and Planning Department, 2014). Although the site is outside the protected inner harbor, the progressively politicized and participatory planning process and the escalating awareness of environmental protection, as discussed in Gurran, Gallent, and Chiu (2016), diminish the effectiveness and efficiency of this new land supply method.

The other land supply choice options – rezoning land use, land resumption from existing land users, rock cavern development, reuse of ex-quarry sites,

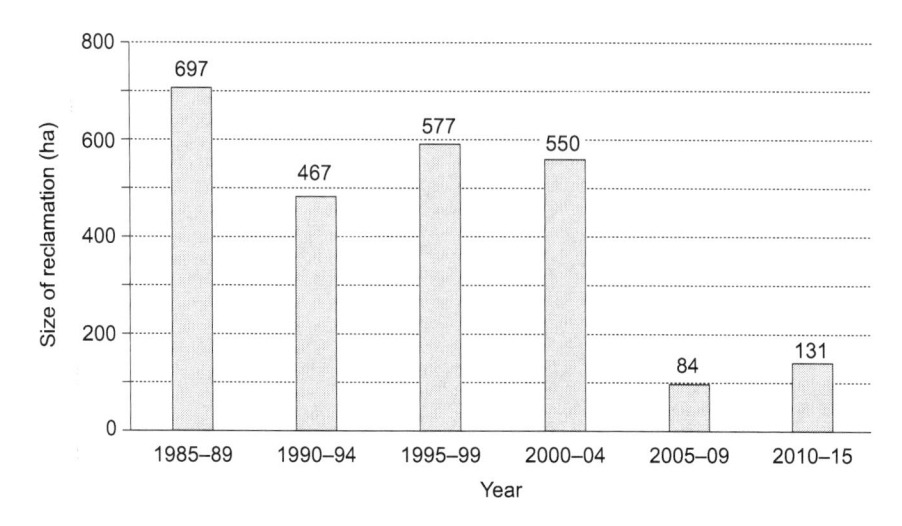

[Figure: Bar chart showing reclaimed land supply by 5-year period: 1985–89: 697 ha; 1990–94: 467 ha; 1995–99: 577 ha; 2000–04: 550 ha; 2005–09: 84 ha; 2010–15: 131 ha. Y-axis: Size of reclamation (ha).]

Figure 10.9 Reclaimed land supply, 1985–2015.
Source: Research Office, Legislative Council Secretariat, 2016.

and redevelopment – are on much smaller scales, or are as thorny, if not more so. Thus, constrained by the natural topography and harnessed by a pluralistic planning decision-making process, Hong Kong faces big challenges securing effective sources of new land supply.

Land management and land policy also impose institutional constraints. Although all land is state owned in Hong Kong, except a small church site (Gurran, Gallent, and Chiu, 2016), once land is leased, it becomes *de facto* private land, with the right of lease renewal except in the rare cases of resumption for significant public purposes. No additional premium payment is required, but an annual rent equivalent to only 3% of ratable value is levied. Thus state-ownership of land has no relevance in enhancing redevelopment as a way to increase housing land supply. In theory, as the only new land supplier, the government can regulate new land supply to mitigate the supply-and-demand imbalance to make housing more affordable. However, the government also has to maximize the financial return of land investment and to maximize the exchange value of land as a public resource. Thus, the dual role of the government as a major market player and the custodian of land as a public resource diminishes the influence of public ownership of land to enhance housing affordability in the market. The execution of these dual roles may sometimes be lopsided, as exemplified in the case of abandoning all new land searches and formation from the mid-2000s until 2010 under depressing market conditions, neglecting the role of the government to build up a land reserve for supplying land readily in the market when housing demand turns around quickly, as was the case in 2011. The neglect of this role was subsequently regarded as the root cause of the breathtaking affordability problems faced by the community today. Thus, balancing its conflicting dual roles as a market player and a public administrator poses another challenge to the Hong Kong government.

3.2 Challenge 2: Limited financial tools to mitigate homeowners' risks

An obvious limitation of the Hong Kong government in regulating market demand is the lack of independent monetary policy and domestic control over interest rates. The low levels of mortgage interest rates in most years since 2002 have resulted from the pegging of the Hong Kong dollar to the US dollar in 1983 in order to stabilize the Hong Kong currency and also in anticipation of possible economic turmoil after the change of government in 1997. Unfortunately, the business and property cycles of Hong Kong and the U.S. have never matched since 1983, and interest rates dictated by U.S. monetary policies have accentuated fluctuations in Hong Kong's housing market. As can be seen in Figure 10.10, although housing prices have skyrocketed since 2010, mortgage interest rates have stayed at the historically low level of about 2.75%, enticing people to purchase housing for consumption or investment purposes in an upward market.

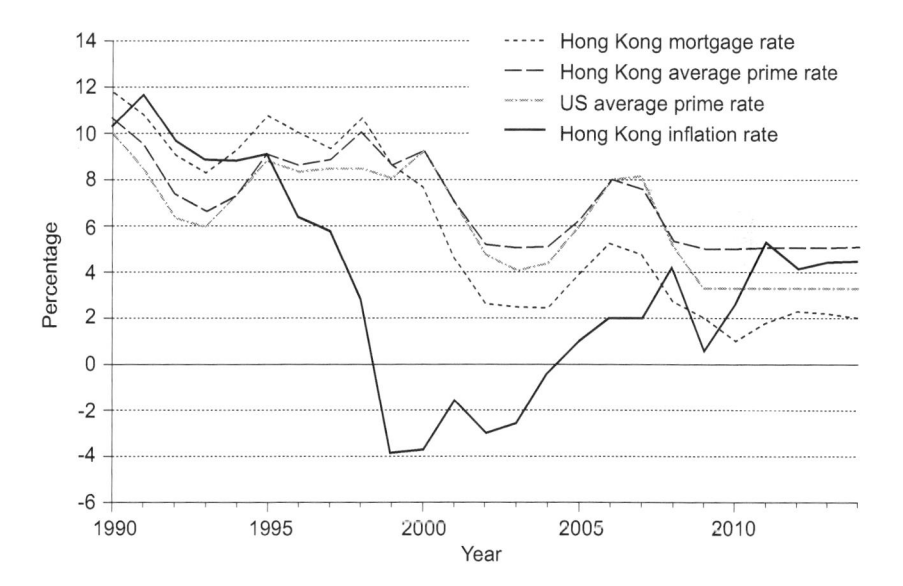

Figure 10.10 Average best lending rates of Hong Kong and the U.S., mortgage rates, and the inflation rates of Hong Kong, 1990–2014.

Note: The mortgage rate reflects the weighted average mortgage rate homeowners may obtain according to the approved mortgage rate of Referral customers.

Source: Based on Census and Statistics Department, 2015; Federal Reserve Board, 2015 and Research Department of Referral, 2015.

It must be noted that mortgage loans in Hong Kong are predominantly floating rate loans indexed to the bank prime lending rate, which implies that it is home purchasers, not the lending institutions, who bear the brunt of the interest rate risks. As the American economy gradually improves, interest rate increases are imminent. Homeowners who have not paid off their mortgage loans, who represented 34% (417,914) of total owners in 2016, may be at risk, especially those who began homeownership in recent years, when housing prices had soared. So far the government has not taken any specific measures to reduce the risks of the homeowners, apart from reminding the public of the possibility of interest rate increases.

Financial regulators, having no control over the interest rate, regard the loan-to-value cap as the main financial tool, together with debt service ratios, that the Hong Kong government can use to modify the housing market conditions (Ahuja and Nabar, 2011). In the early 1990s, the government tightened the loan-to-value ratios, mainly to protect the banking sector from the adverse impact of possible property market crashes incurred by the change of government in 1997. Since 2011, in view of the overheated market conditions, the government has again tightened the loan-to-value ratios.

However, these precautionary measures basically aim to suppress housing demand and to reduce the risks of financial institutions, rather than those of home purchasers, as reflected in a research paper produced by the Hong Kong Monetary Authority: "The [loan-to-value] tool is found to be effective in reducing systemic risk stemming from the boom-and-bust cycle of property markets" (Wong et al., 2011). Nonetheless, the liquidity constraint may have an indirect effect on lowering the purchasers' risk, since the loan amount and therefore the floating monthly mortgage repayment will be reduced. With a greater equity in the property, the owner can also withstand greater shocks in the property cycle. Thus, it was argued that the loan-to-value policy contributed to the low mortgage delinquency ratio of 1.43%, even after the 60% plummet in property prices consequent to the Asian Financial Crisis (Wong et al., 2011), and can be considered as a prudent lending criterion for reducing mortgage default rates (Wood, et al., 2013). However, the launch of the mortgage insurance policy in 1999 to promote homeownership in a depressed economy, which enabled higher loan leverage if home purchasers bought insurance for the additional mortgage loans, has re-exposed homeowners to higher risks. Thus, the mortgage insurance policy diversifies the risks of the financial institutions but not those of the homebuyers.

Further, top-up financing to lure buyers was provided by developers in the primary housing market, as well as by other co-financiers not regulated by the Hong Kong Mortgage Authority, such as finance companies. As shown in Figure 10.11, co-financing was most active during the housing market bust years, exposing homebuyers to additional risks in terms of their ability to sustain monthly income for mortgage repayments during economic decline. The top-up financing hence lowers market entry hurdles but increases purchasers' risks. The Hong Kong Housing Authority, as a direct provider of subsidized owner-occupier housing, also acts as the guarantor of the purchaser's mortgage; but again, it is only a measure to

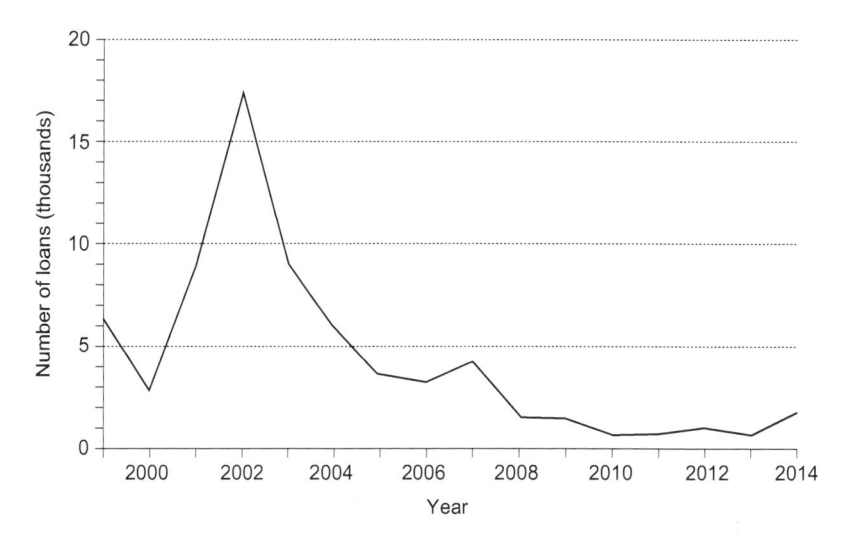

Figure 10.11 Number of loans associated with co-financing schemes, 1999–2014.

Note: Co-financing schemes refer to those schemes that involve the provision of top-up finance by property developer(s) or other co-financier(s) in addition to mortgage loans advanced by authorized institutions.

Source: Based on Hong Kong Monetary Authority, 2015.

ensure the financial institutions' full recovery of the outstanding balance of the mortgage and related costs in case of mortgage default.

Thus, overall, there is no institutional measure in Hong Kong which specifically aims to reduce or protect home purchasers from the risks of property shocks or interest rate variations. So far, no significant risk-sharing financial instruments akin to those suggested in Smith et al. (2009) have been developed in Hong Kong. However, other institutional factors, such as the loan-to-value ratio, have partially counteracted the risks, fostering the low mortgage delinquency ratios.

3.3 Challenge 3: Managing investment demand

An important factor that compounds housing price inflation and fluctuation is the openness of the Hong Kong economy to foreign investment generally, and to property investors specifically (Chiu, 2006). In the aftermath of the Global Financial Crisis, hot money from the West and also from China flowed into Hong Kong's housing market, fueling price hikes. Although the government identified that between 2011 and 2012 an average of 85% of residential property buyers were local people, in the primary housing market, which is regarded as the price setter, an average of 15% were non-local individual buyers; another 17% were local and non-local company buyers (Financial Services and the Treasury Bureau, 2013). Thus, external investors' influence in Hong Kong's property prices is

220 Rebecca L. H. Chiu

not insignificant, especially in the high-priced segment, luring speculators and homeownership churners (Figure 10.3). Although housing price changes may not be as marked as in some other international housing markets, the sheer rise in gross housing price is naturally very attractive to many: an overall rise of 12 times between 1983 and 2013; 3.7 times between 1990 and 1997; 3.8 times between 2003 and 2013; and two times between 2009 and 2013. It is only since February 2013 that targeted tax measures have been levied on foreign property buyers.

Another important factor that fuels the desire of the Hong Kong people to purchase homes as a form of household investment and retirement protection is the currency policy, as discussed above, which deprives the government itself of interest instruments to regulate inflation and housing prices. As shown in Figure 10.10, inflation rates were higher than mortgage rates between 1990 and 1995 and since 2010, when the economy resumed strength. Hence, home purchases and trading up were generally regarded as the best hedges against inflation. This protection function adds an additional "insurance" and "investment" function of homeownership discussed in Smith et al. (2009). There are also homeowners who engage in equity borrowing for income smoothing, alternative investments, or financing buffering. The remortgage volumes shown in Figures 10.12 and 10.10 show that refinancing activities are higher during times of price booms and negative real interest rates; but they decreased after price-cooling measures were introduced in 2011, confirming that manipulation of

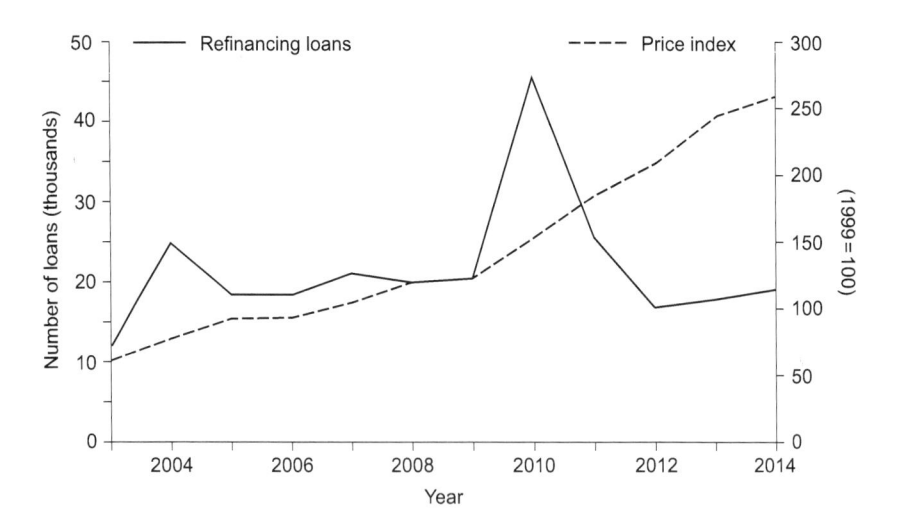

Figure 10.12 Number of housing loans refinanced compared to price index of private housing, 2003–2014.

Note: Refinancing loans include re-mortgage and mortgage transfer. Data of refinancing loan number before 2003 is not available.

Source: Based on Hong Kong Monetary Authority, 2015 and Rating and Valuation Department, 2015.

housing wealth has been happening. Maneuvering homeownership by higher-income households to accumulate household wealth is made possible in Hong Kong mainly because of the frequent market oscillations, the currency policy, and the openness of the housing market.

Housing wealth has a welfare role, as pointed out in Lee (2013) and Smith et al. (2009). This is particularly the case in Hong Kong as the city has no pension system, and the Mandatory Provident Fund, which was only launched in December 2000, currently has a small monthly maximum contribution of US$192 from both employers and employees.

A basic flaw in the 2014 Long Term Housing Strategy is its demographic-based housing demand projection model (Transport and Housing Bureau, 2014), which postulates that housing demand in the next 10 years is based only on household growth under different economic scenarios. Investment demand is considered insignificant, the argument being that housing investors would place their properties in the rental market. Thus, regardless of whether residential properties are purchased for self-use or for investment, they are all available in the market to meet the housing demand. Hence, investment purchase need not be separately accounted for in demand projection, and only household growth needs to be considered (Transport and Housing Bureau, 2014). Figure 10.13 shows that such a simplistic approach in treating housing investment demand is grossly wrong. Vacancy rates for larger housing units, which are prime targets for foreign investors – that is to say the two unit categories of 100–159.9 square meters and above 160 square meters, with some spilling over into the segment 70–99.9 square meters – have generally been much higher than those of smaller units and are well above 10%. It has to be noted that the carrying cost of an unmortgaged

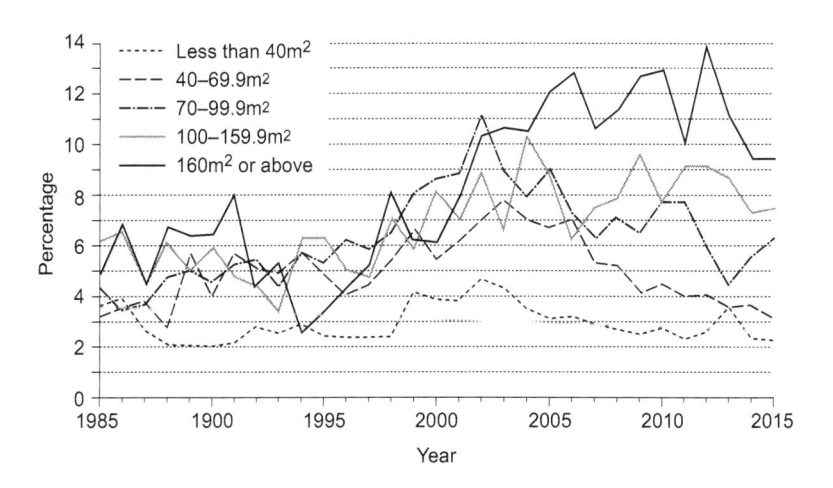

Figure 10.13 Changing housing vacancy ratio by unit size over time.
Source: Rating and Valuation Department, 2016a, 2016b, 2016c.

empty housing unit in Hong Kong is low as it consists only of rates (municipal service charges) and management fees.

The government is unable to explain, let alone solve, the glaring housing affordability problems of recent years (Figure 10.3), when the number of housing units has been consistently higher than the number of households – for instance, at 2,668,000 and 2,449,400, respectively, in 2015 (Census and Statistics Department, 2015; Information Services Department, 2015). The introduction of an extra stamp duty for foreign and local home investors since 2011 has had no significant effect on decreasing housing prices but has only reduced the number of transactions in a given time period (Figure 10.4). Thus, albeit an active player in the property market and the only new land supplier, the Hong Kong government has yet to fully understand the different types of housing demands and their implications for housing supply and affordability. Fundamentally, the government has conveniently mixed up the concepts of housing need and housing demand, the former being demographically driven, and the latter being market-driven and particularly difficult to harness when the market is open to global and regional markets, and when home purchases are used by the local people as a hedge against inflation, to accumulate household wealth, and as a tool for retirement protection.

The challenges confronting Hong Kong's dualistic housing system are severe affordability problems in homeownership, inadequate supply of affordable rental housing and consequent housing mobility bottlenecks, and low housing space standards. These challenges are in fact corollaries of three structural challenges. The first group is the constrained land supply due to hilly terrain, a politicized and pluralistic planning process, and the conflicting role of the government as both landowner and market stabilizer. The second group is the limited financial tools usable by the government to regulate supply and demand imbalances and to minimize risks in sustaining homeownership. The third group is the uncontrolled (until recently) investment demand induced by the openness of Hong Kong's housing market to global property investors funded by the greatly increased flow of international and mainland investment funds seeking higher returns in Asian markets, especially after 2009, and the continued inflation- and retirement-protection function of homeowning to local investors. The recent fiscal measures to restrain investment demand have not been able to cool down housing prices. The impacts of investment demand have not been taken into account by the government's 2014 Long Term Housing Strategy, which adopts a purely demographic-driven approach. This negligence poses the danger that housing affordability problems may continue into the future, and possibly worsen, if the Asian economies continue to grow and attract investment funds to the region.

4. The implications of Hong Kong's housing challenges for Chinese cities

Hong Kong's housing challenges have direct implications for China's prime cities such as Shanghai, Beijing, Guangzhou, and Shenzhen, with their very active housing markets and high housing prices. While Hong Kong is targeted by global and regional property investors, these Chinese cities have to cope with strong

Housing challenges in Hong Kong 223

domestic demand driven by the large influx of intercity migrants and investors, as well as rapidly rising incomes. For instance, population has grown from 10.9 million in 1990 to 21.7 million in 2015 in Beijing (Beijing Municipal Bureau of Statistics, 1991, 2016), from 12.8 million to 24.2 million in Shanghai (Shanghai Municipal Bureau of Statistics, 1991, 2016), and from 2.0 million in 1990 to 11.4 million in 2015 in Shenzhen (Shenzhen Municipal Bureau of Statistics, 1991, 2016).

Whereas Hong Kong uses fiscal measures to dampen price increases, the Chinese city governments use administrative (residency status) and financial regulatory measures (interest rate control and mortgage loan approval by state-owned banks and housing provident fund management offices) to restrict non-local homebuyers and also lower local residents' investment demand. In Hong Kong, as discussed earlier, transaction volumes dropped, but housing prices continue to rise. Figure 10.14 shows that mainland China's price-cooling measures have also been ineffective in the prime cities in spite of tightened purchase restrictions over the years; these measures are reversed quickly to keep the housing market buoyant (as Shenzhen did in 2015). As a free economy, Hong Kong avoids strong direct government interventions because it wants to protect its status as the freest economy and deems fiscal measures to be more appropriate than direct regulatory controls. China, as a marketized socialist system, has fewer concerns of this kind, but the Chinese government's policies still need to consider the dominant role of housing as an investment good because very limited alternative investment options are open to the general public for household wealth accumulation.

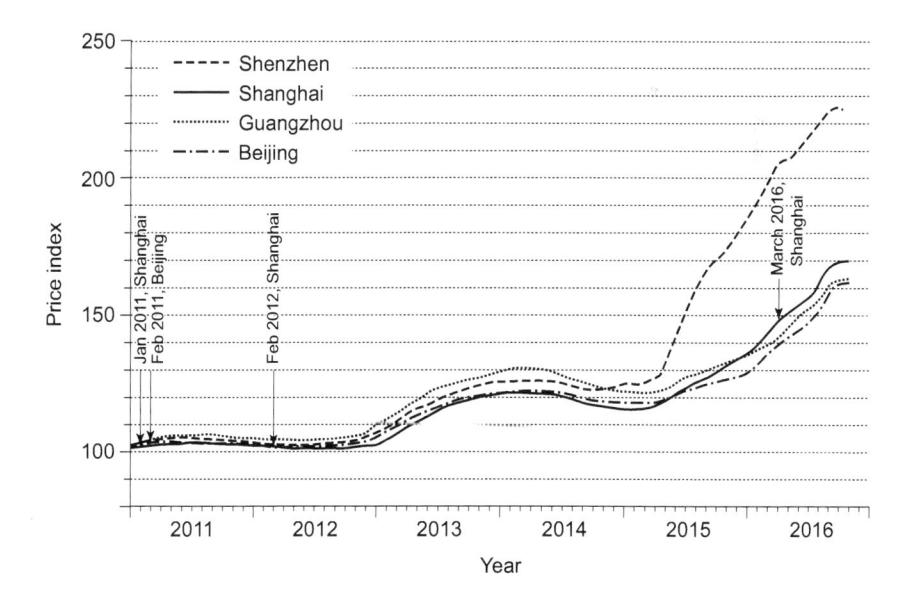

Figure 10.14 Housing price trends of four prime cities in China (2011–2016) with announcement dates of price-cooling measures.
Source: Based on National Bureau of Statistics of China, 2011–2016.

Prime Chinese cities face similar problems to Hong Kong regarding the insufficient supply of affordable rental and owner-occupier housing, for the same reasons: a supply shortage; rapid increase in young households, especially migrants; and a continuous supply of substandard housing by the market to meet the needs and renters' affordability, notably housing in urban villages (He, 2010). In terms of public housing supply, Hong Kong nonetheless operates a much bigger and higher-quality public rental housing program, catering to 46% of the total population, while those of Shanghai and Shenzhen are generally lower than 10%.

While their public housing stock is small, the prime cities in mainland China adopt a mixed mode of housing subsidy schemes. For instance, Shenzhen offers four types of housing subsidies with a special focus on semi-skilled and skilled workers: public rental housing; rental allowances; purchase and relet schemes, whereby the government or employers purchase or rent market housing for reletting to skilled workers at lower prices; and affordable housing schemes, previously the Economical and Suitable Housing Scheme (*Jingjishiyong Zhufang*) and currently the Affordable Commodity Housing Scheme (*Anjuxing Shangpinfang*). The hybrid modes reduce over-subsidization, as in Hong Kong's rental housing sector, where there is inertia in evicting tenants whose income has exceeded the subsidy limit. The drawback of Shenzhen's hybrid mode is the impact of market volatility on the effectiveness of the cash subsidy and on the government's and employers' ability to purchase market housing for relet if housing subsidy resources are housing price inelastic. In mainland China's prime cities, migrant workers pose bigger housing issues, not only because of the large-scale immigration, but also because of the controversy over whether migrant workers should be legitimate recipients of housing subsidy provided by local governments, and whether local governments can afford to take care of such a large population group. It has to be noted that insufficient supplies of both affordable rental and owner-occupier housing of basic or good standards affect livability, upward social mobility, and social stability in Hong Kong and Chinese cities alike.

In terms of the shortage of land supply, unlike Hong Kong, which has a much smaller land area, approximately one-half of Shenzhen, one-sixth of Shanghai, one-seventh of Guangzhou, and one-sixteenth of Beijing are bounded by political borders; the Chinese prime cities all have a rural hinterland. Nonetheless, the control on the expansion of construction land in the rural areas and the keen competition for land from different land users still incur similar problems in finding sufficient land to meet housing needs and demand. What makes new land supply quicker in mainland Chinese cities is the pro-growth and pro-efficiency planning process, which sometimes may even follow instead of lead development. Being owners of urban land in both the case of the mainland and Hong Kong, governments in China face a common issue: how to play the conflicting roles of being both a market participant and the market's social regulator. The implication from Hong Kong is that building a land reserve to meet volatile demand is necessary and feasible but requires government commitment to fulfill its role as a social regulator. As a market player, however, the government is expected to exercise the power to decide land supply in terms of amount, location, and timing in view of market conditions to capture the maximum value of land and to maximize public revenue. This is especially the case for local governments in Chinese cities, as economic growth, to which the housing market contributes, stands as a major criterion to assess the job performance of government leaders.

Finally, the implications from the Hong Kong case for mainland China's prime cities are contingent upon some fundamental issues that are yet to unfold or be resolved. The first is the positioning of housing as a basic consumption good, with and without a strong investment element in relation to investment outlets for household savings, and whether an asset-based welfare policy is to be established through housing investment by households. A second is the implications of the nature and purpose of the chosen form of subsidy provision – i.e., to adopt a direct producer subsidy, or a consumer subsidy approach, or a mixed mode, in the context of a political economy dominated by state capitalism, and with a predominantly marketized housing system. A third issue is whether housing tiers constitute a ladder for social mobility – i.e., the social engineering role of housing in a marketized socialist system, and a marketized housing system. A fourth issue is the role of the government in a state-owned land system and government-dominant planning system.

Acknowledgment

The work described in this chapter was funded by a grant from the Research Grants Council of the Hong Kong Special Administrative Region, China (Project No. 17612215 HKU).

Notes

1 In Hong Kong, land-use zoning has a legal status that must be observed by all landowners and users, who are subject to prosecution if they use the land and properties differently. This status is enforceable in Hong Kong, but less so in Chinese cities subject to temporal, location, and governance factors.
2 HK$7.8 = US$1 on May 3, 2015.
3 A small "Starter Homes" scheme was to be piloted in August 2018, but no sustained supply has been promised yet. A trial scheme of allowing private tenants to buy subsidized owner-occupying housing in the secondary market with the same terms as public tenants, i.e. at a subsidized price, was introduced in 2013, enabling 4,000 private tenants to become homeowners. The scheme was regularized in 2008, with the government promising a supply of 2,500 per year (Transport and Housing Bureau, 2018). These schemes improve the homeownership opportunities of lower-income private tenants, but are still far below the demand level.

References

Ahuja, A. and Nabar, M. (2011). *Safeguarding Banks and Containing Property Booms: Cross-Country Evidence on Macroprudential Policies and Lessons from Hong Kong SAR*, IMF Working Paper WP11/284. Washington: International Monetary Fund.

Beijing Municipal Bureau of Statistics. (1991). *1990 Beijing Statistical Yearbook*. Beijing: China Statistics Press.
Beijing Municipal Bureau of Statistics. (2016). *2015 Beijing Statistical Yearbook*. Beijing: China Statistics Press.
Census and Statistics Department of the Government of the Hong Kong SAR. (2012). 2011 Hong Kong Population Census, www.census2011.gov.hk/en/index.html. Accessed March 20, 2015.
Census and Statistics Department of the Government of the Hong Kong SAR. (2015). All Tables, Publications, Charts and Articles, www.censtatd.gov.hk/hkstat/quicklink/index.jsp. Accessed March 20, 2015.
Census and Statistics Department of the Government of the Hong Kong SAR. (2016). All Tables, Publications, Charts and Articles, www.censtatd.gov.hk/hkstat/quicklink/index.jsp. Accessed March 20, 2017.
Centadata (2014). Centadata, www1.centadata.com/ephome.aspx. Accessed March 20, 2015.
Chiu, R. L. H. (1998). The Home Ownership Drive of Hong Kong Special Administrative Region (HKSAR), *Planning and Development Journal of the Hong Kong Institute of Planners*, 14(1), 12–20.
Chiu, R. L. H. (2005). Housing in the Social Development Perspective. In R. J. Estes (Ed.). *Social Development in Hong Kong: The Unfinished Agenda*, Oxford University Press: London and New York, 157–169.
Chiu, R. L. H. (2006). Globalization and Localization: Economic Performance and the Housing Markets of the Asian Tigers Since the Financial Crisis. *Housing Finance International*, 20(3), 12–17.
Chiu, R. L. H., Tang, B. S., Tse, J. K. H., Lum, T., Chui, E. Y. T., and Kee, T. (2015). *Full Report: A Comprehensive Study of Housing in an Ageing Community*, Contract research report submitted to Hong Kong Housing Society, Hong Kong, Chapter 4.
Development Bureau and Planning Department (2014). Hong Kong 2030+ Towards a Planning Vision and Strategy Transcending 2030; Public engagement. Hong Kong: Government Printer.
Federal Reserve Board. (2015). FRB: H.15 Release—Selected Interest Rates—Historical Data, www.federalreserve.gov/releases/H15/data.htm. Accessed March 25, 2015.
Financial Secretary's Office, HKSAR Government. (2013). Hong Kong Economic Reports. www.hkeconomy.gov.hk/en/reports/archive.htm. Accessed October 20, 2014.
Financial Services and the Treasury Bureau. (2013). The Administration's Responses to Matters Arising from the Meeting of the Bills Committee on Stamp Duty (Amendment) Bill 2013 held on 22 November 2013, www.legco.gov.hk/yr12-13/english/bc/bc05/papers/bc051213cb1-491-2-e.pdf. Accessed April 20, 2015.
Gerlach, S. and Peng, W. (2005). Bank Lending and Property Prices in Hong Kong. *Journal of Banking & Finance*, 29(2), 461–481.
Gurran, N., Gallent, N., and Chiu, R. L. (2016). *Politics, Planning and Housing Supply in Australia, England and Hong Kong*. New York and London: Routledge.
He, S., Wu, F., Webster, C., and Liu, Y. (2010). Poverty Concentration and Determinants in China's Urban Low-income Neighbourhoods and Social Groups. *International Journal of Urban and Regional Research*, 34(2), 328–349.
HKSAR Government Press Releases. (2016). Press Releases, May 10, 2016. www.info.gov.hk/gia/general/201605/10.htm. Accessed June 10, 2016.

Hong Kong Housing Authority of the Government of the Hong Kong SAR. (2014a). 2013/14 Annual Report, www.housingauthority.gov.hk/mini-site/haar1314/en/index.html. Accessed March 10, 2015.

Hong Kong Housing Authority of the Government of the Hong Kong SAR. (2014b). Housing in Figures 2014, www.housingauthority.gov.hk/en/common/pdf/about-us/publications-and-statistics/HIF.pdf. Accessed March 10, 2015.

Hong Kong Housing Authority of the Government of the Hong Kong SAR. (2016a). Housing in Figures 2015, www.housingauthority.gov.hk/en/common/pdf/about-us/publications-and-statistics/HIF.pdf. Accessed March 10, 2017.

Hong Kong Housing Authority of the Government of the Hong Kong SAR. (2016b). 10 May 2016 Press Releases: Ballots drawn for Sale of Home Ownership Scheme Flats 2016, www.info.gov.hk/gia/general/201605/10/P201605100621.htm.

Hong Kong Housing Authority of the Government of the Hong Kong SAR. (2016c). Special Analysis of Housing Situation of General Applicants for PRH as at End-June 2016, www.housingauthority.gov.hk/en/common/pdf/about-us/housing-authority/ha-paper-library/SHC52-16.pdf. Accessed March 10, 2017.

Hong Kong Housing Society. (2014a). All Tables, Publications, Charts and Articles. www.hkhs.com/index.asp?lang=eng.

Hong Kong Housing Society. (2014b). Heya Green Sales Brochure. www.heyagreen.com/download/sales_brochure_files/Heya_Green_Sales_Brochure_Part%20B_20140925.pdf.

Hong Kong Monetary Authority of the Government of the Hong Kong SAR. (2015). Monthly Statistical Bulletin. www.hkma.gov.hk/eng/market-data-and-statistics/monthly-statistical-bulletin. Accessed March 20, 2015.

Information Services Department of the Government of the Hong Kong SAR. (2015). *2015 Hong Kong Yearbook*. Hong Kong: Government Printer.

Lee, J. (2013). Housing Policy and Asset Building: Exploring the Role of Home Ownership in East Asian Social Policy. *China Journal of Social Work*, 6(2), 104–117.

Lee, R. K. B. (2014). Challenges in Hong Kong Land Supply by Reclamation. www.engineeringforum.hk/publications/2014-06-13/4_Challenges%20in%20HK%20Land%20Supply%20by%20Reclamation.pdf. Accessed June 10, 2016.

National Bureau of Statistics of China. (2011–2016). *2011–2016 Monthly Report: Housing Sale Price Changes in 70 Large and Medium Cities*. China: Government Printer.

Panel on Development, Legislative Committee. (2011). CB(1)2293/10-11(04), *Development of Greening Master Plans (GMPs) for the New Territories: Legislative Council Panel on Development on 24 May 2011*. www.legco.gov.hk/yr10-11/english/panels/dev/papers/dev0524cb1-2293-1-e.pdf. Accessed July 20, 2017.

Rating and Valuation Department of the Government of the Hong Kong SAR. (2014). 2014 Hong Kong Property Review. www.rvd.gov.hk/doc/en/hkpr14/PR2014_full.pdf.

Rating and Valuation Department of the Government of the Hong Kong SAR. (2015). Private Domestic – Price Indices by Class (Territory-wide). www.rvd.gov.hk/doc/en/statistics/his_data_4.xls. Accessed March 10, 2015.

Rating and Valuation Department of the Government of the Hong Kong SAR. (2016). Private Domestic – Price Indices by Grade (from 1981). www.rvd.gov.hk/en/property_market_statistics/. Accessed November 10, 2016.

Rating and Valuation Department of the Government of the Hong Kong SAR. (2016). Domestic Sales (from 2002). www.rvd.gov.hk/en/property_market_statistics/. Accessed November 10, 2016.

Rating and Valuation Department of the Government of the Hong Kong SAR. (2016). Private Domestic – Completions, Stock, Vacancy and Take-up. www.rvd.gov.hk/en/property_market_statistics/completions.html. Accessed May 7, 2017.

Research Department of Referral of the Government of the Hong Kong SAR. (2015). Referral Mortgage Brokerage Services. www.mreferral.com/tables-and-charts.html. Accessed March 25, 2015.

Rex, J. and Moore, R. (1971). *Race, Community and Conflict: A Study of Sparkbrook*. London: Oxford University Press.

Saunders, P. (1990). *A Nation of Home Owners*. London: Unwin Hyman.

Smith, S. J., Searle, B. A., and Cook, N. (2009). Rethinking the Risks of Home Ownership. *Journal of Social Policy*, 38(1), 83–102.

Smith, S. J., Searle, B. A., and Powells, G. (2010). Introduction. In S. Smith and B. Searle (Eds.). *The Blackwell Companion to the Economics of Housing: The Housing Wealth of Nations*. New York: John Wiley & Sons, 1–27.

Sing Tao Daily. (2015). March 27, 2015, p. A6.

Sing Tao Daily. (2016a). June 10, 2016, p. A3.

Sing Tao Daily. (2016b). August 12, 2016, p. A8.

Shanghai Housing Security and Administration Bureau. (2016). All Tables, Publications, Charts and Articles. www.shanghai.gov.cn/nw2/nw2314/nw2319/nw14868/index.html. Accessed December 20, 2016.

Shanghai Municipal Bureau of Statistics. (1991). *1990 Shanghai Statistical Yearbook*. China Statistics Press.

Shanghai Municipal Bureau of Statistics. (2016). *2015 Shanghai Statistical Yearbook*. China Statistics Press.

Shenzhen Housing and Construction Bureau. (2016). All Tables, Publications, Charts and Articles. www.szjs.gov.cn/ztfw/zfbz/fpfy/. Accessed December 20, 2016.

Shenzhen Municipal Bureau of Statistics. (1991). *1990 Shenzhen Statistical Yearbook*. China Statistics Press.

Shenzhen Municipal Bureau of Statistics. (2016). *2015 Shenzhen Statistical Yearbook*. China Statistics Press.

Tang, B. S., Zhang, J., and Chiu, R. (2015). In R. L. H. Chiu, B. S. Tang, J. K. H. Tse, T. Lum, E. Y. T. Chui, and T. Kee (Eds.). *Full Report: A Comprehensive Study of Housing in an Ageing Community*, Contract Research Report Submitted to Hong Kong Housing Society, Hong Kong, Chapter 4.

The Government of the Hong Kong SAR. (1997). *Progress Report: The 1997 Policy Address*. Hong Kong: Printing Department.

The Government of the Hong Kong SAR. (2010). *2010–11 Policy Address: Sharing Prosperity for a Caring Society*. Hong Kong: Government Printer.

The Government of the Hong Kong SAR. (2013). *2013 Policy Address: Seek Change Maintain Stability Serve the People with Pragmatism*. Hong Kong: Government Printer.

The Land Registry. (2014). Statistics of Agreements for Sale and Purchase of Building Units. www.landreg.gov.hk/en/monthly/agreement.htm. Accessed March 20, 2015.

Transport and Housing Bureau. (2014). *Long Term Housing Strategy Implementation Milestones as at December 2014*. Hong Kong: Government Printer.

Transport and Housing Bureau. (2018) Hong Kong: The Facts (Housing). www.thb.gov.hk/eng/psp/publications/housing/hongkongthefacts/index.htm. Accessed August 6, 2018.

Wood, G., Smith, S., Ong, R., and Cigdem, M. (2013). *The Edges of Home Ownership*, AHURI Final Report No. 216. Melbourne: Australian Housing and Urban Research Institute.

Wong, E., Fong, T., Li, K., and Choi, H. (2011). *Loan-to-value Ratio as a Macro-prudential Tool – Hong Kong's Experience and Cross-country Evidence*, Working Paper 01/2011. Hong Kong: Hong Kong Monetary Authority.

Yip, N. M., Forrest, R., and La Grange, A. (2007). Cohort Trajectories in Hong Kong's Housing System: 1981–2001. *Housing Studies*, 22(1), 121–136.

11 Building an equitable and inclusive city through housing policies

Singapore's experience

Sock-Yong Phang

SINGAPORE MANAGEMENT UNIVERSITY

> For each of them is very many cities but not a city, as those who play say. There are two, in any case, warring with each other, one of the poor, the other of the rich. And within each of these there are very many.
>
> –Plato (380 B.C.), *The Republic*, Book IV

1. Introduction

Inequality is an age-old concern. In recent years, the rise of income inequality has received worldwide media and policy attention, beginning with the Occupy movement of 2011–2012 and a wave of notable scholastic books such as Stiglitz (2012), Piketty (2014), and Atkinson (2015). Piketty's *Capital in the 21st Century*, an unlikely bestseller, contained a vast amount of data showing that the rich are taking rising shares of income and wealth in the advanced economies. Piketty's approach towards capital and wealth is an aggregative one, and he does not treat real estate or land as a different or distinct form of capital. He deals with neither spatial inequality nor the role of house price inflation in accentuating inequality. An aspatial approach leads to discussions about solutions to inequality focusing on aspatial aspects of higher income and wealth taxes, health and education policies, and labor market interventions such as minimum wages and universal basic incomes (Piketty, 2014; Atkinson, 2015; IMF, 2018).

National inequality data, however, mask considerable variations across cities within the *same* country. The biggest and most dynamic cities in a country also tend to have the highest housing prices, the highest concentration of housing wealth, and the largest income and wealth inequality gaps. Baum-Snow and Pavan (2013), Eeckhout et al. (2014), and Behrens and Robert-Nicoud (2014) have analyzed the link between city size and inequality. In an early review of Piketty's book, a careful study by Rognlie (2015) disaggregates capital and finds that, while the net capital share in developed economies has increased since 1948, this increase comes *entirely* from the *housing* sector. He therefore advocates that observers concerned about the distribution of income should keep an eye on housing costs.

Atkinson suggests that the most valuable asset for the majority of UK homeowners is likely to be their home, and increased housing wealth is a major reason for the rise in popular wealth in the UK (Atkinson, 2015). Bastagli and Hills (2013) also present evidence of the importance of housing in wealth accumulation and wealth distribution for Great Britain. Using British Household Panel Survey data from 1995 to 2005, they show that changes in housing wealth heavily affected changes in wealth, with the highest percentage increase in housing wealth taking place in the middle of the distribution. Phang (2015, 2018) highlights the housing affordability problem in global cities and the role of homeownership policies in Singapore for mitigating income and wealth inequality.

Just as there is no reason the market will deliver a socially desirable distribution of income, there is no reason the market will deliver equitable housing outcomes and inclusive neighborhoods in cities. Households' preferences for local public goods, taxed goods, income, and race of neighbors tend to lead to divided cities: neighborhoods segregated by income (Comandon et al., 2018) and by race (Kucheva and Sander, 2018). Without appropriate policy intervention, we can expect the problems of inequality, housing affordability, unequal housing wealth distribution, and segregation by income and ethnicity to be accentuated and more visible in urban areas and even more so in large metropolises.

In several countries, historical factors and policies explain segregated neighborhoods. Land-use planning agencies and powers could also be overly fragmented and decentralized, resulting in resistance to mixed-use and/or higher-density developments. The not-in-my-backyard (NIMBY) behavior of homeowners and natural preferences for households to live among others with similar preferences and views contribute to income and racial segregation and social tensions in cities. As early as 1971, Nobel winner Thomas Schelling showed how a slight-but-not-malicious individual preference to avoid being in even a slightly minority status would quickly lead to completely segregated neighborhoods (Schelling, 1971). Schelling also studied how a neighborhood's racial composition could change suddenly, "tipping" from one situation to another after a critical mass was reached. More recently, Bishop (2009) and Florida (2017) have called attention to these urban challenges in their books, *The Big Sort* and *The New Urban Crisis*, respectively.

Intergenerational mobility, along with inequality, has received much attention in recent years. Cross-country studies show a well-known negative correlation between income inequality and intergenerational mobility that Krueger (2012) has termed "The Great Gatsby Curve" (Corak, 2013; Narayan et al., 2018). Chetty et al. (2014), however, find that for the US, despite the rise in income inequality in recent decades, overall intergenerational mobility has not changed significantly. Delving deeper, they find substantial variation in intergenerational mobility across geographical areas and identify racial and income segregation at the community level as a major factor that strongly correlates with mobility—i.e., the social prospects of children are dependent on the characteristics of the neighborhood in which they are raised. In particular, upward income mobility is significantly lower in areas with a larger African-American population (Chetty et al., 2018).

The deepening concern about inequality since 2014 has contributed to various international organizations developing global agendas to promote equity and inclusiveness, with a focus on cities. In March 2016, the Organisation for Economic Cooperation and Development (OECD) and Ford Foundation launched an "Inclusive Growth in Cities Campaign" to provide mayors around the world with a platform to exchange solutions to overcome urban inequalities.[1] The United Nations Habitat III conference, held in Quito in October 2016, set a "New Urban Agenda" to create sustainable, equitable cities for all.[2] A key focus of Habitat III was how to stop the widening inequality gap between the richest and poorest in many cities. In 2017, the Urban Land Institute released a new publication, *Building Equitable Cities*, that emphasized the importance of using place-based and people-based strategies to reduce income inequity, increase educational achievement, and reduce racial and ethnic segregation (Bowdler et al., 2017). The recently launched OECD study "Divided Cities: Understanding Intra-Urban Inequalities" (OECD, 2018) reports on how inequality and segregation play out across city neighborhoods by considering multiple cities in an international context.

The Habitat for Humanity CEO, Jonathan Reckford, rightly observed that housing is critical to equitable and inclusive cities: "With 60 percent of people worldwide projected to be concentrated in urban areas by 2030, developing sustainable communities that are inclusive and equitable for all will require creating affordable housing located near job opportunities" (Reckford, 2018). How can a city provide a supply of land and related infrastructure to "create affordable housing located near job opportunities" in a sustainable equitable way?

In the late nineteenth century, the American reformer Henry George proposed a solution to the above question. He observed that the explosive growth of industrial output in nineteenth-century America generated dramatic increases in urban land prices. The windfalls for landowners, in turn, fueled a frenzy of land speculation, real estate bubbles, and volatility. While industrialists, bankers, and landowners amassed enormous wealth in the Gilded Age, there was a simultaneous rise in poverty, inequality, economic crises, and social unrest. In 1879, George published a critique of the capitalist system in *Progress and Poverty: An Inquiry into the Cause of Industrial Depressions and of Increase of Want with Increase of Wealth; The Remedy*. His controversial proposed solution to the problems of inequality and crises was that land should be common property, society should share in any increase in land rents, and the tax on land value increases should be 100%.[3]

Singapore's work in building an equitable and inclusive city is extensive. While Singapore's focus on land is in line with George's views (Phang, 1996), there was no mention of George in the intense policy debates on land policy reform in the 1960s. The policy makers of that era recognized the huge importance of land to jump-start the economy and the housing program. Their observations on the unfairness of unearned land value increments, the negative impacts of land speculation, and the need for the state to capture land values mirrored those of George. However, the policies they implemented differed. This chapter on

Singapore's land and housing policies contributes to the discussions of alternative solutions to the challenges of urban inequality and divided cities.

Singapore is a rather unusual city—a densely populated city of 5.6 million people that happens also to be a country. All the physical spaces that are required for the normal functioning of a country (such as the needs of a military, port, airport, and reservoirs) have to fit into a city with a land area of 720 square km—slightly smaller than New York City and half the size of Greater London.[4] Devan (2018) highlights that 42–43% of land has been allocated for use by the military, and for gateways (airport and port) and manufacturing. At the same time, Singapore is also a global financial center, open to international capital flows and with almost 40% of the population comprising non-citizens (foreigners and permanent residents).

Given the unusual physical setting and land constraints of Singapore, the government has been acutely conscious of the limitations of the market to deal with the multiple complexities and competing demands for land efficiently and equitably. The extreme scarcity explains the obsessive policy attention to land policy and highly interventionist measures in the land, housing, and land transport sectors. Slums and squatter settlements were prevalent in Singapore in the 1960s at the time of its transition from a British colony to independence. The unemployment rate was estimated at 10% in 1965. The majority of the population lived in crammed conditions in pre-war shophouses and attap palm/zinc-roofed housing. It was under such dire conditions that the government made land and housing reforms, with the objective of improving housing conditions being a key priority.

After five decades Singapore's land and housing policies have produced some unusual outcomes compared to other cities of the world: 90% of the land belongs to the government, and 82% of the resident population live in high-rise apartments originally built by government agencies. The wholly state-owned Housing and Development Board (HDB) is Singapore's largest housing developer. As the HDB sells the bulk of its flats (on a 99-year leasehold basis), the resident population homeownership rate is unusually high—at 91% (see Figure 11.1). There are no poor neighborhoods, and ethnic quotas are used to manage the racial composition of blocks and HDB estates. This careful policy attention to land, housing, and social integration has contributed to social and political stability and economic development by creating "good places for Singapore citizens and residents to live and work, setting the infrastructure for people in Singapore to create a poster-child East Asia economic miracle of the late twentieth century" (Sargent, 2018).

The second section of this chapter presents data on income inequality and spatial measures of income and ethnic segregation in Singapore, comparing these statistics with cities in other countries. Section 3 provides an overview of the land, housing, and fiscal policies used to reduce the housing divide and to create mixed income and ethnically diverse neighborhoods. Looking ahead, the chapter concludes in Section 4 with a discussion of the current debates in Singapore on additional measures to mitigate economic inequality and to improve social mobility and integration.

234 Sock-Yong Phang

▨ Before accounting for taxes and transfers   ▧ After accounting for taxes and transfers

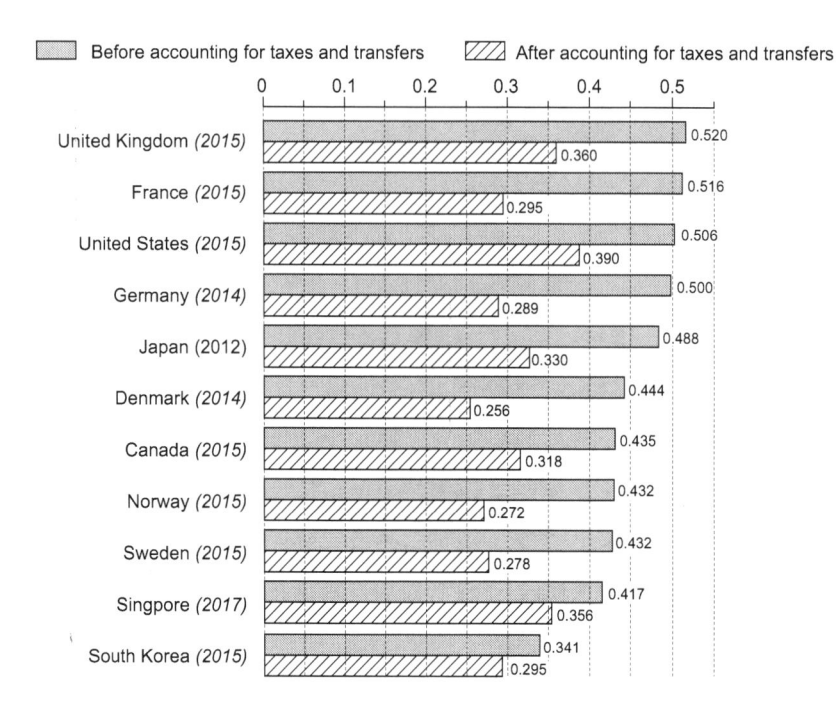

Country	Before	After
United Kingdom (2015)	0.520	0.360
France (2015)	0.516	0.295
United States (2015)	0.506	0.390
Germany (2014)	0.500	0.289
Japan (2012)	0.488	0.330
Denmark (2014)	0.444	0.256
Canada (2015)	0.435	0.318
Norway (2015)	0.432	0.272
Sweden (2015)	0.432	0.278
Singapore (2017)	0.417	0.356
South Korea (2015)	0.341	0.295

Figure 11.1 Homeownership rate.
Source: Singapore Department of Statistics and Housing and Development Board (HDB) websites.

2. Measures of inequality and segregation

2.1 Income inequality and housing wealth distribution

The Gini coefficient index is a widely used measure of income inequality. Most available Gini coefficients report figures over time for a given society before and after accounting for taxes and transfers. However, in using the coefficient for comparative purposes, researchers also have to be aware of the different methods and data used in its calculations. Variations in methodology include use of total income, income from work or wage data, inclusion or exclusion of unemployed and/or non-working households, inclusion or exclusion of foreigners, and use of incomes per household member or the OECD's method of adjustment for household size to account for economies of scale with regard to household expenditures.

Figure 11.2 shows the OECD international comparison of countrywide income Gini coefficients from the latest available data. While Singapore's 2017 Gini coefficient of 0.417, before taxes and transfers, is lower than most OECD countries, the Gini coefficient after taxes and transfers, at 0.356, is higher than most OECD countries (with the exception of the US and the UK). According

Building an equitable and inclusive city 235

to the Singapore Ministry of Finance (2015), the smaller difference between the "before" and "after taxes and transfers" Gini coefficients for Singapore, compared to the OECD countries, reflects the lower tax burden in Singapore and its preference for providing targeted subsidies rather than large social transfers.

However, Gini coefficients for cities are typically much higher than the countrywide figures (Phang, 2015; 2018). Singapore's Gini coefficient (after taxes and transfers) is lower than the Gini coefficients of other global cities such as New York (0.504),[5] London (0.44), Hong Kong (0.40), and Paris (0.37). Devan (2018) suggests that this is in part due to Singapore's substantial manufacturing sector, which contributes 20% of GDP and provides a range of jobs in the middle, compared to other global cities with high-paying jobs in finance and banking at one end and low-paying jobs at the other. Moreover, Gini coefficients do not reflect income-net-of-housing costs, and local policies can make a vast difference to housing affordability, household consumption, and welfare.

Rising income inequality also leads to rising wealth inequality, which in turn drives income inequality. Piketty (2014) notes, "inequality of wealth is always and everywhere much greater than the inequality of income from labor" (pp. 306–307). High-income earners have higher saving rates, leading to growing concentration of wealth that contributes to higher capital incomes. Saez and

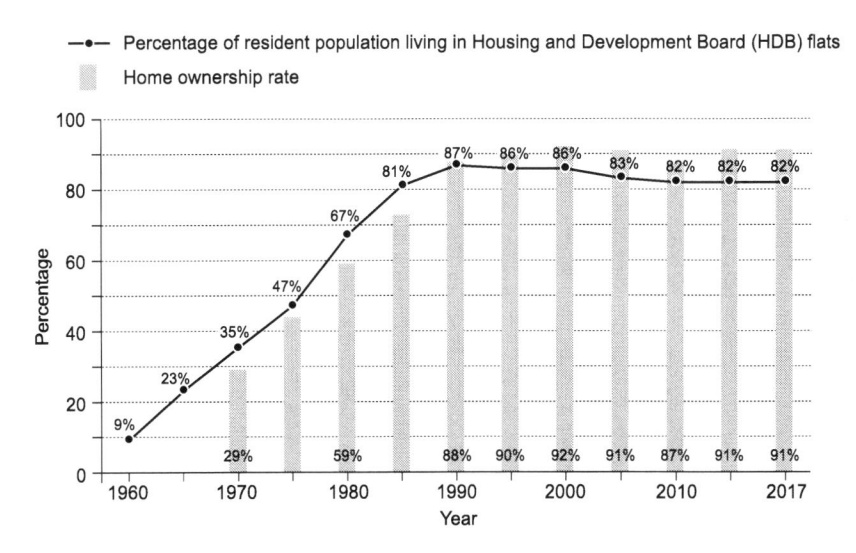

Figure 11.2 International comparison of income Gini coefficients.

Note: Singapore's Gini is based on resident household income from work whereas data on OECD economies is based on income from all sources (which includes non-work income from investments and property).

Source: Singapore Ministry of Finance, Parliamentary Reply on 'Before and After Taxes and Transfers – Singapore's Gini Coefficient', 19 Mar 2018. Access at: www.mof.gov.sg/Newsroom/Parliamentary-Replies/before-and-after-taxesand-transfers---singapore-s-gini-coefficient.

Zucman (2016) find that this "snowballing" effect has dramatically affected the shape of US wealth distribution over the last 30 years—the richest 10% of the population held 77% of the nation's wealth in 2012 compared to 65% in the mid-1980s, leaving less than a quarter of national wealth for the remaining 90% of Americans. Piketty (2014) acknowledges the difficulty of solving the wealth inequality problem:

> To my knowledge, no society has ever existed in which ownership of capital can reasonably be described as 'mildly' inegalitarian, by which I mean a distribution in which the poorest half of society would own a significant share (say one-fifth to one-quarter) of total wealth ... Of course, how one might go about establishing such an 'ideal society'–assuming that such low inequality of wealth is indeed a desirable goal–remains to be seen.
>
> Piketty, 2014, pp. 322–323

In Singapore, half a century of consistent housing policies has resulted in a high homeownership rate and more equitable distribution of housing wealth. House price appreciation over time has benefited homeowners in both the HDB and private housing sectors. Government data on household wealth for 2015 show that 80% of resident households that reside in the HDB sector have a share of 48% of gross housing wealth. Using average market prices by house type, Phang (2015; 2018) estimates the proportion of gross housing wealth owned by the bottom 50% of households in 2015 to be at around 25%.[6] In fact, based on the above estimates, Singapore's homeownership policies have resulted in gross *housing* wealth distribution approximating capital ownership distribution in Piketty's "ideal society."

2.2 Income segregation

Within cities, the spatial concentration of poverty is of increasing policy concern. Reardon and Bischoff (2016) provide evidence that even as income inequality has continued to rise in the US, income segregation in metropolitan areas has followed a similar rising trend since the 1980s. The proportion of families living in poor or affluent neighborhoods, used as one of the measures of income segregation, increased over time from 15.0% in 1970 to 34.3% in 2012. The proportion of families living in middle-income neighborhoods fell from 64.7% in 1970 to 40.5% in 2012. A 2018 study found income segregation to be the highest in the US among OECD countries, and lowest in cities in countries with low levels of overall inequality, such as Australia, New Zealand, Denmark, and the Netherlands (OECD, 2018).

A multiplicity of factors that affect how inclusive or segregated a city's neighborhoods are include zoning laws, housing policies, real estate factors, racial and ethnic composition, historical factors, and migration trends. The neighborhood where a child grows up can have long-term implications for his or her future (Chetty, Hendren, and Katz, 2016; Chetty et al., 2018; Wodtke, Harding, and

Elwert, 2011). Chetty et al. (2016) provide evidence for the US that moving to a lower-poverty neighborhood significantly improves college attendance rates and earnings for children who were young (below age 13) when their families moved. The gains from moving fall with the age at which children move, consistent with recent evidence that the duration of exposure to a better environment during childhood is a key determinant of an individual's long-term outcomes.

I now compare Singapore's income segregation patterns with US metropolitan areas to contrast the differences in the income composition of their neighborhoods. I use the dataset provided by the Singapore Department of Statistics for 28 planning areas that classify resident households into 18 income categories.[7]

First, following the classification used by the Stanford Center on Poverty and Inequality and by Reardon and Bischoff (2016), I compute the nationwide ratio of the mid-point of an income category to the 2015 median household income (S$8,666).[8] Based on this ratio, I classify each income category into six groups: affluent ($r \geq 1.50$); high income ($1.25 \leq r < 1.50$); high-middle income ($1.00 \leq r < 1.25$); low-middle income ($0.80 \leq r < 1.00$); low income ($0.67 \leq r < 0.80$); or poor ($r < 0.67$). Table 11.1 shows the distribution by household income in 2015 according to this classification. Thirty percent of households are classified as "affluent," with incomes at least one-and-a-half times greater than the median household income. Thirty-four percent of households are classified as "poor," with incomes less than two-thirds of the median.

I then compute for each of the 28 planning areas the ratio (r) of the area's median household income to that of Singapore's median household income. Based on this ratio, I classify each planning area as affluent ($r \geq 1.50$); high income ($1.25 \leq r < 1.50$); high-middle income ($1.00 \leq r < 1.25$); low-middle income ($0.80 \leq r < 1.00$); low income ($0.67 \leq r < 0.80$); or poor ($r < 0.67$). An affluent neighborhood thus defined therefore has more than half of the households with incomes at least one-and-a-half times greater than Singapore's median household income. A poor neighborhood has more than half of the households with incomes less than two-thirds of Singapore's median household income.

Table 11.2 shows the proportion of households residing in planning areas defined by their income for Singapore and contrasts these figures with those for US metropolitan areas (117 metropolitan areas with a population more than half a million). For 2015, although poor households comprise 34% of the Singaporean resident population, no planning area is categorized as "poor," and there is only one "low-income" planning area out of 28. While, nationwide, 30% of the population are classified as "affluent," there are only two "affluent" planning areas. This results in a low proportion of the population living in either "poor" planning areas or "affluent" planning areas: only 2.6% compared to the 34.3% figure for US metropolitan areas.[9]

There is a strong correlation between house types and household incomes in Singapore. Table 11.3 shows the average monthly household income from work for resident employed households by dwelling type for 2017. The composition of house types at the local level determines to a large degree the integration of

238 Sock-Yong Phang

Table 11.1 Distribution of households by household income categories in 2015.

Income category (r = household income from work/median household ratio)	Number of employed resident households (1,000s)	% of total
Poor (r < 0.67)	373.9	33.8%
Low income (0.67 ≤ r < 0.80)	70.8	6.4%
Low-middle income (0.80 ≤ r < 1.00)	130.4	11.8%
High-middle income (1.00 ≤ r < 1.25)	111.3	10.1%
High income (1.25 ≤ r < 1.50)	90.7	8.2%
Affluent (r ≥ 1.50)	329.4	29.8%
Total	1,106.5	100.0%

Source: Singapore Department of Statistics, General Household Survey 2015. Access at: www.singstat.gov.sg/publications/ghs/ghs2015content

Table 11.2 Proportion of families/households in low-, middle-, and high-income neighborhoods: US metropolitan areas > 500,000 and Singapore.

Income category	US metropolitan areas		Singapore
	2000	2012	2015[a]
Poor	15.2%	18.6%	0.0%
Low income	11.9%	11.0%	0.7%
Low-middle income	23.2%	18.9%	73.2%
High-middle income	23.9%	21.6%	18.7%
High income	13.1%	14.2%	4.9%
Affluent	12.7%	15.7%	2.6%
Middle income	47.1%	40.5%	91.9%
Poor + Affluent	27.9%	34.3%	2.6%

Note: Singapore 2015 numbers based on resident household income and household data for 28 planning areas.

Source: Reardon and Bischoff (2016, p. 7) for the US using family data; author's estimation using dataset from Singapore Department of Statistics, General Household Survey 2015. Access dataset at: www.singstat.gov.sg/publications/ghs/ghs2015content

households of different incomes. As the planning area is too large to capture local variations, I use 2017 population and house type data that are available at the subzone level. For 180 subzones (with populations above 2,000), I classify each subzone according to the median house type and compute the proportion of the population living in subzones as defined by the median house type.

While 20.8% of the resident population reside in private housing (see Table 11.3), 10.0% or less than half reside in predominantly private housing neighborhoods (see Table 11.4). There is only one neighborhood where the median house type is the HDB one- or two-room flat. Only 7.5% of the population reside in neighborhoods where the median house type is a three-room HDB flat. The majority (82.3%) of the resident population reside in "middle-income"

Table 11.3 Singapore: relationship between average household income and house type, 2017.

House type	Proportion of resident households	Average monthly household income from work (S$)	Ratio of household average income to median income of S$9,023
HDB 1- and 2-room flats	5.8%	$2,748	0.30
HDB 3-room flats	17.8%	$6,450	0.71
HDB 4-room flats	31.8%	$9,260	1.03
HDB 5-room and executive flats	23.5%	$12,554	1.39
Private housing: condominiums and other apartments	15.6%	$20,491	2.27
Private housing: landed properties	5.2%	$26,701	2.96
	100%	$12,027	1.33

Source: Singapore Department of Statistics (2017), Key Household Income Trends 2017, p. 34. Access at: www.singstat.gov.sg/find-data/search-by-theme/households/household-income/latest-data

neighborhoods where the median house type is either the HDB four- or five-room flat. Using "subzone" housing type data rather than "planning area" income data increases the degree of spatial segregation of "poor" and "affluent" to a score of 17.7%.

2.3 Ethnic segregation

In 2017, of Singapore's population of 5.61 million, 3.44 million were citizens, 0.53 million were permanent residents (PRs), and 1.64 million were foreigners. Singapore's resident population is a multi-racial, multi-cultural, multi-religious mix of Malay, Chinese, Indian, and Other. The ethnic composition of the resident population in 2015 was 13.3% Malays, 74.3% Chinese, 9.1% Indians, and 3.3% Other. The "Other" category included Filipinos, Caucasians, Eurasians, Arabs, Thais, and Japanese. In the 2015 General Household Survey, Buddhism was the most important religion, professed by 33.2% of the population; 18.5% of the population indicated they had no religion. Almost all Malays (99%) were Muslims, while the two main religions professed by the Chinese were Buddhism/Taoism (55.2%) and Christianity (20.9%), with 23.3% professing no religion. The religions professed by Indians included Hinduism (59.9%), Islam (21.3%), Christianity (12.1%), and Sikhism (4.3%).

Table 11.5 compares the socio-economic characteristics of the three main ethnic groups for 2015. The data indicate that Malay households had lower median household incomes and the lowest proportion of household heads with a university qualification (6%). A disproportionate number of Malay households (14.6%) reside in one- to two-room HDB flats, compared to the national average of 5.6%. While Indian households had a lower homeownership rate (84.1%), the proportion of household heads with university qualifications was significantly

Table 11.4 Proportion of population residing in low-, middle- and high-income neighborhoods, 2017.

Median house type	Number of subzones	Proportion of resident population living in subzone defined by median house type
HDB 1- and 2-room flats	1	0.2%
HDB 3-room flats	19	7.5%
HDB 4-room flats	77	54.5%
HDB 5-room and executive flats	36	27.8%
Private housing: condominiums and other apartments	33	7.1%
Private housing: landed properties	14	2.9%
Total	180	100%
Middle-income (HDB 4–5-room and executive flats)	113	82.3%
Poor (HDB 1- to 3-room flats) + Affluent (private housing)	47	17.7%

Source: Singapore Department of Statistics, Population Trends 2017. Access at: www.singstat.gov.sg/publications/population-trends

Table 11.5 Socio-economic characteristics by ethnic group, 2015.

	Population	Chinese	Malay	Indians
Proportion of resident households		74.3%	13.3%	9.1%
Median monthly household income (S$)	$8,666	$8,000–$8,999	$5,000–$5,999	$8,000–$8,999
Proportion of households with head of household with university education	28.2%	28.5%	6.0%	37.9%
Homeownership rate	90.8%	93.2%	87.0%	84.1%
Proportion of households residing in 1- or 2-room HDB flats	5.6%	4.4%	14.6%	6.6%
Proportion of households residing in 4- or 5-room and executive HDB flats	56.2%	56.0%	61.8%	55.8%
Proportion of households residing in private housing	19.5%	21.1%	2.5%	19.1%
Proportion of households with incomes S$20,000 and over	13.4%	14.3%	2.5%	13.6%

Source: Data from Singapore Department of Statistics, General Household Survey 2015. Access at: www.singstat.gov.sg/publications/ghs/ghs2015content

higher (at 37.9%) than the national average (of 28.2%). For all three races, the HDB four-room flat was the median house type as well as the most common house type, and more than half of households for all three races reside in middle-income housing defined as four- and five-room HDB flat types.

The presence of residentially segregated neighborhoods is a common feature of housing patterns in cities with ethnically diverse populations and large minority groups. In Singapore, however, other than in the historical areas of Kampong Glam, Chinatown, and Little India that have been designated as conservation areas, there are no obvious concentrations of minority groups in housing patterns.

The dissimilarity index of Massey and Denton (1988) can measure the spatial segregation of any one group from any other mutually exclusive group across the geographic units that make up a larger geographic entity. This index has a minimum value of zero, and the maximum value is 100. In a completely racially segregated city, the index would be 100; in a city where all neighborhoods have a racial composition identical to that of the entire city, the index will be zero. The formula for the index of dissimilarity is:

$$\frac{1}{2}\sum_{i=1}^{N}\left|\frac{a_i}{A}-\frac{b_i}{B}\right|$$

where:
 a_i = the population of group A (e.g., Chinese) in the ith subzone
 A = the total population in group A in Singapore (e.g., Chinese in Singapore)
 b_i = the population of group B in the ith subzone (e.g., non-Chinese)
 B = the total population in group B in Singapore (e.g., non-Chinese population in Singapore)

The summation is over the number of geographical units. The value of the index indicates the proportion of individuals from one of the specific ethnic populations (e.g., Chinese) who would need to relocate in order to obtain an even distribution of that population across all geographical areas.

For Singapore, I used data from the 2015 General Household Survey for which data on the ethnicity of resident population by 224 subzones are available. On average, each subzone has about 17,000 residents. I estimate the dissimilarity index to be 16.5% for the Chinese, 26.2% for the Malays, and 10.2% for the Indians—i.e., Indian households are the most dispersed, and Malay households are the most segregated. These levels of segregation are low compared to those of large cities in the US. For example, using census tract data, the 2010 segregation indices for New York-Northern New Jersey-Long Island were 78.0% for Black-White, 62.0% for Hispanic-White, and 51.9% for Asian-White.[10]

3. Land and housing policies for an inclusive singapore

US Supreme Court Justice Oliver Wendell Holmes, Jr. opined, "taxes are what we pay for civilized society, including the chance to insure."[11] To build an inclusive city, Singapore has gone significantly beyond taxes; it has utilized a suite of inter-related policies that encompasses land and housing legislation, land-use planning, land value capture, state housing developer and housing finance institutions, housing market regulations, taxes, and subsidies.[12]

It is not surprising that among a set of 33 advanced, emerging, and newly industrialized economies, the International Monetary Fund Index of Government Participation in Housing Finance Markets ranks Singapore number one on the list (IMF, 2011). Several of Singapore's land and housing policies are redistributive in their intent and impact. In this section, I provide a summary list of 15 policies that have contributed to a sustainable affordable housing framework and enhanced social inclusion.

1. *Government as dominant housing developer*: The HDB, established in 1960, is Singapore's largest housing developer. Seventy-three percent of Singapore's entire housing stock in 2017 is classified as HDB housing, and 94% of government-built flats have been sold at subsidized prices on a 99-year-leasehold basis to eligible households. Owners of subsidized flats may sublet bedrooms, and after a minimum period of occupancy, may rent or sell the flat for which there are active (albeit regulated) rental and resale markets. A resale levy amount of between S$15,000 and S$55,000 depending on flat type is payable to the HDB only if the owner buys a second subsidized flat from the HDB or a unit under the Executive Condominium Scheme.
2. *State land for housing*: The government has the power to acquire land from private landowners for economic development purposes, including for residential purposes. It undertakes reclamation of land from the surrounding seas, and the physical land area of Singapore has increased by about 24% since the 1960s. State land, as a proportion of total land area, grew from 44% in 1960 to 76% by 1985 and is estimated to be around 90% currently.
3. *Housing Provident Fund*: Singapore's pension system is based on defined contributions rather than defined benefits. Employees' enforced contributions and accumulated savings in their personal accounts with the Central Provident Fund (CPF) can be used for housing-related down payments and mortgage payments (but not for rental payments). In 2016, members (through their employers) deposited the equivalent of 8.7% of GDP into the CPF; withdrawals from the CPF for housing comprised 46% of contributions or about 4% of GDP (Phang, 2018). The CPF makes mortgage payments on behalf of members, thus reducing mortgage default risk.
4. *Availability of mortgage loans*: The government provides mortgage loans through the HDB for HDB flat purchasers at an affordable interest rate of 2.6%, regardless of borrower default risk—i.e., while there are subprime mortgages, the HDB does not charge subprime mortgage rates. HDB mortgage loans are for up to 25 years, with a loan-to-value ratio at 90%, and monthly installments capped at 30% of applicants' monthly income. Since 2003, commercial banks have been allowed to provide loans to buyers of HDB flats, and this has reduced the HDB's share of mortgage loans outstanding from 59% in 2000 to 17% by 2015. However, the HDB continues to be the most important lender in terms of number of mortgage loans. In 2016, of the 1.002 million CPF members who withdrew savings for housing mortgage payments, 52% had loans from the HDB (Phang, 2018).

5. *House price affordability*: The HDB prices its new flats for sale below the prices fetched by similar older units in the resale market. The government monitors housing affordability measures such as the price-to-income and mortgage payment-to-income ratios closely. To enhance housing affordability for lower-income households, the government varies the effective subsidy for each household according to household characteristics using housing grants. For a low-income household with monthly income below S$1,500, housing grants to purchase an HDB flat can be as high as S$80,000.

   The HDB's pricing has a redistributive impact as larger flat types purchased by higher-income households have higher per-square-meter prices (Phang, 2018). In 2017, the house price-to-median annual household income ratio for HDB flats in Sengkang Town was 1.9 for two-room flats (45 sq. m), 3.7 for three-room flats (65 sq. m), 4.7 for four-room flats (90 sq. m), and 5.0 for five-room flats (110 sq. m). The monthly mortgage payment-to-household-income ratio ranged from 7% for two-room flats to 25% for five-room flats. Price and mortgage affordability have resulted in high homeownership rates of 84% for the bottom 10% of the resident population (i.e., citizens and PRs) and 87% for the bottom 20%.

6. *Facilitating private housing supply*: The government revises the Concept Plan and the Master Plan for land use every 10 and five years, respectively. The process has allowed for changes and intensification of land use with economic and population growth (Phang, 2018). Under the Government Land Sales (GLS) program, the government sells land to private-sector developers with conditions tailored to achieve its planning and private housing supply objectives. Land prices are determined by a competitive tender process, with revenue from sales channeled into government reserves. In addition to private housing, the government also sells land for the public–private hybrid "Executive Condominium Scheme," which provides housing for couples with monthly household incomes below S$14,000. To facilitate redevelopment to higher densities, legislation was amended in 1999 to allow private collective (en bloc) sales for new housing developments to proceed without the need for 100% of owners to agree. The legal threshold percentage that will allow for collective sale for redevelopment is 90% for properties less than 10 years old and 80% for developments 10 years or older.

7. *Measures to prevent land speculation and hoarding:* The Singapore Land Authority has put in place measures to prevent the use of GLS sites for land speculation and hoarding. It specifies a project completion period as part of the tender conditions, and developers are required to pay a premium to extend the completion period. For non-GLS developments, foreign and listed developers are required to finish building their projects within five years of acquiring a site, and to sell all units within two years from project completion. To extend the deadlines would require payment of punitive charges at 8%, 16%, and 24% of the land purchase price for the first, second, and subsequent years, respectively, pro-rated to the proportion of unsold units.

8. *Segmentation of housing markets:* Residency status, household income, marital status, age, and current and previous ownership of properties determine eligibility to purchase specific types of housing (Phang, 2018). This set of criteria ensures that those of higher income and wealth do not "crowd out" households with lesser means. Foreigners are not eligible to purchase HDB housing, and a PR household with PR status for at least three years may buy an HDB resale flat. For citizen and PR households, each household is limited to owning only one HDB flat. The household income (monthly) ceiling is currently S$1,500 for a flat under the Public Rental Scheme; S$6,000 for a new two- or three-room HDB flat; and S$12,000 for a new four- or five-room HDB flat, for buying a resale flat with the CPF Housing Grant, or for obtaining an HDB concessionary loan. The income ceiling is S$14,000 to purchase a new Executive Condominium unit. There is no income ceiling for the purchase of an HDB resale flat.
9. *Regulating housing demand:* Since 2011, the government has acted to dampen demand for private housing by investors and foreigners through hefty buyer stamp duties that also serve as anti-speculation taxes. The latest July 6, 2018 round of demand "cooling" measures brings the Additional Buyer Stamp Duty (ABSD) payable for residential properties to 25% for corporate entities and 20% for foreigners. PRs pay 5% ABSD for their first property and 15% for their second and subsequent properties; citizens pay 12% for their second property and 15% for their second and subsequent properties. The ABSD is in addition to buyer stamp duty (of up to 4%) that is required on all transactions; there is a seller stamp duty (of up to 12%) for holding periods of less than three years to discourage speculation. There are also caps on loan-to-value, mortgage service, and total debt service ratios, as well as tenure of housing loans offered by commercial banks.[13]
10. *Progressive property taxes:* Property taxes in Singapore are a percentage of annual rental values. The residential property tax regime is progressive, with rates for owner-occupied housing starting at 0% (for annual values up to S$8,000) and rising to 16% (for annual values above S$130,000). Owner-occupiers of one- and two-room HDB flats are exempt from property taxes, and the rate for other HDB owner-occupiers is at the lowest tier of 4%. In 2017, three-room HDB flat owner-occupiers paid annual property taxes of up to S$18.40; four-room HDB flat owners paid between S$52 and S$100; five-room HDB flat owners paid between S$83.20 and S$131.20; and executive flat owners paid S$95.20 to S$143.20 (*The Straits Times*, November 28, 2016b). The property tax rate for non-owner-occupied housing ranges from 10% (first S$30,000) to 20% (above S$90,000).
11. *Transfers according to house types:* In 1994, the government began using house type as a proxy for household income and wealth for purposes of targeted transfers. The amount of rebates to offset the Goods and Services Tax (a consumption tax), waivers on HDB rents, and for HDB service and conservancy charges depends on house type. Other fiscal transfers such as CPF

top-ups and cash vouchers also use house type or house value to determine the amount of transfers.

12. *Integration of house types inside housing estates*: Spatially, the HDB has been able to integrate different groups across the income spectrum in its estates and new towns as it is the dominant housing supplier (Phang, 2018). With 80% of the resident population residing in HDB housing, there is no social stigma attached to public housing in Singapore. This fact also makes for more stable HDB communities as residents do not view a need to upgrade "out" of the HDB sector as incomes increase. There is a deliberate policy of providing for all flat types (generally ranging from one room to five rooms) in public housing estates. Blocks of different room types are mixed together when planning a precinct of about 2.5–3.5 hectares, considered of sufficient scale for different house types. Within each block, there is also mixing of successive unit types: two- and three-room, three- and four-room, and four- and five-room— i.e., between "socio-economic levels that are likely to be compatible with each other" (Liu and Tuminez, 2015; Wong and Yeh, 1985, for block plans). Private housing developments that are located within predominantly HDB areas also help to integrate residents in the two distinct housing segments.

13. *Ethnic integration policies*: Residential areas were highly segregated by race and Chinese dialect groups in colonial Singapore (Phang, 2018). In the 1970s, the HDB allocated new flats in a manner that would mix the different races in the new housing estates. However, when a trend of Malay ethnic regrouping through the resale market became evident in the 1980s, the government implemented an Ethnic Integration Policy in 1989, under which racial limits were set for the HDB blocks and neighborhoods.[14] These quotas (see Table 11.6) limit the proportion of flats in a block and in a neighborhood that can be owned by a particular race. When these limits are reached (the information is available online), those wishing to sell their HDB flats in the particular block or neighborhood are constrained to sell to another household of the same ethnic group. With the increase in non-citizens residing in the HDB sector over time, PR quotas were introduced in 2010, and non-citizen tenant quotas were implemented from 2014.

14. *Accessibility to facilities and public transport*: In the development of HDB towns, in addition to residential use, the HDB's planners plan simultaneously for the provision of comprehensive public facilities. HDB towns are transit-oriented developments; land-use planning provides for public transportation, bus terminals, and Mass Rapid Transit stations. The general land-use allocation in a new town includes playgrounds, car parks, shopping facilities, schools, parks and gardens, and sports and recreation. Institutional facilities include the town's administrative office, clinics, hospitals, community centers, and places of worship. In addition, space is allocated for fresh produce markets, cooked food centers, and commercial and industrial use in order to provide necessary amenities and employment for residents (Wong and Yeh, 1985; Cheong, 2017).

Table 11.6 Ethnic limits under the HDB's Ethnic Integration Policy.

	Resident population 2017	Neighborhood quota	Block quota
Chinese	74.3%	84%	87%
Malay	13.3%	22%	25%
Indian + Other	9.1% + 3.2%	12%	15%
PR homeowners		5%	8%
Non-citizen and PR tenants		8%	11%

Source: Singapore government and HDB websites. Access at: www.hdb.gov.sg/cs/infoweb/residential/buying-a-flat/resale/ethnic-integration-policy-and-spr-quota
www20.hdb.gov.sg/fi10/fi10296p.nsf/PressReleases/51A6512F14F32CC848257C6200226A1E?OpenDocument

15. *Access to schools*: Access to good schools is one of the most powerful factors for housing location choice in the cities of most countries. In the US, funding of local schools from local property taxes results in wide disparities in school quality that further reinforce patterns of income segregation.

Singapore's education system enjoys a good international reputation, and students do well in global benchmarking tests for mathematics and science (*The Straits Times*, November 29, 2016a). In contrast to the US system, schools and educational institutions at all levels are heavily subsidized from the national budget. Citizens are required to enroll in a national primary school (at age six) for six years of free compulsory education, unless granted an exemption. The system ensures free access to high-quality primary education for citizens, regardless of income and housing location. A majority of primary schools thus admit children from all socio-economic and ethnic backgrounds, with 80% of primary schools having at least 5% of students from the top and bottom socio-economic quintiles (Ong, 2018). However, some primary schools, due to their history, alumni, and location in private housing estates, do have large proportions of students from higher-income groups. A national "Primary School Leaving Examination" sorts 12-year-olds into secondary schools based not on housing location, but on preferences, examination results, and primary school affiliation.

4. Current Singaporean debates for building a more inclusive society

Singapore has allocated an unusual amount of policy attention and resources to homeownership that has led to more equitable housing distribution, enhanced social stability, and reduced economic inequality. However, no system is perfect, and no society is static. There is always room for improvement. This concluding section therefore discusses some of the ongoing debates, concerns, suggestions, and recent policies to mitigate income inequality and promote social inclusion in Singapore.

First, Singapore's welfare policies have a strong bias towards using housing as the dominant channel of support to lower-income groups. This bias is apparent when compared with Australia, New Zealand, and countries in Western Europe that have more comprehensive welfare systems. However, while the provision of housing can encourage—and, in Singapore's case, has resulted in no small measure in—a more equitable and inclusive society, it is only one policy. Put simply, it may have to be complemented by reforms in other spheres. For example, there is no across-the-board legislated minimum wage and no unemployment benefits in Singapore. Instead, the government provides targeted public assistance based on assessment of eligibility and only when other forms of help have been exhausted.[15] Over the past decade, though, there has been a shift in government policies towards increased social expenditures in healthcare, education, and skills upgrading. Under the Progressive Wage Model, skill-based minimum wages were introduced in 2012 for citizens and PRs for the cleaning, security, and landscape sectors. Resisting calls for unemployment insurance, the government instead funds programs for training, skills upgrading, and lifelong learning, as well as providing cash and CPF payouts for older lower-wage workers.

Second, the government has identified aging and slowing social mobility as two big challenges facing Singapore. With one in four Singaporeans expected to be more than 65 years old by 2030, a growing number of households with a large proportion of their assets in illiquid housing equity will need to monetize their housing wealth for retirement financing. The government has put in place a number of schemes to facilitate this withdrawal. The subletting of rooms and renting out the HDB flat appears to be the most popular option. Downsizing to a smaller or shorter lease flat (elderly studio or two-room Flexi) presents another option. The Lease Buyback Scheme, under which the HDB buys back the tail end of the property's lease, and which allows for the elderly to age in place, has seen a low take-up rate of around 1% of eligible households (Phang, 2018). There is a need to explore additional options for housing wealth monetization, as a lack of channels may cause an increase in consumption inequality among the elderly.

Third, the evidence on intergenerational mobility in Singapore is mixed. Ng (2015) finds intergenerational mobility to be moderately low in Singapore and postulates that mobility will be increasingly challenging given that certain characteristics of the current education system (streaming at the secondary school level and the costs of university education) reinforce immobility. In contrast, Yip (2012) and a 2015 study by the Ministry of Finance judge intergenerational income mobility to be moderate to relatively high compared to other countries such as the US, Japan, the UK, and Denmark. In the past year, Singapore's president, as well as several high-ranking public policy makers, have highlighted the need for policies to mitigate inequality and improve social mobility, with education and building inclusive neighborhoods identified as important strategies.[16]

Fourth, within the housing system, the homeownership bias has resulted in a marginalized public *rental* housing scheme. To discourage rental as a tenure choice, the HDB has limited rental flats available to smaller flat types and has not changed the low monthly income ceiling of S$1,500 since 2003. HDB neighborhoods have rental and owner flats located in separate blocks, and there were several years (1982 to 2006) in which no rental blocks were built (Ng, 2018a). In 2015, 3.7% of resident households resided in HDB one- and two-room rental flats, of which 32.1% comprised Malay households. In contrast to homeowners, tenants have not benefited from the across-the-board housing price appreciation over the decades. Vasoo (2018) has noted the re-emergence of ethnic ghettoes in one- to two-room HDB neighborhoods as a problem that needs to be addressed. In a fundamental policy change, the HDB began integrating rental and sold flats within the same block in its new building projects from 2014 (*The Straits Times*, May 17, 2018a).

Fifth, neighborhood diversity, may not, in and of itself, translate into social cohesion.[17] Other than the HDB rental–owner housing divide, a 2017 survey on social networks highlighted the divide between those from "elite" and "neighborhood" schools, and between residents of public and private housing.[18] Respondents in the survey were able to name network members from different racial, religious, and nationality groups. However, social networks across "elite" and "neighborhood" schools and across public and private housing were significantly weaker than the networks within groups.[19] To promote social mixing across housing types, the prime minister announced the creation of a new form of grassroots organization, the Residents' Network, in July 2018. Existing local Residents' Committees that currently serve HDB estates, and Neighborhood Committees that cater to private housing estates, may continue as they are or rename themselves, or neighboring committees can merge to form a single Resident Network (*The Straits Times*, July 21, 2018c).

Finally, within Singapore, a few groups do not benefit from the housing welfare system. The HDB's definition of a family nucleus, for purposes of eligibility for its housing schemes, excludes single, unwed mothers and their children. Although the number of children born out of wedlock is not high (2.2% of resident births in 2015), this deliberate policy of exclusion reflects Singapore society's preference to preserve the traditional norm of "parenthood within marriage" (*The Straits Times*, November 29, 2017b).

On another front, official statistics on income, wealth, households, and housing provide data on the resident population (citizens and PRs), but they exclude the large number of foreigners. In 2017, there were 1.37 million foreigners working in Singapore, comprising 37.4% or more than one-third of the total labor force. Foreigners working in Singapore are categorized by type of work pass and the sector they work in. Of the 1.37 million foreigners in the labor force, 14% held an employment pass and 14% a mid-level skilled pass (S-pass). The employment pass holders mostly reside in private housing, and Singapore ranks among one of the most expensive cities in the world for expatriates (*The Straits Times*, June 26, 2018b).

At the other end of the income spectrum, almost one million foreigners working in Singapore hold work permits that allow for neither dependent privileges nor the prospect of future residency. In 2017, 18% of foreigners working in Singapore were female foreign domestic workers; 53% worked in the construction, manufacturing, marine shipyard, process, or services sectors. While domestic workers reside with their employers, other work permit workers (more than 700,000) reside in purpose-built dormitories, approved workers' quarters in industrial areas, and as multiple occupancy tenants in HDB and private housing. The increased visibility of low-wage foreign workers from several different countries residing in local residential areas has given rise to some concerns among resident households.[20] In 2014, the HDB introduced non-citizen tenant quotas in its housing estates; there are also regulations on rental occupancy limits and on eligibility of work permit workers to rent HDB housing based on nationality (Malaysian/non-Malaysian) and employment sector.[21] At the same time, the government regularly reviews housing standards and conducts inspections across all types of foreign worker accommodations (Ng and Neo, 2018).

The problems and new challenges discussed in this section, however, should not detract from Singapore's significant core achievements in making homeownership affordable and in building an inclusive society for the vast majority of its resident population.

Notes

1 Information on OECD inclusive growth agenda may be found at: www.oecd.org/inclusive-growth/
2 Information on Habitat III's "New Urban Agenda" may be found at: http://habitat3.org/the-new-urban-agenda/
3 O'Donnell (2015) provides an account of the ideas of Henry George and shows how these remain relevant to our present times.
4 Singapore's land area was 582 square km in 1960. Through expensive reclamation of land from surrounding waters, the land area has increased by 24% since the 1960s.
5 Florida (2017, p. 83) provides data on Gini coefficients for US metropolitan areas: New York-Northern New Jersey-Long Island (0.504); San Francisco-Oakland-Fremont (0.475); Chicago-Naperville-Joliet (0.468).
6 The analysis of housing wealth distribution in Singapore was motivated by Piketty's concern that "inequality of wealth is always and everywhere much greater than the inequality of income from labour." Piketty presents data for the US, which shows that in the early 2010s the top decile owned 72% of America's wealth, while the bottom half's claim was just 2%. In France, the richest 10% owned around 62% of national wealth, and the poorest 50% owned less than 4% (p. 257). In his view, this unequal ownership of capital is a prime driver of income disparities.
7 Singapore Department of Statistics General Household Survey 2015, Table 152, Resident Households by Planning Area and Monthly Household Income from Work. Access at: www.singstat.gov.sg/find-data/search-by-theme/population/geographic-distribution/latest-data
8 References to dollars in this chapter refer to the Singapore Dollar (S$). The exchange rate in July 2018 was approximately US $ 0.73 to S$1.
9 Low income: Outram; high income: Bishan, Marine Parade, Novena; affluent: Bukit Timah and Tanglin; the other 22 planning areas are classified as either low-middle income or high-middle income.

250 Sock-Yong Phang

10 University of Michigan Population Studies Centre Racial Residential Segregation Measurement Project. Access at: www.psc.isr.umich.edu/dis/census/segregation2010.html
11 The quote is from Justice Oliver Wendell Holmes, Jr.'s dissenting opinion in the 1927 court case of Compañía General de Tabacos de Filipinas v. Collector of Internal Revenue.
12 Refer to Phang (2018) for detailed discussion and analyses.
13 For details of measures and changes over time, refer to www.srx.com.sg/cooling-measures
14 The Singapore government acknowledged the influence of Thomas Schelling in the formulation of the Ethnic Integration Policy (Dodge, 2006, p.142).
15 Ng (2018b) describes Singapore as a "residual welfare model" that creates barriers to access with huge administrative costs, while Haskins (2011) regards Singapore's social policy as "a crucible of individual responsibility."
16 Singapore's president, prime minister, deputy prime minister, and minister for education have given speeches or interviews on inequality and social mobility in 2018.
17 Putnam (2007) provides evidence for the US that trust (even of one's own race) is lower, altruism and community cooperation rarer, and friends fewer in ethnically diverse neighborhoods.
18 The Institute of Policy Studies at the National University of Singapore conducted a face-to-face survey in 2016 of 3,000 respondents to measure social capital. The methodology and findings may be accessed at: http://lkyspp2.nus.edu.sg/ips/wp-content/uploads/sites/2/2017/11/Study-of-Social-Capital-in-Singapore_281217.pdf
19 See Ong (2018) for Education Minister Ong's proposals for education policies in the next phase to address the inequality challenge.
20 In 2008, residents of a private housing estate petitioned their Members of Parliament to appeal against a proposed plan by the government to convert an unused school in their neighborhood into a foreign worker dormitory facility. See Shaw and Ismail (2010) and Baey (2010) for discussions of the Serangoon Gardens "uproar." De Koninck (2017, p. 46) identifies the locations of 42 of these foreign worker dormitories, with the largest of these nearly all-male "proletarian towns" providing more than 10,000 beds each.
21 Since November 7, 2006, non-Malaysian work permit holders from the construction sector have not been able to sublet HDB flats or rooms. This regulation was extended to the marine and process sectors on May 1, 2015. Since January 2017, non-Malaysian work permit holders from the manufacturing sector are not eligible to rent a whole HDB flat and can only rent rooms. Only non-Malaysian work permit holders in the service sector can rent whole flats (*The Straits Times*, January 24, 2017a).

References

Atkinson, A. B. (2015). *Inequality: What can be done?* Cambridge, MA: Harvard University Press.

Baey, G. H. Y. (2010). *Borders and the Exclusion of Migrant Bodies in Singapore's Global City-State*. Master of Arts Thesis, Kingston, Ontario, Canada: Department of Geography, Queen's University.

Bastagli, F. and Hills, J. (2013). Wealth Accumulation, Aging, and House Prices. In J. Hills, F. Bastagli, F. Cowell, H. Glennerster, E. Karagiannaki, and A. McKnight (Eds.). *Wealth in the UK: Distribution, Accumulation, and Policy*. Oxford: Oxford University Press, pp. 63–91.

Baum-Snow, N. and Pavan, R. (2013). Inequality and City Size. *Review of Economics and Statistics*, 95(5), 1535–1548.

Behrens, K. and Robert-Nicoud, F. (2014). Survival of the Fittest in Cities: Urbanization and Inequality. *Economic Journal*, 124(581), 1371–1400.

Bishop, B. with Cushing, R. G. (2009). *The Big Sort: Why the Clustering of Like-Minded America is Tearing Us Apart*. New York: Houghton Mifflin Harcourt.

Bowdler, J., Cisneros, H., Lubell, J., and Phillips, P. L. (2017). *Building Equitable Cities: How to Drive Economic Mobility and Regional Growth*. Washington, D.C.: Urban Land Institute.

Cheong, K. H. (2017). The Evolution of HDB Towns. In C. K. Heng (Ed.). *50 Years of Urban Planning in Singapore*. Singapore: World-Scientific, pp. 101–126.

Chetty, R., Hendren, N., Kline, P., Saez, E., and Turner, N. (2014). Is the United States Still a Land of Opportunity? Recent Trends in Intergenerational Mobility. *American Economic Review Papers and Proceedings*, 104(5), 141–147.

———, Hendren, N., and Katz, L. (2016). The Effects of Exposure to Better Neighborhoods on Children: New Evidence from the Moving to Opportunity Experiment. *American Economic Review*, 106(4), 855–902.

———, Hendren, N., Jones, M. R., and Porter, S. R. (2018). Race and Economic Opportunity in the United States: An Intergenerational Perspective. NBER Working Paper No. 24441.

Comandon, A., Daams, M., Garcia-Lopez, M.-A., and Veneri, P. (2018). Divided Cities: Understanding Income Segregation in OECD Metropolitan Areas. In OECD (2018), *Divided Cities: Understanding Intra-Urban Inequalities*. Paris: OECD Publishing, pp. 19–51.

Corak, M. (2013). Income Inequality, Equality of Opportunity, and Intergenerational Mobility. *Journal of Economic Perspectives*, 27(3), 79–102.

De Koninck, R. (2017). *Singapore's Permanent Territorial Revolution: Fifty Years in Fifty Maps*. Singapore: NUS Press.

Devan, J. (2018). Preface. In C. Gee, Y. Arivalagan, and F. Chao (Eds.). *Singapore Perspectives 2018: Together*. Singapore: World Scientific, vii–xii.

Dodge, R. (2006). *The Strategist: The Life and Times of Thomas Schelling*. Singapore: Marshall Cavendish.

Eeckhout, J., Pinheiro, R., and Schmidhein, K. (2014). Spatial Sorting. *Journal of Political Economy*, 122(3), 554–620.

Florida, R. (2017). *The New Urban Crisis: How Our Cities Are Increasing Inequality, Deepening Segregation, and Failing the Middle Class—and What We Can Do About It*. New York: Basic Books.

George, H. (1879/1971). *Progress and Poverty: An Inquiry into the Cause of Industrial Depressions and of the Increase of Want with Increase of Wealth ... The Remedy*. San Francisco: Author's Proof Edition. New York: Robert Schalkenbach Foundation.

Haskins, R. (2011). Social Policy in Singapore: A Crucible of Individual Responsibility. *ETHOS*, 9(June), 73–80.

Institute of Policy Studies, Singapore. (2017). A Study of Social Capital in Singapore. http://lkyspp2.nus.edu.sg/ips/wp-content/uploads/sites/2/2017/11/Study-of-Social-Capital-in-Singapore_281217.pdf

International Monetary Fund. (2011). *Global Financial Stability Report*. Washington, D.C.: International Monetary Fund.

———. (2018). *How to Operationalize Inequality Issues in Country Work*. IMF Policy Paper. Washington, D.C.: International Monetary Fund.

Krueger, A. (2012). The Rise and Consequences of Inequality. Presentation at the Center for American Progress, January 12, 2012.

Kucheva, Y. and Sander, R. (2018). Structural versus Ethnic Dimensions of Housing Segregation. *Journal of Urban Affairs*, 40(3), 329–348.

Liu, T. K., and Tuminez, A. S. (2015). The Social Dimension of Urban Planning in Singapore. In D. Chan (Ed.). *50 years of Social Issues in Singapore*. Singapore: World Scientific, pp. 97–116.

Massey, D. S. and Denton, N. A. (1988). The Dimensions of Residential Segregation. *Social Forces*, 67(2), 281–315.

Narayan, A., Van der Weide, R., Cojocaru, A., Lakner, C., Redaelli, S., Mahler, D. G., Ramasubbaiah, R. G. N., and Thewissen, S. (2018). *Fair Progress?: Economic Mobility Across Generations Around the World*. Equity and Development. Washington, D.C.: World Bank.

Ng, I. (2015). Education and Social Mobility. In F. B. Yahya (Ed.). *Inequality Singapore*. Singapore: Institute of Policy Studies and World Scientific, pp. 25–50.

Ng, K. H. (2018a). Direct Interventions, not just Social Mixing, Needed to Narrow Housing Inequality. *Channel NewsAsia*, June 23, 2018.

———. (2018b). To Tackle Inequality, Policy Mind-sets Must Change. *The Straits Times*, May 11, 2018.

Ng, K. H. and Neo, Y. W. (2018). HDB Should Apply its Own Occupancy Rules to Rental Housing. *The Straits Times*, February 16, 2018.

O'Donnell, E. T. (2015). *Henry George and the Crisis of Inequality*. New York: Columbia University Press.

OECD. (2018). *Divided Cities: Understanding Intra-Urban Inequalities*. Paris: OECD Publishing.

Ong, Y. K. (2018). The Unfinished Business of Tackling Inequality. Speech delivered by the Minister for Education in Parliament. May 2015. www.moe.gov.sg/news/speeches/speech-by-mr-ong-ye-kung--minister-for-education--at-the-debate-of-presidents-address--the-unfinished-business-of-tackling-inequality

Phang, S.-Y. (1996). Economic Development and the Distribution of Land Rents in Singapore: A Georgist implementation. *American Journal of Economics and Sociology*, 55, 489–501.

———. (2015). Superstar Cities, Inequality and Housing Policies. Celia Moh Professorial Chair Lecture, delivered at the Singapore Management University, April 23, 2015. A summary of the lecture was published as Home Prices and Inequality: Singapore versus Other Global Superstar Cities in *The Straits Times*, April 4, 2015.

———. (2018). *Policy Innovations for Affordability Housing in Singapore: From Colony to Global City*. Cham, Switzerland: Palgrave Macmillan.

Piketty, T. (2014). *Capital in the Twenty-first Century*. Cambridge, MA: The Belknap Press of Harvard University Press.

Putnam, R. D. (2007). *E Pluribus Unum*: Diversity and Community in the Twenty-first Century; The 2006 Johan Skytte Prize Lecture. *Scandinavian Political Studies*, 30(2), 137–174.

Reardon, S. F. and Bischoff, K. (2016). *The Continuing Increase in Income Segregation, 2007–2012*. Stanford Center for Education Policy Analysis. Access at: http://cepa.stanford.edu/content/continuing-increase-income-segregation-2007–2012

Reckford, J. (2018). *Housing for Inclusive and Equitable Cities*. Wrigley Lecture at Arizona State University, January 30, 2018.

Rognlie, M. (2015). Deciphering the Fall and Rise in the Net Capital Share: Accumulation or Scarcity? Brookings Papers on Economic Activity.

Saez, E. and Zucman, G. (2016). Wealth Inequality in the United States since 1913: Evidence from Capitalized Income Tax Data. *The Quarterly Journal of Economics*, 131(2), 519–578.

Sargent, T. (2018). Foreword. In S.-Y. Phang, *Policy Innovations for Affordability Housing in Singapore: From Colony to Global City*. Cham, Switzerland: Palgrave Macmillan.

Schelling, T. C. (1971). Dynamic Models of Segregation. *Journal of Mathematical Sociology*, 1(2), 143–186. Reproduced in Schelling, T. C. (1978). *Micromotives and Macrobehavior*. New York: Norton.

Shaw, B. J. and Ismail, R. (2010). 'Good Fences Make Good Neighbours'? Geographies of Marginalization: Housing Singapore's Foreign Workers. Paper presented to the SEAGA International Conference, November 2010.

Singapore Department of Statistics. (2017). Population Trends 2017.

Singapore Ministry of Finance. (2015). Income Growth, Inequality and Mobility Trends in Singapore. Occasional Paper.

Stiglitz, J. (2012). *The Price of Inequality: How Today's Divided Society Endangers Our Future*. New York: W. W. Norton & Company.

The Straits Times. (2016a). Lower property tax for HDB flat owners in 2017. November 28.

The Straits Times. (2016b). Singapore students top global achievement test in mathematics and science. November 29.

The Straits Times. (2017a). New HDB rental rules for work permit holders. January 24.

The Straits Times. (2017b). MND rejects petition to reform public housing policies for single parents. November 29.

The Straits Times. (2018a). Parliament: HDB to go one step further to integrate rental and sold flats in same block: Lawrence Wong. May 17.

The Straits Times. (2018b). Singapore is 4th most expensive city in the world for expats, with Hong Kong costliest: Mercer. June 26.

The Straits Times. (2018c). Residents' Network to help spur social mixing in estates. July 21.

Vasoo, S. (2018). Investments for Social Sector to Tackle Some Key Social Issues. In S. Vasoo and B. Singh (Eds.). *Critical Issues in Asset Building in Singapore's Development*. Singapore: World Scientific, pp. 21–35.

Wodtke, G., Harding, D., and Elwert, F. (2011). Neighborhood Effects in Temporal Perspective: The Impact of Long-Term Exposure to Concentrated Disadvantage on High School Graduation. *American Sociological Review*, 76(5), 713–736.

Wong, A. and Yeh, S. H. K. (Eds.). (1985). *Housing a Nation: 25 Years of Public Housing in Singapore*. Singapore: Maruzen Asia for Housing & Development Board.

Yip, C. S. (2012). *Intergenerational Income Mobility in Singapore*. Singapore: Ministry of Finance. www.mof.gov.sg/Portals/0/Feature%20Articles/IG%20Income%20Mobility%20In%20Spore.pdf

Part 3
Urban housing development and outcomes in China

12 The long-term dynamics of housing in China

Bertrand Renaud

HOMER HOYT INSTITUTE, U.S.

1. Introduction: four decades of profound change

China has undergone epochal changes since market-oriented reforms were launched in 1978 under the leadership of Deng Xiaoping. The most basic indicators tell us that China's socio-economic transformation has been extremely rapid and, arguably, the most profound in China's entire history. Relying on International Monetary Fund (IMF) data, between 1980 and 2015 China's total population has increased by 387 million to 1.374 billion people, which is a population increase larger than the entire US population in 2015.[1]

The share of China's urban population has risen steeply from 19.6% to 54.8%. In constant RMB, national GDP expanded 25.9 times and per capita GDP rose 18.6 times. Taken together, these population and GDP indicators yield the striking result that the size of the Chinese *urban economy* has expanded massively by some 72.4 times in a mere 35 years. The impact of this powerful urban transformation has been very large on China itself and on the rest of the world as well.

Internationally, China's share of world GDP has risen from 2.3% in 1980 to 17.1% by 2015. If China and the US maintain their respective level of nominal GDP growth rates, the Chinese economy will become larger than that of the US in less than a decade. Already, in PPP (purchasing Power Parity) dollars terms, China was the largest economy in the world in 2017. The financial system has facilitated this high growth and has emerged as one of the world's largest, with total financial assets equal to 470% of GDP (IMF, December 2017). This system is playing a critical role in the growth and stability of housing and real estate.

The housing sector itself has grown into a major component of the national economy during the more than 72 times expansion of the Chinese urban economy. Housing and the other types of real estate are quintessential non-traded goods that have long received limited attention from development economists, and if at all as a social issue. Yet, by providing services to local businesses and households, the real estate sector's performance in terms of prices, levels of output and volatility is an integral part of the performance of the overall economy.

Housing is by far the largest component of the entire real estate sector. Housing represents approximately 80% of the total value of all real estate assets in a high-income, fully urban economy like the US.[2] Spatially, housing occupies the largest share of total urban land use in cities. It is a major contributor to new land consumption away from rural use because, in addition to a city's population increase, the demand for housing space rises with household incomes.

At early stages of development and low levels of urbanization, housing is of limited significance for national economic management. Trade and industrialization are dominant policy concerns. However, the cumulative weight of housing in the national economy increases continuously as a country's industrialization, urbanization, and per capita income rise. Eventually, housing becomes the single largest asset class in high-income economies. For instance, in 2015 the total value of US housing assets was $26 trillion, which was larger than the stock market, but not all non-bank financial assets. Meanwhile, the 2015 US GDP was approaching $18 trillion. At a high-income stage, the mismanagement of the housing sector and of its two-way interactions with the wider economy can have very costly consequences, as shown by the experiences of the US and several other Western, high-income countries during the Great Recession of 2008–2009, and Japan in the 1990s.

Chapter 12 offers a short macro-history of China's housing and urban development. In particular, it explains the transformation of urban housing during China's 35-year growth take-off between 1978 and today's "growth transition." The analysis has four parts. In Part 2, three economic models of quite different vintages give us complementary insights into the urbanization of China, and into the scale and location of new urban housing. The non-spatial Lewis model (1954) clarifies the meaning of China's present "growth transition." The von Thünen model (1826) gives us insights into the slow urbanization process in pre-industrial China. Then the major new synthesis of trade and economic geography of the Krugman model (1991, 1995) explains the key factors that have driven the movement of labor and capital and transformed China's very large system of more than 660 cities during its 35-year growth take-off. Looking at housing proper, the Burns-Grebler hypothesis shows the long-term pattern of change in the share of housing investment in annual GDP over the course of China's urbanization.

A short Part 3 focuses on the "growth transition," which China has just entered. It summarizes the actual development records of countries around the world since the end of WW2. Then it compares China's rapid economic growth take-off with those of five East Asian economies (Japan, Korea, Hong Kong, Singapore, and Taiwan), which are the *only* five societies to have successfully managed their economic growth take-off after WW2, then have avoided falling into the "middle-income trap" during their own "growth transition," and are now high-income economies.

Part 4 focuses directly on China's powerful new housing system that has emerged during China's growth take-off decades of 1980–2015. It stresses how the historical housing privatizations of 1998–2003 have altered the course of China's long-term development, and it analyzes the main features of the Great Housing Boom of 1998–2013, which has been the first market-based housing cycle in China's "socialist market economy."

Closing this macro-history of Chinese housing, Part 5 points to a cluster of needed reforms that China faces to achieve a higher quality of urbanization and to better house the one billion urban residents anticipated by 2030.

2. Dynamics of urbanization and of housing during development

No single economic model can account for the transformation of China from a pre-industrial society of the past to the post-industrial society of the future. However, three models of very different vintages together give important insights into the long-term dynamics of the economic development, and its urban dimensions. As Rodrik stressed recently, economics is a social science and economic models do not provide universal laws in the way physical sciences do: "Knowledge accumulates in economics, but horizontally, with newer models explaining aspects of social outcomes that were unaddressed earlier. Fresh models do not replace older ones, but they bring a new dimension that may be more relevant in some settings ... In economics context is all" (Rodrik, 2015 p.67). The Lewis, von Thünen, and Krugman models clarify the urban dynamics of China at different stages of its development.

2.1 The three main stages of long-term economic development

Based on international experience, development economics finds that a country typically grows through three main stages over lengths of time that can vary significantly (Rodrik, 2005). First, there is a take-off stage of initial industrialization, often marked by very rapid economic growth accompanied by progressively accelerating urbanization. Then the economy enters a growth transition: as it reaches a middle-income level, its share of manufacturing employment peaks and rates of urbanization are also at their highest. This stage is called a "growth transition" because major economic adjustments are needed to shift the economy upwards from basic industrialization to higher value-added activities. In terms of causal ordering, substantial political and institutional adjustments are prerequisite to the full implementation of these economic adjustments. If these adjustments are successful, the country progresses to its third stage of sustained development on a relatively steady path to becoming a fully urbanized, much larger and more diversified advanced economy. Various indicators, labor market indicators in particular, show that China has just entered its own growth transition around the period 2011–2014.

The Lewis dual-sector model of development (Lewis, 1954) can account for these three stages of economic development. The Lewis model focuses on the transfer of underemployed surplus labor from the subsistence agricultural sector to an emerging modern industrial sector implicitly concentrated in cities, but the Lewis model does not have a spatial dimension. Over time, the growth of the industrial sector absorbs the surplus rural labor, promotes urbanization, and stimulates sustained development and rising incomes.

At the start of the Lewis development process, the country is a pre-industrial society and its subsistence agricultural sector is characterized by an abundance of labor of low productivity, low wages, and disguised unemployment. As the rural population keeps increasing, the marginal productivity of farm labor falls toward zero because rural land is fixed and farm technology is low and static. Lewis calls the farm workers whose marginal productivity is so low "surplus labor." In contrast, the small (urban-based) manufacturing sector is defined by higher wage rates compared to the subsistence sector, and has a higher marginal productivity. The manufacturing sector uses production processes that are relatively capital intensive. Investment and capital formation in the manufacturing sector are possible over time as profits are reinvested in the capital stock. At the start of industrialization, improvements in the low marginal productivity of the farm sector remain low because national investment priorities go toward the capital stock of the manufacturing sector (and its supporting infrastructure).

When capital accumulation in manufacturing catches up with population, the economy reaches a point when the "excess supply of labor" at very low wages is exhausted. Lewis called this point a turning point. The "Lewis turning point" coincides with rising manufacturing wages following the exhaustion of surplus labor for several reasons. An increase in the size of the industrial sector compared to the subsistence sector may turn the terms of trade against manufacturers (whom Lewis calls "capitalists") and force them to pay workers a higher percentage of their product in order to keep their income constant. The rural subsistence sector may also be adopting improved techniques of production, and this may raise the level of subsistence wages, which in turn forces higher manufacturing wages. With the growth of cities, the cost of living of workers also increases, due to growing congestion and rising housing prices, which increases pressure on the manufacturing wage after the Lewis turning point. The rate of capital accumulation in labor-intensive manufacturing will decline and a restructuring of the economy will be needed. The Lewis turning point is an important indicator that the country has reached a growth transition.

Dale Jorgenson (1961) has shown that the economic incentives driving the transfer of labor in the Lewis model from the subsistence rural sector to the urban manufacturing sector formally yield a logistic curve. Figure 12.1 shows how the Lewis model accounts for the three stages of long-term development observed empirically. The growth take-off stage is on the left side of the curve; the side of low per capita income and low levels of urbanization. As the country moves away from its pre-industrial conditions, the rate of growth of the economy accelerates. Rates of population transfer and of urbanization reach their peaks as the economy moves up toward the inflexion point, which corresponds to the growth transition stage. Mathematically, the Lewis turning Point occurs around the inflexion point, when the country's population becomes 50% urban. Beyond that point the economy moves to its third developmental stage on a path toward full urbanization and a high per capita income.

The logistic curve implies that the growth rates of urbanization will accelerate as the economy approaches its inflexion point (its growth transition). Past the

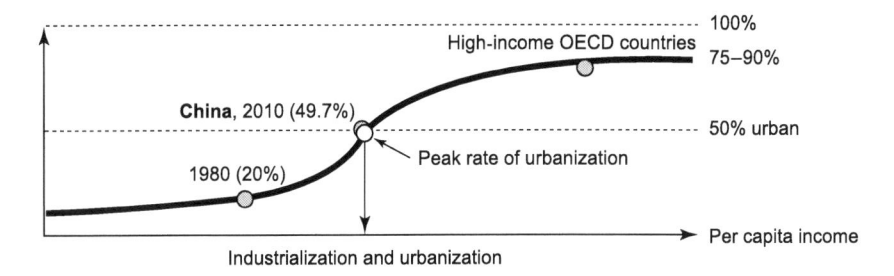

Figure 12.1 The three stages of long-term economic development.

inflexion point, during the third stage of development, steady growth will tend to be slower than during the take-off stage. However, a slower annual GDP growth rate now applies to a very much larger GDP than during the take-off decades, which also means a very much larger annual output from the economy, including housing output. As already noted, the Chinese urban economy expanded 72 times during the 35 years of its growth take-off. At the start of its development, *urban* housing was a small share of a small national economy. By the time of the growth transition, urban housing has become a leading sector of a much larger economy. During the third stage of development the housing sector becomes larger and larger as the level of urbanization reaches a saturation level close to but less than 100%.

The scale of the horizontal axis of Figure 12.1 is measured in decades. For China, the take-off stage alone has taken about 35 years (1978–2014). South Korea, which has had the fastest speed of development on record, took only five decades to travel through the *entire* growth curve. Singapore has had an even faster economic growth trajectory than Korea, but it is a city economy without a reservoir of rural labor to empty out and the Lewis model does not apply to it. However, it remains that Singapore is a very valuable empirical model of the *internal* development of a major global city during its own stages of economic development.

Much research has taken place since the seminal Lewis model of 1954. Findings about the strong correlation between industrialization and urbanization are consistent across time periods and across countries. "The relationship between GDP and urbanization is tight"; but one ambiguity remains: what is cause and what is consequence (Duranton, 2014; 2015) Can urbanization generate economic growth, as some government officials have asserted during China's growth slowdown? Does such a causal effect exist? What is undeniable is that the efficiency of cities impacts the productivity of the economy. The quality of governance and financial management play a central role in a city's efficiency, and the local management of housing is part of it.

Many countries have experienced high economic growth rates during their growth take-off, but most have failed to manage their growth transition stage successfully. They have fallen into the "middle-income trap." Their urban population continues to expand due to rural-urban migration push factors, but their cities include large and even dominant low-productivity informal sectors, as is the case for countries in Latin America and other regions of the world. In countries that have politically and economically mismanaged the growth transition stage, the growth of these economies becomes unsteady and volatile (Pritchett, 2000). In such volatile contexts, the insights of the Lewis model become inappropriate and the logistic curve of development of Figure 12.1 is no longer representative. In contrast, the Lewis model gives a good account of the entire development experiences of Japan, South Korea, and Taiwan, and of their urbanization. The open question today is whether China will also manage its risky growth transition stage successfully.

2.2 Slow pre-industrial "von Thünen" urbanization

The Lewis model does not address the spatial distribution of economic activities across the cities of a country. On the other hand, the von Thünen and the Krugman models have a major geographic dimension. They are complementary models that provide important insights into the development of China (and other East Asian economies) before and during industrialization.

The von Thünen model (1826) focuses on pre-industrial market cities and on the spatial organization of agricultural activities around these cities *before* industrialization. During the pre-industrial era, cities functioned as administrative centers and market cities trading agricultural products. Demographic and economic growth rates were slow. The von Thünen urban dynamics explain how cities located in the richest agricultural regions had their growth constrained by high transportation costs and limited economies of scale in traditional manufactures.[3] Administrative cities were dependent on their government's capacity to raise taxes nationwide and locally to provide security, infrastructure, and other public services. Major cities combined administrative and market functions.

Urban systems have a long history in East Asia, where very large cities, some with populations over one million, existed in China, Japan, and Korea a millennium before the start of the first Industrial Revolution in Europe in the late eighteenth century. Such pre-industrial urban systems have been analyzed for China by G. William Skinner (1964 and 1977) and reinterpreted in depth by Yu (2015). In pre-industrial China and the other East Asian societies, there were large cities already, but the total share of cities was well under 10% of the total population. These pre-industrial East Asian systems of cities have been the matrix within which rapid industrialization took place after WW2. In China, the beginnings of modern industrialization in cities were stirring during the period 1912–1937, but they were obliterated by the war with Japan and then the civil war during the period 1937–1949 (Naughton, 2007).

The nineteenth-century von Thünen model was the first economic model to explain the location of economic activities around cities. Von Thünen's two major analytical contributions to the field of economics are the concepts of land rent and marginal cost. His model has laid the foundations of spatial economics and economic geography, but it says nothing about the forces that drive the transfer of labor to cities, which is the central concern of the Lewis model.

2.3 "Krugman" industrialization and urbanization during the growth take-off stage

The Krugman model (1991, 1995) is a major conceptual and analytical breakthrough that finally brings into an integrated framework core elements of trade theory, which was until then completely non-spatial, together with core concepts of economic geography concerned with the location of economic activities. A central element of the Krugman model is the analysis of the origins and growth of cities, which is good because most economic geography models have an odd theoretical hole at their core by taking the existence of cities as given.

The Krugman model is a very valuable tool to think about the interactions among the economic factors behind the massive transformation of China from a small, closed economy in 1980 into a large manufacturing economy integrated into the global economy by 2015. Most valuable, the analytical power and relevance of the Krugman model is not limited to the growth take-off stage; we can use it to think about the three stages of long-term development.

The model gives analytical insights into why the vastly complex actual urban system of a country behaves the way it does.[4] In Krugman's own words, the model is a framework to understand "the dynamics of an economy as a self-organizing system of cities over the long-term development of a country." What makes the model particularly attractive is its explanations of how interactions between economic forces associated with transportation, information, and technological innovations drive the continuously changing urban and regional structure of an economy as it becomes better integrated into the world economy over time. Why do cities exist and grow? How do these cities interact? How do economies of scale permit some cities and regions to create a lasting comparative advantage? How do incomes differ across cities and regions and shape labor flows across cities? Why does regional specialization increase and then decrease over time?

The case of the US, which is one of the few that we can compare with China in geographic size, provides the opportunity to see the long-term effects of progressive integration on the spatial organization of an economy of continental scale. There exists a quantitative analysis of the evolution of the regional structure of the US between 1860 and 1987, from the time when the fragmented young American republic was facing large physical and political obstacles to its economic integration, to the time when the US economy had already experienced several decades of rapid income convergence and economic integration across the continent (Sukkou Kim, 1998). In between, the US regional and urban system went through a peak of industrial specialization and economic divergence across regions in the 1920s. During the 1920s, the US crossed its 50% urban population marker and the share of manufacturing employment in the total labor force reached its peak. In China, both events occurred in the 2010s.

264 Bertrand Renaud

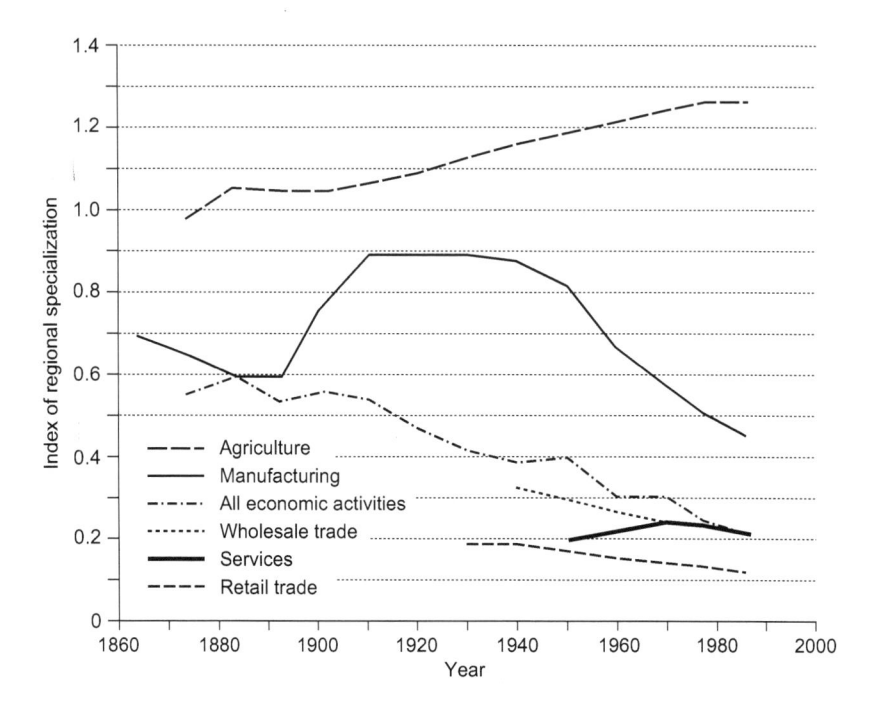

Figure 12.2 US regional specialization in manufacturing and other activities, 1860–1987.
Source: Sukkou Kim, 1998, Economic Integration and Convergence, J. of Econ. History V.58, N.3, p.666.

The process of regional divergence and convergence in the US is summarized in Figure 12.2, borrowed from Kim (1998). The parallels with the spatial history of the People's Republic of China since 1949 are very suggestive. Within Figure 12.2, the most telling curve traces the evolution of an index of manufacturing specialization. An index value of zero means that there is no specialization across regions, a value of two means manufacturing concentration in one region. In 1860, the US manufacturing sector was small and spread across a fragmented continent, so regional specialization was limited. The share of the US economy in the world economy was also very small: the US economy did not become larger than that of France until 1900 (Maddison, 2001). The US spatial economy in 1860 might be compared with the small, closed economy of the PRC that was fragmented into autarkic, poorly connected regions until the start of the market reforms in 1978.

Technological innovations in transportation and communication steadily improved the mobility of labor, goods, factors of production, and information during the US growth take-off, which also led to a rapid divergence in industrialization, incomes, and urban population across regions. At the center was the fast-growing, rich, manufacturing core of the US economy, located in its northeastern "manufacturing belt." At the periphery you had the poor regions of the South. Without going through the details of the Krugman model, the central intuition explaining the emergence of the manufacturing belt is that economies of scale in manufacturing were large enough compared to transport costs to cause a geographical concentration of manufacturing activity and population favoring the larger market (Krugman, 1998). Here again, there is a parallel with the emergence of the coastal manufacturing belt of China after 1980, which left behind hinterland provinces as a source of migrant labor.

The American growth transition did not go particularly well, with the stock market crash of 1929, the Smoot-Hawley Tariff Act of 1930, the banking crisis, and the Great Depression that spread globally and further lowered international trade.[5] The level of concentration of US manufacturing did not change until the end of WW2. Then, with more technological innovations in manufacturing and other activities, the US economy became rapidly integrated with the increasing mobility of labor and capital and more location choices for business across the entire urban system. By 1970 the decentralization of manufacturing activities was already greater across regions and cities of the US than it had been in 1860. The obvious hypothesis is that the Chinese system of cities will experience a similar spread of economic activities, income convergence, and spatial integration if and when the rebalancing policies of the growth transition succeed.

The Krugman model is directly helpful to our understanding of the macro-dynamics of Chinese housing across China's urban system of now more than 660 cities. The link between housing and urbanization is direct: every urban dweller lives somewhere. In terms of value, housing makes up the dominant share of the total existing assets and of new annual investments in a city. In terms of space, housing typically uses more than 65% of the total urban land of a city.

2.4 Urban dimensions of long-term economic development: a restatement

Figure 12.3 is just an urban relabeling of Figure 12.1. It emphasizes the long-term development of an economy from its pre-industrial urban state (at the start of the logistic curve) to its post-industrial state, when its urbanization is high and reaches saturation at its upper end.

Comparative research shows that a coincident indicator of the growth transition is the level of per capita GDP estimated in Purchasing Power Parity dollars ranging from PPP $10,000 to $13,000 (Brülhart and Sbergami, 2009; Eichengreen, Park, and Shin, 2013) – hence the moniker of "middle-income trap" given to the growth transition stage. This is the income level that China reached between 2011 and 2014.

2.5 Evolution of the ratio of housing investment to GDP: the Burns-Grebler hypothesis

Housing conditions and annual investments change along the logistic curve of urbanization. Figure 12.4 illustrates the Burns and Grebler cross-country

266 Bertrand Renaud

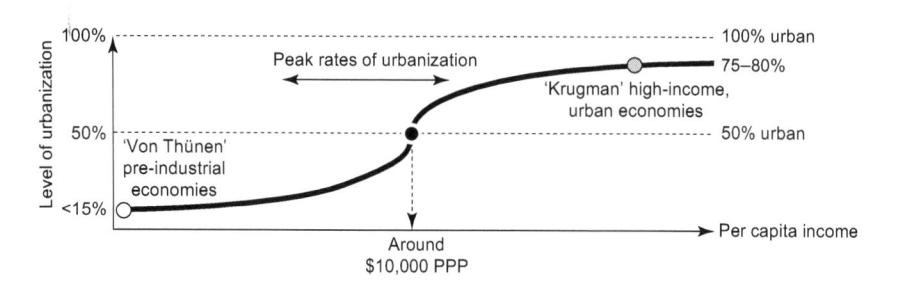

Figure 12.3 Urban dimensions of long-term economic development: a restatement.

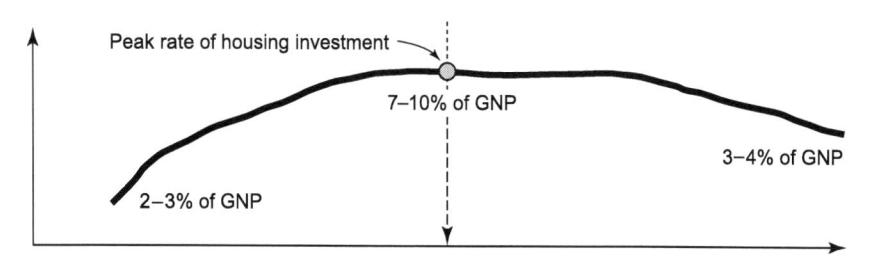

Figure 12.4 Housing investment and level of urbanization.

hypothesis that the ratio of housing investment to GDP follows an inverted U-shaped pattern that is linked to the stage of economic development (Burns and Grebler, 1978). This hypothesis has been confirmed empirically on cross-section studies of countries (Malpezzi, 1999). It has also been tested across Chinese regions.

At low levels of economic development, the ratio of housing investment to GDP is also low. The largest share of public investment in cities goes to public infrastructure such as roads, water, and sanitation systems, and other public goods that facilitate industrialization. The share of new housing investment in annual GDP is low and of the order of 2–3% of GDP.

During the growth take-off stage, as industrial activities and urban population expand, the share of housing investment in GDP rises to peaks of around 8% as the economy reaches the 50% inflexion point. In city states like Singapore, which is known to have overinvested in housing prior to 1985, the share of annual housing investment in GDP peaked at values higher than 10%.

Then the relative share of housing investment in GDP slows down to around 3–4% at the upper end of the logistic curve, when urbanization has reached its saturation level in high-income economies. By then, housing assets form one of the two largest asset classes in the national economy. At the high-income stage, when an economy is fully urban, a lower ratio of housing investment to GDP still means a much larger absolute volume of annual housing investment. The actual time path of the share of housing investment in GDP in a given country is a curve jagged by macro-economic shocks and recessions.

The growth take-off stage is a period of crucial transformation of housing systems on their way to the growth transition. It is during this stage that many of the lasting features of large, organized mass housing systems fall into place and the real estate sector itself becomes a major component of the national economy. The growth take-off decades are the period when basic legal, regulatory, and information systems and the organization of the industry are put in place. Because the path dependency of housing institutions is quite durable, at high income levels, housing systems still bear a strong relation to the modern mass housing systems that emerged during the growth take-off stage as seen in the case study of Korea by Cho, Kim et al. in this volume. Institutional changes tend to become more marginal.

3. The actual development record since WW2 and the middle-income trap

How does the actual record of economic development compare with the three models of development and urbanization we are using? In a two-year study of the lessons of economic development over seven decades, the International Commission on Growth and Development (2008) found that only 13 out of more than 150 developing countries of all sizes have experienced sustained high growth, which the Commission defined as 7% real GDP growth per year for 25 years or longer.[6] These 13 economies are: Botswana, Brazil, the People's Republic of China, the Hong Kong Special Administrative Region (SAR) Indonesia, Japan, the Republic of Korea, Malaysia, Malta, Oman, Singapore, Taiwan, and Thailand. The six East Asian economies of China, Japan, South Korea, Taiwan, Hong Kong and Singapore are part of this small elite group of 13 development take-off successes.

Then, out of these 13 economies that experienced a sustained period of high growth, only five economies have now become advanced, high-income economies. All of them are East Asian. Other economies have been caught in a middle-income trap of slow and volatile growth following their successful take-off decades. Can China's own growth transition match the performances of its five East Asian predecessors? What do these East Asian societies have in common? What factors might explain their outstanding development performance? Are their respective housing systems very similar, or do they differ?

Every growth transition is a difficult period that combines risky political and economic adjustments. Economically, the five East Asian countries had outgrown the coordinating benefits of the developmental state for infrastructure investments and labor-intensive manufacturing. They had to develop a different and much more sophisticated growth model. Politically, this economic transition significantly affected the interests of the elites that had emerged during the take-off and whose incentives had been aligned with the needs of the country to achieve rapid growth. In particular, these elites had benefited in important ways from the directed credit policies of the government relying on financial repression policies. Now, financial policies needed to be liberalized in order to serve efficiently the needs of the new post-transition economy and new competing elites – and small and medium enterprises (SMEs) in particular. In addition, East Asian growth transitions typically occurred under considerable domestic and external uncertainties. This is again the case for China today, where domestic reforms are needed when the world environment is becoming much more unpredictable and risky.

In terms of causal ordering, the record shows that the decisive factors in a country's successful crossing of its growth transition are the adjustments to the political system that was in place during the take-off stage. These political adjustments are necessary to achieve the flexible and open system needed for long-term sustainable growth. Without them, needed economic reforms may not be implemented, or only partially. This is what Japan, Korea, Taiwan, and Singapore achieved, each in its own way. On the other hand, middle-income countries such as Argentina, Brazil, and Mexico in Latin America, Indonesia, Malaysia, and Thailand in Southeast Asia, or Iran, Saudi Arabia, and Turkey in the Middle East have not succeeded so far. They are still mired in their individual middle-income traps. China's economic plans for the rebalancing of the economy were spelled out in details at the start of the Xi Jinping administration. It is the political transition that will determine how well economic reforms will take place.

3.1 China has reached its growth transition and housing is a major sector of its economy

China officially crossed its 50% level of population urbanization in 2011. This 50% urban population marker compares with the historical peaks in other East Asian countries, where the shares of manufacturing employment in total employment peaked during that period. This was also the case for the US in 1920, as noted earlier.

Figure 12.5 shows the historical sequence of growth take-offs and peak rates of urbanization in Japan, Taiwan, South Korea, and, eventually in China, several decades after the other East Asian economies. Figure 12.5 reports only Japan's *second* crossing of the 50% marker. Japan, which was the first East Asian country to industrialize after the nineteenth century Meiji reforms, had already reached this marker before WW2, but the devastations of the war drove Japan's economy backward and its urban population had fallen back to 35% in 1944. Japanese urbanization crossed the 50% marker after WW2 for the second time in the 1950s.

During the take-off stages of China, Japan, South Korea, and Taiwan, rapid urbanization was led by labor-intensive, export-oriented industrialization that tended to concentrate in the cities of the new and rapidly growing core industrial region, most vividly illustrated in China by the new city of Shenzhen. The tempo of urbanization, which reflects the speed of urban concentration in China, peaked at 4.78% per year during the period 1980–1985.

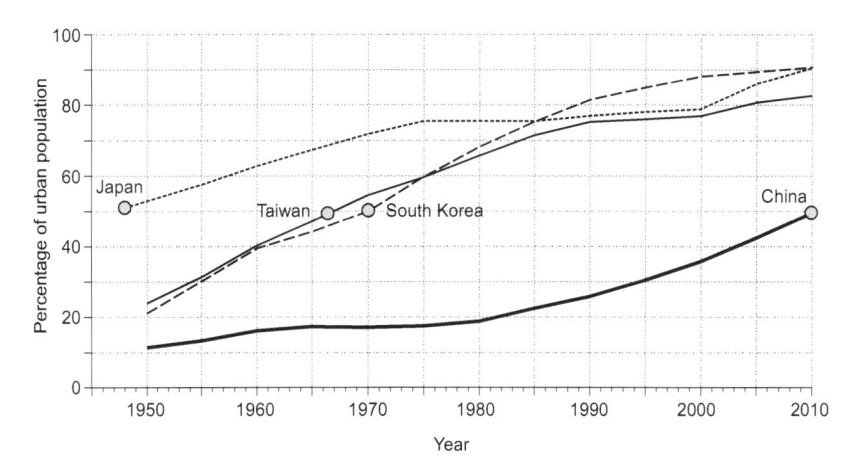

Figure 12.5 Growth transition and 50% level of urbanization in China, Japan, South Korea, and Taiwan.
Note: The UN figures for Korea have been edited (i.e replaced) by the official ROK-MLTI Annual report on Land 2013.
Source: ROK Ministry of Land Transport and Infrastructure, Annual Report on Land 2013.

Because of its continental scale, China's regional labor markets differ across the country. Findings about the time when the Chinese labor markets reached their Lewis turning point also differ (Knight et al., 2010). Population aging is accelerating the effects of the Lewis turning point. China's National Bureau of Statistics reported that the size of the Chinese population of working age 15–59 already fell by 3.5 million in 2012. Real wages began to rise more rapidly. Manufacturing production is switching to higher value-added goods. The service sector is expanding in Chinese cities, not only because of the growth of consumer demand coming from an expanding urban population with rising incomes. More importantly, with the transition from labor-intensive manufacturing to higher value-added manufacturing, the service sector also expands as new manufacturing goods include a rising share of services inputs, as occurred in the US after WW2.

Past the Lewis turning point, interactions between housing and the macroeconomy become stronger through the urban labor market channel that is expanding both in size and income level. Rapidly rising housing prices put pressures on urban wages, whose rise in turn weakens exports. In this manner, the non-traded housing sector can have an indirect impact on the performance of exports. When the Lewis turning point was reached in Taiwan and Korea around the 50% urban population marker, wages also rose rapidly. Yet, rural-urban migration continued as different industries producing higher value-added goods were developing. The pressures of rising incomes and an expanding urban population, together with expectations that such trends will continue, have a powerful impact on expectations for the rate of increase of land and housing prices. The situation had been similar in Japan during the 1960s and 1970s, decades of extremely high Japanese growth which fed the costly myth of ever-rising land prices known in Japan as *toshi shinwa* or "land fairy tale." Figure 12.5 shows the year when each Asian economy crossed the 50% urbanization level. These East Asian milestones are spread over six decades.

Spatially, the third stage of sustainable growth differs from the urban concentration observed during the take-off stage. The spatial decentralization of manufacturing progresses across the urban system, as occurred in the US after the 1930s (Krugman, 1998; Sukkoo Kim, 1998). During this final development stage, housing assets continue to accumulate and become the largest asset class in the economy. Housing cycles begin to play a leading role in the stability of the economy, as shown in the US case, where Leamer argued just before the global Great Recession of 2008–2009 that "housing IS the cycle" (Leamer, 2007).

4. China's development from an East Asian perspective

4.1 Distinguishing features that China shares with other East Asian economies

China and the other East Asian economies share distinguishing features that, combined together, set these economies apart from other economies of the world.[7]

East Asian societies have some of the highest population densities in the world. East Asian rural population densities have been higher than densities considered "urban" in some countries of Latin America and Africa. Under the right institutions and policies, high initial population densities can be favorable to the diffusion of the Krugman urban dynamics.

These countries' shared Sinitic cultural heritage has had a lasting impact on the governance of public and private institutions, as well as on prevailing norms of public and private behavior. There are multiple social ramifications to this Sinitic heritage. Among these are a deep tradition of learning that was fostered by the meritocratic selection of government officials. East Asian levels of human capital have been much higher than in other countries with similar per capita GDP.

After WW2, East Asian "developmental states" successfully implemented pragmatic and flexible strategies that proved to be highly consistent with the Krugman urban dynamics and resulted in the highest economic growth rates in the world, as well as extremely rapid growth of cities and their housing sectors.

Western-centric views dominated perspectives on the institutions of economic development after WW2 because this was where advanced economies were concentrated. It is therefore worth elaborating briefly on political order and the role of the state in East Asian history. In his two-volume study of the emergence of modern political order in world history, Francis Fukuyama (2011; 2014) focuses on "the process by which political institutions emerge, evolve and eventually decay" (2014, p.7). He points out that "China was the first civilization to establish a non-patrimonial modern state, which it did some eighteen centuries before political units appeared in Europe" (2014, p.10). In one of three chapters on "the strong Asian state" and the modern Chinese state, Fukuyama finds that

"East Asia's problem is not state weakness but the inability to constrain the state" (2014, Chapter 23). During the growth take-off stage, strong state effectiveness is a significant factor in the management of the infrastructure.

4.2 Post-WW2 East Asian development policies supported the Krugman urban dynamics

In addition to initial conditions favorable to urbanization and industrialization, East Asian governments implemented policies that proved supportive of the Krugman urban dynamics. So did China after Deng Xiaoping's pragmatic market-oriented reforms started in 1978, adopting policies favorable to efficient urbanization:

- Proactive demographic policies captured the dividends of the demographic transition in terms of employment, savings, and education.
- Successful rural land reforms implemented after WW2 in Japan, South Korea, and Taiwan had multiple benefits for wealth distribution and for the growth of an organized real estate sector. However, Communist China went the opposite way. After 1949, the PRC nationalized land, with consequences that are still being felt today.
- A policy emphasis on industrialization through manufacturing exports sped up innovations and favored employment concentration in large cities and the emergence of core industrial regions. These labor-intensive manufacturing exports played a critical role in financing more domestic investments at a time when domestic savings were still low (Rodrik, 1994).
- Manufacturing exports sped up the rate of domestic innovation in order to meet the higher standards of international demand from high-income countries.
- Concentration of manufacturing in major urban centers facilitated faster knowledge spillovers, labor market pooling, and specialized suppliers; it also facilitated infrastructure investments.
- Access to the large US market for Japan, Korea, Hong Kong, and Taiwan during the 1960s and 1970s was strategically important when international trade was still fragmented and supply chains were short.
- Spatial infrastructure plans were coordinated with national economic plans. These plans often *anticipated* urbanization needs. In addition to transport, energy, and communications investments, these plans included the creation of industrial estates and special export zones.
- "Flexible rigidities" (Ronald P. Dore) in policy monitoring meant that successful policies were sustained, while unsuccessful ones were quickly discarded.
- The governance of East Asian cities and their financial management was considerably better than in most other developing countries because of the long tradition of strong government in East Asia. A substantial body of urban laws developed rapidly. Central governments also monitored the performance of individual cities through ministries of home affairs or their national equivalents.

With these policies, East Asian societies achieved some of the fastest rates of economic growth in the world during their take-off decades. Their rapidly growing urban systems also experienced the highest "tempos of urbanization" in the world, possibly thanks to their high rural densities. The tempo of urbanization is measured by the difference between the urban population growth rate and the rural population growth rate observed during a given period of time, usually five years. A high tempo of urbanization means a rapid rate of urban population concentration, which will go preferentially to cities where new, trade-oriented manufacturing is growing. In East Asia, those were the largest coastal cities. South Korea appears to hold the world record for a tempo of urbanization, but China's tempos have not been far behind during the past two decades.

4.3 Similar growth objectives across East Asia, but four different housing strategies

Important shared initial conditions at the start of the development growth take-off did not mean identical housing systems. National economic policy choices led to different policy priorities for housing. Over time, the supply of housing has ended up being organized quite differently across the six East Asian economies.

The results of incremental housing policy choices can be grouped into four main types of East Asian housing strategy, which differ remarkably. Given the durable path dependency of housing institutions, East Asian housing systems still bear today the imprint of the policy choices made during the early years of the economic take-off of each country. A shared feature of East Asian housing strategies is the frequency and scope of government interventions. These government housing interventions have occurred in one or more of the four main components of a housing system: land use and urban planning; the direct construction of public housing; housing finance with a reliance on special housing circuits; and direct and indirect housing taxation and subsidies.

Looking back, four different types of *de facto* housing strategy emerged across East Asia after WW2. First, China (and North Korea) took a central planning path to housing provision modeled after the housing policies of the socialist Soviet system, where housing was viewed as a social good to be distributed by the state.

Second, in Japan, the redirection of household savings toward industrial investment, which started in 1925, also reflected the management of the economy as a war economy. Japanese post-war reconstruction after 1945 saw the continuance of many of these policies, as national priority was given to rapid industrialization (Lockwood, 1954; Mills and Ohta, 1976). During the reconstruction following the 1950–1953 Korean War, South Korea followed a similar strategy of redirecting demand away from consumption in order to speed up industrialization and infrastructure investment.

In a third strategy, the small, open economies of Hong Kong and Singapore adopted a non-ideological, pragmatic use of public housing on a very large scale for the social support of their labor force. Yet the dynamics and the outcomes have been very different. In Hong Kong, the public housing sector remains among the largest in the world and the government remained a landlord for 46% of the population in 2016 (see Chiu, Chapter 10). In Singapore, large-scale public housing production was conceived from the start as a strategic step to create a nation of owner-occupants. By 2017, 91% of Singaporeans were owner-occupants, but the housing stock originally built by the public Housing and Development Board (HDB) was 73% of the total stock. The public rental sector is very small (see Phang, Chapter 11).

Fourth, Taiwan's housing strategy stands out from the rest of East Asia where frequent direct government interventions in housing remain the rule. The non-interventionist housing policies of Taiwan relied on market mechanisms.

The sole East Asian exception to a long-term continuity in housing strategy since WW2 is China, where the ideological housing policies of the central planning era had left the housing sector in a disastrous state by the time the Deng reforms started in 1978. With the historical housing privatizations of 1998–2003, China completed a 180-degree turn away from its administrative-command past. The significance and scale of the housing privatization reforms of 1998–2003 for China's current growth transition stage are clear when compared with China's wrong-headed initial housing strategy adopted in 1949.

4.4 Structural distortions specific to Chinese housing

China is not a usual market economy like the other five developed East Asian economies. It prefers to define itself officially as a "socialist market economy." This is not a trivial choice for the institutional structure of China's housing system.[8] Important legacies from the socialist central planning era still affect the structure and dynamics of China's urban system and of housing. Among these legacies, three in particular deeply affect the quality of Chinese urbanization and housing. These are:

- The duality of land property rights between rural and urban land, which has had a distorting effect on the expansion of Chinese cities and has lowered their spatial efficiency;
- The discriminatory "*hukou*" registration system that has created *de facto* two classes of Chinese citizens who have very different access to the benefits of industrialization and urbanization; and
- The distorting fiscal incentives facing local governments, which have driven them to overinvest in urban infrastructure and real estate, and to maximize the returns from sales of the urban land over which they have a monopoly control. The share of land revenues in local government finance from real estate activities has become so large that local government finance in China has been nicknamed "land public finance" (Man and Hong, 2011). In addition, preferential access to state financing expanded the "soft budget constraints" (Kornai, 1992) of local governments, which explains the large share of local government debt in China's national debt in 2016.

These institutional distortions and others create major policy challenges for China's housing and urban system during the growth transition and beyond. Recommended policy principles for the growth transition are presented in the second half of Chapter 16, which provides the roadmap for better housing policies, internationally and in China.

5. Emergence of a new Chinese housing system during the take-off decades

The new housing system that has emerged during China's growth take-off decades has had five distinguishing features. First, China's investment-led growth policies have had direct and indirect impacts on housing investment behavior by households. Second, contrary to Japan and South Korea, China has *overinvested* in housing in spite of its financial repression policies, the drivers of this overinvestment being local governments. Third, by their scale and their completeness, the historical housing privatizations of 1998–2003 have permanently altered the course of China's long-term development. Fourth, these privatizations have played a major role in China's Great Housing Boom of 1998–2013. Fifth, this housing boom has spread unevenly across city sizes, and between coastal regions and China's hinterland. We consider these five features in turn.

5.1 China's investment-led growth compared with earlier East Asian experiences

During the take-off decades, strong East Asian governments regulated the growth of the financial system. They made financial capital subordinate to industrial capital. East Asian financial systems have been bank based, which is typical of financial systems at early stages of economic development, and not capital markets based, like high-income Anglo-Saxon financial systems. The exception was the British Crown Colony of Hong Kong, where financial policies and regulations were shaped by the UK.

To channel financing to industrialization, Japan, South Korea, and Singapore adopted "financial repression" policies, which had a deep impact on the flow of household savings and on the development of the housing finance system.[9] So did China, with the added complexity that it was transforming from a central planning system, where all investment decisions had been made by the state, into a market economy. Unique in East Asia, Taiwan has had realistic interest rates reflecting the true cost of capital since 1962, but its bank regulations still constrained bank residential mortgage lending until 1985 (Renaud, Kim, and Cho 2016, Chapter 5).

Under central planning, China's financial system had been reduced to a state-run mono-banking system. After 1978, China had to recreate entirely a genuine financial system, starting with a new central bank at the core of what remained an unfinished financial sector transformation until the present decade (Hardy, 1998). In practice, the Chinese government has had even greater control over investment patterns than in the other East Asian market economies, where investment decisions by large private business groups such as *zaibatsus* in Japan and *chaebols* in Korea played an increasingly influential role.

Aware of earlier East Asian experiences, Chinese policy makers chose to finance very high rates of domestic investment in indirect ways through resource transfers from the household sector in favor of government investments and state-owned enterprises (SOEs). These resource transfers have operated through three macro-economic policies. The systematic undervaluation of the currency, favored corporate exporters at the expense of domestic household consumers. The rate of growth of wages was kept slower than the rate of growth of labor productivity, which was facilitated by the facts that China was operating in a labor-surplus economy, had not reached its Lewis turning point, and labor unions were state controlled; and "financial repression policies," with credit allocation directed by the government, with the central regulation of deposit and lending rates significantly below the true opportunity cost of capital and its equilibrium level in the economy.

Of these three macro-economic policies, financial repression has caused the largest transfers away from households to "priority investments." The size of these transfers has fluctuated from year to year. Their magnitude depended on the gap between the true current cost of capital in the economy and the repressed low interest rates offered by Chinese banks on deposits. These resource transfers can be quite large and they have ranged between 4% and 8% of GDP (Lee, Syed, and Liu, 2012).

Resource transfer policies away from households have been politically sustainable in China because the real growth rate of household incomes remained quite high, by any international standard, thanks to even higher rates of growth of the Chinese economy: rates of real household income growth were of the order of 7–9% per year, when GDP growth rates were 11% or better during the growth take-off decades. Between 1998, when the housing privatization reforms started, and 2008, when the Global Financial Crisis occurred, the average rate of growth of Chinese real wages was 9% per year, when the average GDP growth rate was even higher at 11.2%.

The degree of financial repression has varied over time. Household deposit rates in banks received low but positive real interest rates during the 1990s, but they fell below inflation during the real estate boom decade. Negative real deposit rates drove Chinese households, with their rapidly rising real incomes and saving rates, to invest most of their funds into housing because it was one of the rare assets available to them offering positive and rising returns. The volatile and inadequately regulated stock markets of Shanghai and Shenzhen suffered from insider trading and were an investment alternative mostly for the highest household income groups, who also dominated housing investments, especially in booming Eastern cities.

The emergence of China's mortgage finance system has been conditioned by the pace of overall financial reforms. The first experimental residential mortgage loans were made in 1986 by China's Construction Bank. During the Great

Housing Boom of 1993–2013, China's new housing finance system has grown extremely rapidly (Deng and Fei, 2008). The depth of China's housing finance system measured by the year-end ratio of mortgage loans outstanding to GDP has grown from zero in 1986 to well past the 25% mark. Household debt has been growing at a very rapid pace and has become a concern for financial regulators. Most of that debt is housing loans.

Real rates of returns on housing ownership have been quite high during the East Asian take-off years. These rates of returns were the highest in the new industrial urban core of the economy, where city population growth reached high annual rates above 7% per year. Housing rates of return were consistently higher than real GDP growth rates. Access to housing finance being limited by financial policies, achieving homeownership has been constrained by the need for large amounts of prior equity. Since the mid-1990s, China has not been different. In China and across East Asia, the housing sector has been a very significant channel of wealth redistribution to higher-income groups. The only East Asian exception is Singapore, where housing policies of five decades have led to the best housing wealth distribution by a factor of five compared to the average of OECD (Organisation for Economic Co-operation and Development) countries (Sock-Yong Phang, Chapter 11).

5.2 How China overinvested in housing and real estate after 1994

Opposite to the experiences of Japan and Korea that had *underinvested* in housing in favor of industrial investments under national financial repression policies, China has *overinvested* in housing and real estate during its peak growth years, especially between 2008 and 2014, in spite of its own financial repression policies. The main channel for this unexpected outcome has been the flow of funds to local governments.

Chinese urban municipal governments have a monopoly over the urban land-use rights recreated by the constitutional amendments of 1988. A key factor in local government behavior has been the 1994 intergovernmental fiscal reforms that initially left local governments with 77% of all public expenditure responsibilities but only 46% of public revenues. Such fiscal imbalances have made local governments heavily dependent on revenues from land transactions and given them a strong incentive to fuel land prices and promote the real estate sector (Man and Hong, 2011). In many Chinese cities, local governments still maintain formal or informal links with real estate firms (You Tien Hsing, 2010). Drawing on the implicit or explicit guarantees of local governments, these real estate firms have had an incentive to overinvest in housing. "Moral hazard" and socialist "soft budget constraints" (Kornai, 1992) have been major incentives to local overinvestment in China.

Officially, Chinese local governments could not borrow, or issue bonds or guarantees until 2014. In practice, local governments were bypassing these regulations by creating autonomous municipal investment companies or *Chengtou* to borrow for infrastructure, real estate, and other projects.[10] Local government debt increased tenfold between 1998 and 2008. Then local government debts have accelerated sharply after the stimulus package of 2009 against the global Great Recession. Together with local governments, real estate is now one the three major contributors to China's rapidly growing national debt level, with a debt-to-GDP ratio that has expanded from 150% in 2008 to 260% in 2016. A third contributor to the rapidly increasing fragility of China's financial system is the still-expanding growth of SOEs' debt.

5.3 China's historical housing privatizations between 1998 and 2003

Internal government debates on housing privatization and also city-level experiments started after the 1988 property rights amendments of the 1982 Chinese Constitution. These amendments had restored the trading of land-use rights in cities after a four-decade freeze on land-use transactions in cities. The decision to privatize the urban housing stock nationwide was made in 1998 by Prime Minister Zhu Rongji (Zhu Rongji, 2013). Urban housing privatizations reached their full potential when restrictions on the resale of privatized units were abolished nationwide in 2003.

The massive housing privatizations of 1998–2003 are unique to China's urban development. Their scale and speed are unmatched anywhere else. These privatizations were a sharp break with China's central planning past. Their impact on the transformation of China's economy and society compares in importance to the decollectivization of agriculture and the implementation of the rural "household responsibility system" between 1978 and 1984 at the start of the transitions from plan to market. The impact of housing privatization on household expectations and the demand for housing was compounded by other reforms that were occurring at the same time. In turn, these domestic reforms were amplified by the global economic boom prior to 2008. Taken together, these reforms resulted in a unique decade in China's urban development, which is unlikely to be repeated.

From a macro-economic perspective, the housing privatizations were expected to have multiple long-term benefits. They would relieve SOEs from the social burden of housing their workers, and prepare them for more intensive international competition as China was negotiating concurrently its entry into the World Trade Organization (WTO) in 2001. The parallel restructuring of the industrial sector with the privatization of a large number of non-strategic SOEs, which also began in 1998, would also facilitate labor mobility and the restructuring of the labor system. The wealth transfer of housing assets from SOEs and other public employers (*danwei*) would improve the autonomy and economic security of many households. Finally, these housing privatizations were consistent with the deepening of financial reforms. The immediate macro-economic impact of housing privatization was quite large and caused a historical shift in the development trajectory of China.

The 1998–2003 housing privatizations were carried out at artificially low prices. There was no choice: cash wages were still quite low and there was only a nascent mortgage market. However, concurrently, the 1998 privatization of SOEs drastically expanded urban *private*-sector employment from its low share of 3% in 1981 to 82% in 2012 (Lardy, 2014, Chapter 3). In particular, the SOE reforms gave rise to a new class of high-net-worth individuals who have had a major impact on the composition of the new housing demand. A powerful housing boom followed the lifting of the 3–5-year resale restrictions in 2003. Millions of households who had gained significant housing equity through the massive housing wealth transfer from the state could now use that equity to upgrade their housing units or buy additional properties. Moreover, the financial costs of owning in China are very low by international standards in middle-income economies, and the economic "user cost" of housing often turned negative. The mortgage finance system dominated by the four large state-owned banks grew extremely rapidly from its zero base in 1996 and fueled the housing boom. These state banks offered hybrid housing loans by merging funding from their own deposits with the individual household savings contracts from the municipal Housing Provident Funds created since 1991 across China.

5.4 The great housing boom of 1998–2013

From a cyclical perspective, the first decade of the twenty-first century is the period when the real estate sector became a major non-state component of China's national economy. For the first time, Chinese housing was now driven by household choices, thanks to the ownership reforms of 1998, together with the generalized reliance on market wages following the prohibition of the direct provision of housing by urban work units (*danwei*) that until then had been a major part of the total compensation of *danwei* employees.

Extremely rapid growth was transforming Chinese urban housing markets. During the peak decade of the growth take-off, China became the world's largest construction market. It added more than two billion m^2 of new floor area every year, which was half of the global real estate supply. About half of that new floor space was for residential use, which was almost evenly divided between urban and rural housing.

Part of this new residential supply was needed to meet demographic growth and urbanization, but a larger part reflected the dramatic rise in housing standards in terms of residential space per capita. China's total population had increased by about a third, from one billion in 2002 to 1.33 billion in 2010, when China crossed its 50% urbanization marker. At the same time, residential space per capita in cities quadrupled from 6.7 m^2 in 1978 to 28.3 m^2 by 2007. At the same time, residential space tripled in rural areas from 9.4 m^2 to about 29 m^2. China's official average per capita floor space now exceeds European averages and those of Japan and most of East Asia. This space average is not expected to rise much further, as increasing housing quality standards are becoming the leading dimension of new supply in response to rising incomes.

As also occurred during the earlier growth take-offs of Japan, Korea, and Taiwan, China's urban housing boom was driven by the massive labor reallocation and rapid industrialization reflected in the Lewis model. In addition, the housing privatization reforms of 1998 and the aggressive restructuring of the SOEs through privatization since 1997 had a major impact in changing the *composition* of housing demand across income deciles, while the percentage of urban private-sector employment rose fourfold between 1998 and 2005.

During the boom, housing divides grew wider. Housing prices rose faster than household incomes in most Chinese cities. In major cities like Beijing and Shanghai, housing prices rose two to three times faster than the average real growth rate of disposable income, which was itself growing at 9% per year between 1998 and 2007. The gap between the real rate of growth of housing prices and the real rate of growth of wages was more than seven percentage points. Therefore, housing price-income ratios kept rising and affordability kept deteriorating for the average urban population. As already noted, this was China's first household demand-driven, as opposed to state-driven, housing cycle. Housing prices were now determined as the margin by the highest income (and high-net-worth wealth) households. These groups included a substantial share of the new private entrepreneurs whose firms were benefiting from the large labor reallocation that was taking place through the concurrent industrial reforms.

The housing price indices (HPI) of the Chinese National Bureau of Statistics were under revision because they had a large, widely known downward bias in reporting housing price movements during the boom. Technically, the best housing price index was from Tsinghua University (TU-HPI) based on the sale of newly built housing units in 35 major Chinese cities (Wu, Deng, and Liu, 2011). The national TU-HPI index shows a real price increase rate of 17% per year between 2006-Q1 and 2010-Q4. If the transient negative external price impact of the 2008 financial crisis is removed, the resulting trend for China's real average rate of return on housing investment was more than 20%, which was well above China's implicit cost of capital. "User cost of capital" analyses show that the carrying cost of owning a housing unit in China has been very low (Chen and Wen, 2014; Renaud, Kim, and Cho, 2016, Chapter 9).

The immediate driver of housing prices was the rapid rise of land values fueled by local governments. Nationwide, real constant-quality land values increased by more than 16% per year between 2000-Q1 and 2013-Q2. Land price increases were much higher in major Eastern cities. The average share of land in total housing costs rose from 37% before 2008 to over 60% after 2010 in Beijing (Wu, Gyourko, and Deng, 2010). Such high rates of land price inflation are not unique to China. Land price increases of comparable magnitudes also occurred across major East Asian cities, when peak rates of urbanization and the top growth years of the take-off coincided (Renaud, Kim, and Cho, 2016, Chapter 2).

An important inefficiency and inequity feature of China's housing boom has been the continuously rising and extremely high vacancy rates in major cities. As Chen and Wen state: "According to China's Household Finance Survey of 2014, in 2013 the average vacancy rate in the first-, second- and third-tier cities in China was 21.2%, 21.8% and 23.2% respectively"…" 35.1% of entrepreneurial households own vacant houses, and the majority (29.9%) own several vacant apartments. Among non-entrepreneurial households, a lower ratio of 22.6%

owns vacant units. Overall, the proportion of households holding vacant housing units rises with their income. In the top 10th income decile, 39.7% of households own several vacant houses, which is about 22 percentage points higher than for households in the bottom income quartile." (Chen and Wen, 2014) The reasons for high vacancy rates in mainland China are similar to those behind very high vacancy rates in Taiwan.

The first decade of the twenty-first century has been exceptional: "the average real rate of return to housing capital in China has been on average 20% between 1998 and 2012. In particular, it increased steadily from 18% in 2001 to 26% before the financial crisis year of 2008" (Chen and Wen, 2014). A leading cause of these exceptional returns to capital has already been mentioned: it was the massive reallocation of labor from SOEs to new private firms that benefited from the associated low wage rates and enjoyed high productivity, especially after 1998 when the government kept pressing for faster privatization of SOEs. The share of private employment in total urban employment rose sharply from 16% in 1998 to 65% in 2008, a fourfold rise in a single decade (Chen and Yi, 2014).

The large-scale reallocation of labor to higher-productivity urban jobs is slowing down rapidly now that China has entered its growth transition, and regional labor markets across China have reached their Lewis turning points. Manufacturing wage rates are rising rapidly and a mean reversion of rates of return to capital is also taking place. Another contributing factor to China's labor market and economic growth transitions is the reversal of the demographic dividend with China's aging population. These processes are similar to the structural transitions that occurred decades earlier in Japan, Taiwan, and South Korea, where the end of the massive labor reallocation also marked the end of their very high GDP growth rates. However, slower GDP growth rates should not be confused for a problem: a 1% GDP increase of the Chinese economy in 2018 is very large because it applies to real GDP, meaning a national output that is 31.6 times larger than in 1980. Moreover, since 2012, China's labor force has started declining, which means that a slower GDP growth rate is much less threatening to social stability.

During the Great Housing Boom, the share of total real estate investment (including non-residential real estate) that was 4.2% of GDP in the starting year of housing privatizations had doubled to 8.2% of GDP in 2004, and almost tripled to 12.0% by 2010. With the massive 2009 credit stimulus equal to 12.5% of GDP (and the largest peacetime stimulus in world history), the weight of real estate rose further to 13% of GDP in 2011 and 13.8% in 2012. Such ratios were higher than the peak ratios to GDP observed in Spain when its finance-driven real estate boom turned into a bubble. Also, within total Chinese real estate investment, the share of new housing rose from 65% at the start of the decade to above 70% by 2012.

5.5 Spatial dimensions of the housing boom across the Chinese system of cities

China has overinvested in housing during the period 1998–2014 on an aggregate basis. However, the relationship between housing investment and city economies has varied across the country. The Chinese economy is of continental scale, very diverse, and its spatial integration through massive infrastructure investments only occurred during the housing boom period. Media perceptions of a generalized Chinese real estate bubble were caused by their focus on a selective group of mostly large coastal cities, whereas China's urban system has some 660 cities.

China's spatial diversity offers the possibility to test at the country level the Burns and Grebler cross-country hypothesis that the ratio of housing investment to GDP follows an inverted U-shaped pattern that is linked to the stage of economic development (see Figure 12.3). When the Chinese economy is partitioned into its three main regions, the Burns-Grebler hypothesis holds between the lagging Western provinces, the rapidly growing Central provinces and the much richer and more mature Eastern provinces where core industrialization began. At the national level, the Granger causality between housing investment and GDP is bidirectional. However, leads and lags between housing investment, non-housing investment, and provincial GDPs during the period 1999-Q1 and 2007-Q4 become strikingly different across these three national regions (Chen, Jie, and Aiyong Zhu, 2008).

Until 2006–2007, *national* aggregate data did not definitely show China overinvesting in housing. However, investment patterns differed markedly across the three regions. Powerful housing price booms and quantitative overinvestments occurred in the large, high-income cities of the Eastern region, where a high share of GDP still originates. Meanwhile, the most rapid rates of urban growth were occurring in the medium-size cities of the country, at least until 2000 (Chan, Henderson, and Tsui, 2008).

Evidence at the end of the take-off period suggests a housing price diffusion pattern from large to small cities and from east to west (Chan, Henderson, and Tsui, 2008). Large national developers were playing a significant role in this price diffusion, as they were promoting new types of (higher-income) units across still-fragmented local housing markets. At the end of the housing boom, around 2013, among China's 100 largest banks, regional banks who lent heavily to real estate were growing at very high rates. They were also reporting much higher returns on assets and high profit rates (ROA) than national SOE banks (Cooper, 2012). After 2009, supply–demand imbalances grew in third-tier cities, as local supply was driven by large national and regional developers. Given the overheated state of the real estate sector, the actual strength of the balance sheets of these regional banks is a major issue for bank regulators because this segment of the financial system has become an important factor in the stability of China's urban economy. Spatially, the beginning of a housing correction is revealing a significant excess housing supply in smaller third-tier cities by national and regional developers.

6. Conclusion: urban and housing issues during China's growth transition

The achievements of China over the past four decades during its multiple transitions from plan to market are truly remarkable. The greatest economic and social transformation in China's entire history has occurred during these decades. At the birth of the PRC, China's population was 594.4 million in the 1953 census. The total population expanded from 1,000.2 million in 1982 to 1,334 million by 2010. In 1982, 210.7 million lived in cities (20.9%) and China was considered "under-urbanized" by global norms (Chan, Henderson, and Tsui, 2008). The urbanization process has accelerated during the growth take-off. During the past decade, China's urban system was expanding by more than 15 million newcomers per year through combined increases in city scale, boundary changes, and migrant settlements.

The urban challenge during China's growth transition is that the pressures on cities are rising further at the same time that major changes in the economy are needed. "All current demographic projections agree that China's population will continue to grow for at least another decade" and "most recent projections envisage a peak population of around 1.45 to 1.5 billion, a figure that will be reached around 2025 to 2030" (Peng Xizhe, 2011).

The reservoir of urban expansion remains deep. "If current trends hold, the urban population will expand from about 665 million in 2010 to 926 million in 2025, and [China's urban system] will hit the 1 billion mark by 2030," and China will be 70% urban (Peng, 2011). This means that China's cities will add 335 million people in the present two decades, which is more than the entire population of the United States. About 240–260 million of these new urban residents will be rural-urban migrants. It is projected that there will be 220 cities with more than one million residents by 2025, compared with just 35 in Europe today. Moreover, 24 of these cities will have more than five million residents. The annual demand for new housing to serve new city dwellers is likely to be of the order of 800 million square meters, or about 80% of annual housing output during the years of the Great Housing Boom.

On a purely aggregate national basis, one could assume that this new round of urban population growth will absorb the housing overinvestment of the past decade and lower exceedingly high vacancy rates in the process. However, the magnitude of the mismatches between household incomes and unit characteristics and prices is likely to be large, but it is not known. Also, the scale of the spatial mismatches between the stock and the annual supply of new housing units, and the housing demand across the household income distribution is also unknown. The likelihood of spontaneous, gradual housing market corrections could be lowered because local governments have important ownership stakes or control lower-tier banks. In addition, local government-affiliated entities also own non-bank subsidiaries which are rarely transparent. Incentives to disguise local losses and thereby prevent needed adjustments can be high. Widespread implicit guarantees add to the risks of financial system instability during the risky growth transition (IMF, December 2017).

Moving to a higher quality of urbanization implies improving the moving parts of an already very large urban system whose size is projected to expand further and to reach one billion people by 2030. The challenge is that, since 2009, the quality of Chinese urbanization has declined visibly in three ways. First, its efficiency in terms of allocation of land, labor, and capital is declining as scarce urban resources are increasingly misallocated (through government reliance on credit-driven growth). Second, the benefits of urbanization are not accessible equally among all citizens, and China now has the most unequal income distribution in Asia, and one of the worst in the world (IMF, 2017b). Third, environmental sustainability in cities and surrounding rural areas is now a national concern. The air, water, and solid wastes generated by cities grow steadily, and a growing number of citizens are exposed to environmental hazards.

To end a growing resource misallocation, reduce inequality of access to urban benefits, and achieve environmental sustainability, China faces the three clusters of issues identified in Section 4.4. As if these clusters of issues were not enough policy challenges by themselves, there remains the question of what can be done to improve housing performance, nationally and locally. Fortunately, these (separable) housing reforms are the focus of the roadmap for housing policy discussed in Chapter 16, which is the centerpiece of this book.

Notes

1 International Monetary Fund data come from the "Article IV Consultations," which are statutory reviews of IMF member countries' economic performance. Further vetting of data with Chinese officials occurs during discussions of additional IMF reports on China.
2 A rare estimation of the composition of real estate asset value for the US in 1989 was: housing, 79%; offices, 8.5%; retail, 7%: warehousing and industrial, 5.5% (Hartzell et al., 1994). In spite of the larger role that the share of land plays in real estate values in China and the rest of East Asia, we can think of a notional composition of real estate assets in high-income East Asia as being: 80% residential, 15% office and retail real estates, and 5% other real estate.
3 As Krugman (1995, Chapter 2) reminds us, von Thünen started a great German tradition of spatial analysis and economic geography that lasted until the mid-twentieth century. However, this work was totally ignored by the dominant Anglo-Saxon school of economics, for which space did not exist. It took until 1966 – i.e. 140 years – for von Thünen's masterwork to be translated into English. Fortunately, studies of Chinese cities have drawn on these German models, especially on Walter Christaller's central place theory of 1933.
4 Two valuable references to understand the intuitions and the analytical approach behind Krugman's model, for which he was awarded a Nobel Prize in 2008, are his Nobel Prize Lecture on "The Increasing Returns Revolution in Trade and Geography" of December 2008, and his earlier 1995 monograph on *Development, Geography and Economic Theory*, which details the analytical antecedents of his model and also includes an appendix explaining the model itself: the basic approach; the assumptions; short-term equilibrium; how the model deals with centripetal and centrifugal location forces and the conditions for the sustainability of industrial concentration; and the dynamics of the multi-location model resulting in the economy as a self-organizing system of cities.
5 There is a parallel between the US in the late 1920s at the end of its high growth take-off years and the rapidly rising concerns about the stability of China's financial

system at the end of its own high growth take-off stage. China's financial system has facilitated these high growth rates, especially during the decade of credit-driven growth since 2009. Like the US financial system in the 1920s, China's financial system has developed rapidly. The concerns are about substantial credit expansion resulting in high corporate debt, indebted local governments, and rapidly rising household indebtedness (IMF, 2017; 2018). Credit intensity has increased "above levels consistent with a high probability of financial distress" (Figure 6, p. 20, IMF, 2017).

6 This study is also known as the Spence Report after the name of its director, Economics Nobel laureate Michael Spence.
7 Vietnam is the only other country belonging to the East Asian civilization, but it could not be included in the present comparative analysis for lack of data.
8 Chinese society went through multiple transitions from plan to market during the take-off decades 1980–2015. These very successful decades followed the instability of the administrative-command decades of 1949–1978 (Naughton, 2007). These rapid transitions generated intense debates between conservatives and reformers about the future of China's economy and society. In 1992, economist Wu Jinglian proposed that China be officially defined as a "socialist market economy", which was accepted and continues today (Wu, Jinglian, 2013, Chapter 19).
9 For a review of recurring features of financial repression, see Beim and Calomiris (2001, Chapter 2).
10 These seemingly "arms-length" municipal investment companies are variously called in English "local government financing vehicles" (LGFV), "local government investment vehicles" (LGFIV), or "local government financing platforms" (LGFP).

References

Beim, D. O. and Calomiris, C. (2001). *Emerging Financial Markets*. New York: McGraw-Hill.
Burns, L. and Grebler, L. (1978). *The Housing of Nations: Analysis and Policy in a Comparative Framework*. New York: John Wiley and Sons.
Brülhart, M. and Sbergami, F. (2009). Agglomeration and Growth: Cross-Country Evidence. *Journal of Urban Economics*, 66(1), 48–63.
Chan, K., Henderson, V., and Tsui, K. (2008). Spatial Dimensions of Chinese Economic Development. In L. Brandt and T. Rawksi (Eds.). *China's Great Economic Transformation*. Cambridge University Press, pp. 776–828.
Chen, J. and Zhu, A. (2008). *The Relationship between Housing Investment and Economic Growth in China: A Panel Analysis Using Quarterly Provincial Data*, WP 2008/17. Uppsala: Uppsala University, www.nek.uu.se/
Chen, K. and Wen, Y. (2014). *The Great Housing Boom of China*, Working Paper 2104–022A. St. Louis, MO: Federal Reserve Bank of St. Louis.
Cooper, J. (2012). Are China's Banks Heading for a Crisis? *The Banker*, April 2, London.
Deng, Y. and Fei, P. (2008) The Emerging Mortgage Market in China. In D. Ben-Shahar, C. K. Y. Leung, and S. E. Ong (Eds.). *Mortgage Markets Worldwide*. Chichester, Sussex, UK: Blackwell Publishing.
Duranton, G. (2014). *The Urbanization and Development Puzzle*. In S. Yusuf (Ed.). *The Buzz in Cities: New Economic Thinking*. Washington DC: The Growth Dialogue. www.growthdialogue.org
———. (2015). Growing Through Cities in Developing Countries. *World Bank Research Observer*, 30(1), February, pp. 39–73.

Eichengreen, B., Park, D., and Shin, K. (2013). Growth slowdowns redux: new evidence on the middle-income trap. NBER Working Paper 8673. Cambridge, MA, www.nber.org/papers/w18673.

Eichengreen, B. (2014). The Tyranny of Economic Growth, http://Caixin.com, May 9.

Fukuyama, F. (2011).The *Origins of Political Order: From Prehuman Times to the French Revolution*. New York: Farrar, Strauss and Giroux.

———. (2014). *Political Order and Political Decay: From the Industrial Revolution to the Globalization of Democracy*. New York: Farrar, Strauss and Giroux.

Hardy, N. (1998). *China's Unfinished Economic Revolution*. Washington D.C.: Brookings Institution Press.

Hartzell, D., Pittman, R., and Downs, D. (1994). An Updated Look at the Size of the U.S. Real Estate Market Portfolio. *Journal of Real Estate Research*, 9(2), 197–212.

Hsing, Y. (2010). *The Great Urban Transformation: Politics of Land and Property in China*. Oxford University Press.

International Commission on Growth and Development. (2008). *The Growth Report: Strategies for Sustained Growth and Inclusive Development*, Washington D.C.: World Bank.

International Monetary Fund. (2017), People's Republic of China, Financial System Stability Assessment, IMF Country Report cr17/358, December 6, 2017.

———, (2018), People's Republic China, 2018 Article IV Consultation, IMF Country Report cr18/240, June 28, 2018.

Jorgenson, D. (1961). The Development of a Dual Economy. *Economic Journal*, 71, 309–334.

Kim, S. (1998). Economic Integration and Convergence: US Regions, 1840–1987. *Journal of Economic History*, 58(3).

Knight, J., Deng, Q., and Li, S. (2010). The Puzzle of Migrant Labor Shortage and Rural Labor Surplus in China. Oxford University Department of Economics, Discussion Paper 494.

Kornai, J. (1992). *The Socialist System: The Political Economy of Communism*. New Jersey: Princeton University Press.

Krugman, P. (1991) Increasing Returns and Economic Geography. *Journal of Political Economy*, 999: 183-199.

——— (1995). *Development, Geography, and Economic Theory (Ohlin Lectures)*. Cambridge, MA: MIT Press.

——— (1998). *The Increasing Returns Revolution in Trade and Geography*. Nobel Prize Lecture, Stockholm.

———, Masahisa Fujita and Anthony Venables. (1999). *The Spatial Economy – Cities, Regions and International Trade*. Cambridge, MA: MIT Press.

Lardy, N. (1998). *China's Unfinished Revolution*. Washington D.C.: Brookings Institution Press.

——— (2014). *Markets over Mao: The Rise of Private Business in China*. Washington D.C.: Peterson Institute of International Economics Press.

Leamer, E. E. (2007). Housing is the Business Cycle. NBER Working Paper No. 13428, September. www.nber.org/papers/w13428

Lee, I., Syed, M., and Liu, X. (2012). *Is China Over-investing and Does it Matter?* IMF Working Paper WP/12/277. Washington, D.C.

Lewis, A. (1954). Economic Development with Unlimited Supplies of Labor. *The Manchester School*. 22(2), 139–191.

Lockwood, W. (1954). *The Economic Development of Japan, Growth and Structural Change 1868–1938*. Princeton New Jersey: Princeton University Press.

Maddison, A. (2001). *The World Economy: A Millennial Perspective*. Paris: OECD. www.oecd.org/dev/developmentcentrestudiestheworldeconomyamillennialperspective.htm

Malpezzi, S. (1999). Economic Analysis of Housing Markets in Developing and Transition Economies, section 2.3.1 on early cross-country research, p. 1806 in Peter Nijkamp, Edwin S. Mills, P.C. Cheshire (1999). *Handbook of Regional and Urban Economic, Vol. 3 Applied Urban Economics*. North Holland Publishing.

Man, J. Y. and Hong, Y. (Eds.). (2011). *China's Local Public Finance in Transition*. Cambridge, MA: Lincoln Institute of Land Policy Press.

Mills, E. S. and Ohta, K. (1976). Urbanization and Urban Problems. In H. Patrick and H. Rosowski (Eds). *Asia's New Giant: How the Japanese Economy Works*. Washington DC: The Brookings Institution.

Muellbauer, J. (2012). When is a Housing Market Overheated enough to Threaten Stability? Conference on Property Markets and Financial Stability, Reserve Bank of Australia, Sydney, August.

Naughton, B. (2007). *The Chinese Economy: Transitions and Growth*. Cambridge, MA: MIT Press.

Peng, X. (2011). China's Demographic History and Future Challenges. *Science*, Vol. 333, July 29, 581–587.

Phang, S. (2014). *Superstar Cities, Inequality and Housing Policy*. Celia Moh Lecture, Singapore Management University. www.smu.edu.sg/news/2015/03/26/income-inequality-and-housing-policies-superstar-cities

Pritchett, L. (2000). Understanding Patterns of Economic Growth: Searching for Hills among Plateaus, Mountains and Plains. *World Bank Economic Review*, 14(2), 221–250.

Renaud, B., Kim, K., and Cho, M. (2016). *Dynamics of Housing in East Asia*. Oxford: John Wiley/Blackwell.

Rodrik, D. (1994), Getting interventions right: how South Korea and Taiwan grew rich. Cambridge, MA: NBER Working Paper 4964.

———. (2005). Growth strategies. In P. Aghion and S. Durlauf (Eds.). *Handbook of Economic Growth*. Amsterdam: Elsevier.

———. (2015). *Economics Rules: The Rights and Wrongs of the Dismal Science*. New York: W.W. Norton and Company.

Skinner, G. W. (1964). Markets and Social Structure in China. *Journal of Asian Studies*, 24(1), 3–43.

———, (Ed.) (1977). *The City in Late Imperial China*. Stanford, CA: Stanford University Press.

Stiglitz, J, Sen, A., and Fitoussi, J. P. (2014). *Mismeasuring Our Lives: Why GDP Doesn't Add Up*. Paris: Commission on the Measurement of Economic Performance and Social Progress.

Vogel, E.. (2011). *Deng Xiaoping and the Transformation of China*. Cambridge, MA: Harvard University Press.

Von Thünen, J. (1826). *Der Isolierte Staat in Beziehung auf Landwirtschaft und Nationalökonomie*. (Translated by Carla M. Wartenberg in 1966). New York: Pergamon Press.

Wade, R. (2004). *Governing the Market* (second paperback edition of 1990 original). Princeton, NJ: Princeton University Press.

Wu, J. (1992). A Suggestion that we Adopt the Authorized Expression (*Tifa*) 'Socialist Market Economy'. In *Wu Jinglian, Voice of Reform*, edited with introductions by Barry Naughton, MIT Press, 2013.

Wu, J., Gyourko, J., and Deng, Y. (2010). Evaluating Conditions in Major Chinese Housing Markets. NBER Working Paper 16189.

Wu, J., Deng, Y., and Liu, H. (2011). House price index construction in a nascent housing market: the case of China. IRES Working Paper IRES 2011–2017. Singapore: National University of Singapore.

Yu, W. G. (2015). *The Chinese Market Economy, 1000–1500*. New York: SUNY Press.

Zhu, R. (2013). *Zhu Rongji on the Record, the Road to Reform 1991–1997*. Washington D.C.: Brookings Institution Press, Documents 47, 48, 53, 54.

13 Urban housing challenges in mainland China

Zhi Liu

LINCOLN INSTITUTE OF LAND POLICY AND PEKING UNIVERSITY-LINCOLN INSTITUTE CENTER FOR URBAN DEVELOPMENT AND LAND POLICY

Yan Liu

PEKING UNIVERSITY-LINCOLN INSTITUTE CENTER FOR URBAN DEVELOPMENT AND LAND POLICY AND PEKING UNIVERSITY

1. Introduction

After the implementation of a series of step-by-step reform actions during 1988–1998, China completed a transition of urban housing supply from a welfare provision system to a predominantly market-based system (Wang, 2011; Yang and Chen, 2014). Since then, China has experienced extraordinarily rapid growth in urban housing stock. According to the National Bureau of Statistics of China, gross housing floor area per capita in urban areas nearly doubled during 1998–2015, from 18.7 square meters in 1998 to 37 square meters in 2017. This level is higher than the average for middle-income countries (20 square meters) and close to the average of 28 European Union countries (42.6 square meters in 2011).[1] Today, 90 percent of the Chinese urban households who have an urban *hukou* (rural household registration) own one or more housing units. The quality of public infrastructure services in most urban residential neighborhoods can be considered adequate to high, according to the international standards.

However, the rapid growth of the housing market eventually got out of balance and resulted in a number of unprecedented economic, financial, and social problems. Hidden under the above numbers is a significant disparity in housing consumption between the high- and low-income groups. Soaring housing prices in the commodity housing markets in the first-tier[2] and some of the second-tier cities have attracted housing investment from better-off households both locally and nationally, making commodity housing increasingly unaffordable for local young and new families, not to mention migrant households. The commodity housing market experience in the third-tier and fourth-tier cities exhibits a rather different pattern. Over-supply of commodity housing units and other reasons led to excessive stock and even ghost towns, wasting significant amounts of housing resources, including land, capital, energy, and construction materials. Across urban China, high rates of housing vacancies are widespread.

While multiple-homeowners are concerned with the formation of a housing price bubble, those who could not afford homeownership feel desperate about their future prospects of purchasing a home.

The urban affordable housing programs generally do not cover the migrant workers due to their rural *hukou* status. Most of them live in urban villages (i.e., former rural settlements that are surrounded by urban development), crowded dormitories, underground spaces (Kim, 2014), or rural villages at the urban fringes. Small property rights housing (discussed later), illegally supplied by rural communities for urban residents or migrant workers, has become an alternative for many urban homebuyers despite incomplete property rights.

To understand the critical housing issues and challenges faced by the Chinese government and the urban residents today, this chapter provides a comprehensive overview of the urban housing trajectory, related policy changes, and residents' housing experiences in mainland China from 1998 to 2017. Specifically, it reviews the cycles of the urban commodity housing market, rental housing market, affordable housing policy, and small property rights housing since the 1998 housing marketization reform. It highlights the factors contributing to the rapid growth of commodity housing markets (including economic sector policy, land policy, and municipal finance), and the factors contributing to the housing demand. Finally, it discusses the urban housing challenges to be addressed in future housing policy reform.

2. The golden period and its aftermath

The housing policy reform of 1998 created momentum for the rapid growth of urban commodity housing markets across China. In 2003 the momentum was further fueled by a sector policy that designated the real estate sector as one of the industrial pillars of the national economy. The policy marked the beginning of a 10-year "golden period" or housing boom from 2004 to 2014. This period was also a period of high GDP growth in China (averaging over 10 percent a year), which was followed by a new period known as the New Normal, with the national GDP growth rate declining to below 7 percent a year.

Sector policy is a policy instrument that the central government uses to drive economic growth. The government selects a production sector that has strong linkages to other production and service sectors and develops it into a pillar of the national economy, in order to drive the development of a large number of sectors. In 1994, for example, the central government designated the automobile industry as a key industry and issued policy directives to encourage households to purchase private cars. The policy triggered the rapid development of the automobile manufacturing industry, as well as car loan and insurance businesses.

The real estate sector policy greatly encouraged the local governments to support and facilitate real estate development through easy credit, lax land supply, and faster approval. Between 2004 and 2014, the share of residential housing investment in GDP rose from less than 2 percent to 10 percent (Figure 13.1), the annual housing gross floor area sold increased from 340 million square meters to

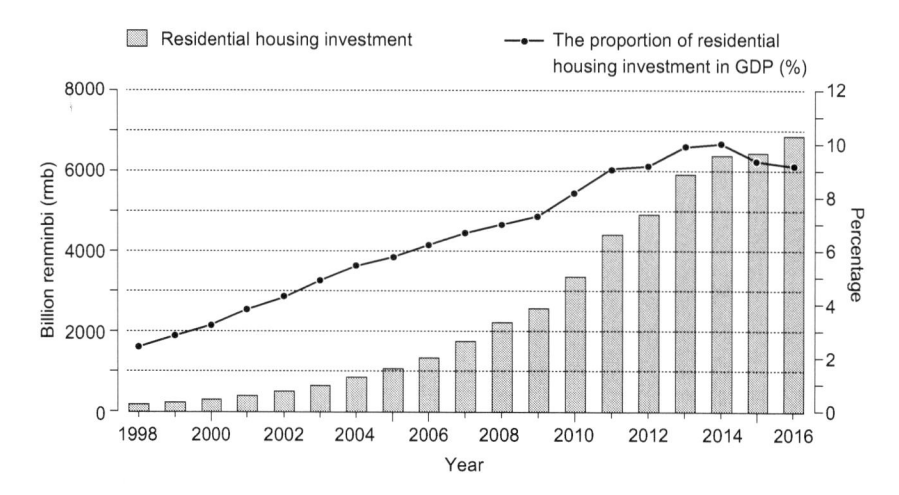

Figure 13.1 Residential housing investment in China, 1998–2016.
Source: China Statistical Yearbook, various years.

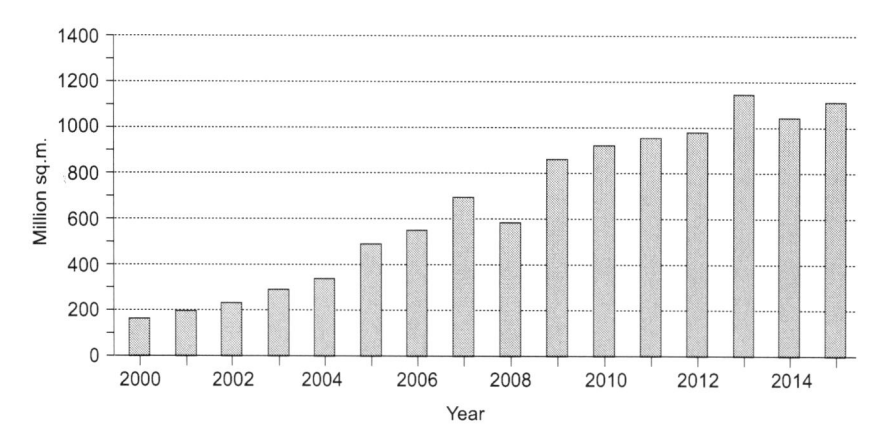

Figure 13.2 Sales of residential housing in square meters, 2000–2015.
Source: China Statistical Yearbook, various years.

over one billion square meters (Figure 13.2), and the average gross housing floor area per capita increased from 26 to 34 square meters.[3]

However, two major problems—rising housing prices and growing housing vacancies—followed. Average housing price indexes for all cities increased by 137 percent during 2004–2014, averaging 9 percent a year (Figure 13.3). Fearing the formation of housing price bubbles from time to time, the central government directed the local governments to implement a series of so-called macro-adjustment

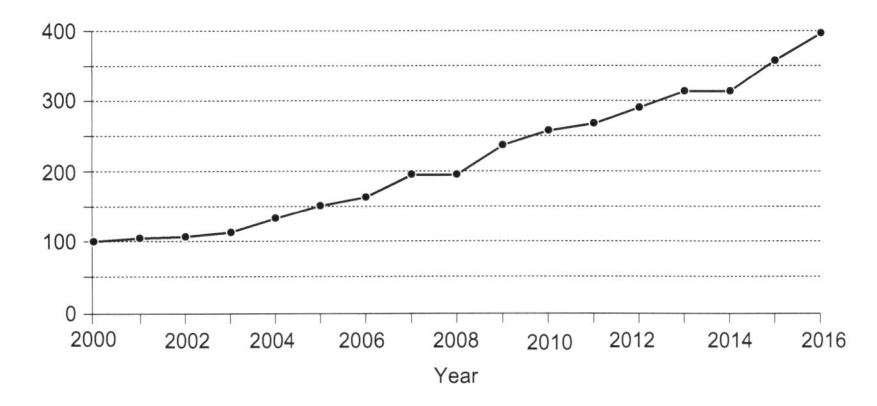

Figure 13.3 Average residential housing price index for 35 major cities, 2000–2016.
Source: China Statistical Yearbook, various years.

measures—such as limiting the supply of land and credit, increasing down-payment requirements, or limiting the eligibility of housing buyers—to cool down the market.[4] When the housing market deflated, the central government responded by directing the local governments to relax the supply of credit, lower the required share of down payments, or lift the eligibility limits of buyers. This kind of back-and-forth adjustment characterizes the real estate market development over the 10-year golden period. Despite the macro-adjustments from time to time, the commodity housing market and prices have grown rapidly.

The average price-to-income ratio for 35 major cities remained at a moderate level in the early 2000s; then it experienced a dramatic rise and reached a high point of 9.2 in 2010, before leveling off around 8.7 (Figure 13.4). The increases in the price-to-income ratio were particularly sharp in the first-tier cities and some of the second-tier cities. The ratios in some of the first- and second-tier cities are reported to have reached over 20 in 2016.[5] This trend significantly reduced housing affordability. In response, the government planned to provide 36 million housing units to the low-income households over the 12th Five-Year Plan period (2011–2015).

China has not carried out a housing census since 1984; thus, no census data exist for housing vacancies. However, available random surveys reveal high levels of vacancies. A nationwide household survey indicated that the urban housing vacancy rate reached 22.4 percent in 2013;[6] many households owned more than one housing unit, and many of these additional units remained vacant (Gan, Yin, and Tan, 2014). Another nationwide housing market survey, carried out by the IT firm Tencent in May 2015, found the housing vacancy rates among the major cities to be 22–26 percent (Tencent Housing, 2015). As China does not yet levy a property tax on the ownership of private residential properties, the carrying cost for a vacant housing property is minimal.

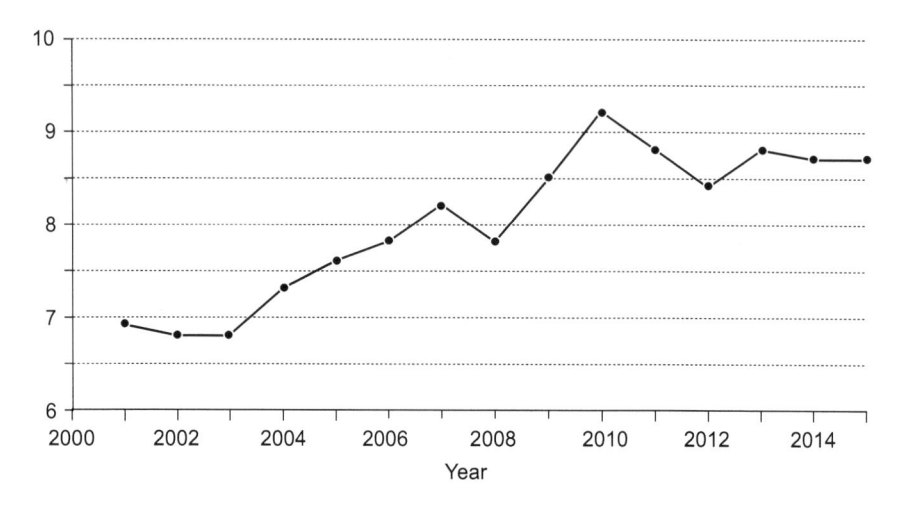

Figure 13.4 Average price-to-income ratio for 35 major cities, 2001–2015.
Source: E-house Real Estate Institute (2015) and China Statistical Yearbook, 2016.

Table 13.1 Housing for sale in China, 2011–2015.

	Area for sale (million square meters)					Increase over the previous year (%)				
	2011	2012	2013	2014	2015	2011	2012	2013	2014	2015
Residential housing	181	236	324	407	452	35.8	39.7	37.2	25.6	11.2
Office building	13	16	20	26	33	10.8	26.4	25.7	34.4	24.7
Commercial business housing	61	71	93	118	147	11.9	20.9	31.1	26	24.6

Source: National Bureau of Statistics of China, various years.

The golden period came to an end in 2014. Excessive new housing supply (i.e., newly built but unsold housing units) became a widespread phenomenon across China. Table 13.1 reports the real estate stocks (including residential housing, office buildings, and commercial housing) according to the data released by the National Bureau of Statistics of China. The annual residential housing floor area for sales increased substantially during 2012–2015, suggesting a rapid accumulation of housing stock. Many cities struggled to absorb the excessive supply via demand-stimulation measures.

But the commodity housing markets have also become increasingly polarized. While housing demand in the first-tier cities remains strong and the prices have continued to increase strongly over the last three years, demand in the lower-tier

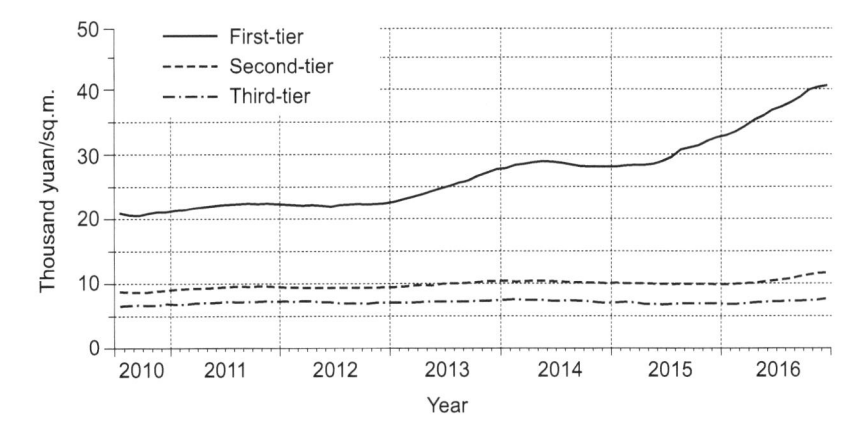

Figure 13.5 Residential housing prices for 100 cities since mid-2010.
Source: Wind Database based on data released by China Index Academy (http://fdc.fang.com/index/BaiChengIndex.html).

cities grows little, and so do the prices (see Figure 13.5). The continuing increases of housing prices in the first-tier cities could be explained by the fact that these cities are the most economically vibrant cities in China, with superior public services and other urban amenities that attract household investment not only from the residents of the cities themselves but also from other cities. In comparison, the third- and fourth-tier cities attract much less household investment.

3. Impacts of land-based finance on housing markets

The housing market development is highly related to China's land policy and municipal finance (Liu and Wang, 2014). By the Constitution, urban land is state owned, and rural land is collectively owned by the villages. There is no private ownership of land. The Land Administration Law further stipulates that the conversion of rural land for urban use requires the transfer of land ownership from the village collectives to the state, and only the state has the power to convert rural land to urban land through expropriation. At the city level, the state is represented by the municipal governments.

For urban expansion, a municipal government would expropriate rural land, service the land with infrastructure, and sell the use rights of the land for residential and commercial real estate development, or allocate the land for public use. The compensation to the villages for the farmland taken is based on its agricultural value. The sales of use rights for residential and commercial land are carried out through competitive tendering. The sale prices are often many times higher than the compensated values of rural land, providing a great source of revenue to the municipal governments.

The Chinese municipal governments are responsible not only for the provision and management of municipal services, but also for local economic development. In fact, the latter is always given a top priority, as GDP growth has long been a primary performance indicator of the governments at all levels. Public investment in urban expansion and infrastructure becomes an important tool for the municipal governments to use to promote local economic growth.

The fiscal means available to municipal governments are limited, however. The current tax-sharing system was established in 1994. The central government determines all tax rates and tax assignments (i.e., which taxes belong to the center, which belong to the local governments, and which are shared between the central and local levels). Local governments essentially possess no tax autonomy. And until recently they were not allowed to borrow directly from commercial banks. However, municipal governments found various ways to raise additional funds, including, most significantly, sales of land-use rights for residential and commercial real estate development, and borrowing from commercial banks indirectly through municipal government-owned urban development investment corporations, often using land as collateral.

The sales of land-use rights generate a significant amount of revenue to municipal governments. As the data in Figure 13.6 indicate, revenue from land concessions often amounted to 40–60 percent of the tax revenue of the local governments during 2001–2016. Most (if not all) of the land concession revenue was spent for capital investment in urban infrastructure. The ability of municipal governments to mobilize land-based finance, which includes land concession revenue and commercial borrowing with land as collateral, is a key factor behind the rapid urban development and redevelopment in China over the last 20 years.

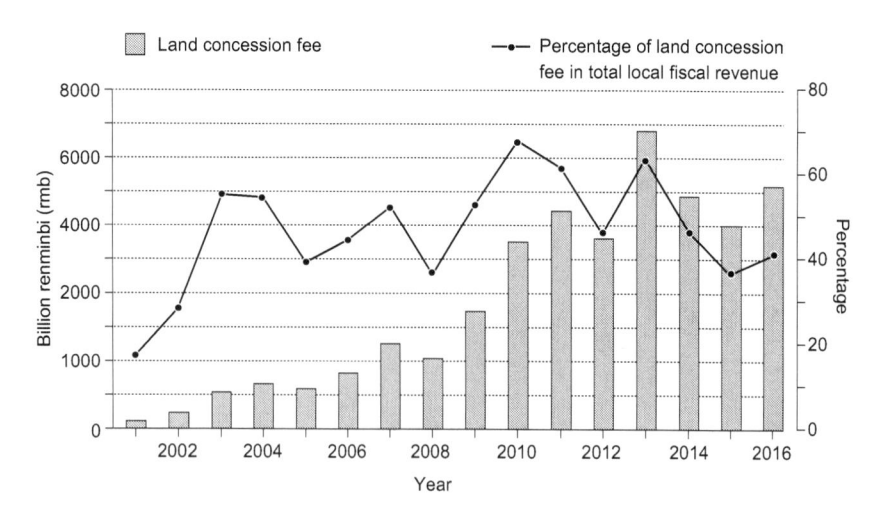

Figure 13.6 Land concession revenue and local tax revenue, 2001–2016.
Source: China Statistical Yearbook, various years.

The sales of land-use rights for residential and commercial real estate development are only part of the land operations by cities. In competition for investment and employment, most cities offer sizable amounts of industrial land at very low prices, or even free of charge, to attract manufacturing firms. Municipal governments believe that the newly attracted manufacturing firms would bring in jobs, raise GDP, and increase local tax revenues, and that the new jobs would generate new demand for housing and services, the expanded supply of which would generate further local GDP and local tax revenues. The costs of subsidizing industrial land are covered by the revenue from sales of land-use rights for residential and commercial real estate development. The prices of commercial and residential land are many times (ranging from 5 to 15) higher than the industrial land prices (Liu and Wang, 2014).

Land operations at the city level are also heavily constrained by the national policy for farmland preservation. The policy requires that a national total of 1.8 billion *mu* of basic farmland be preserved for food security purposes.[7] Under the policy, each province must maintain a fixed amount of basic farmland. When a certain amount of basic farmland is taken for urban development, the province must generate the same amount of basic farmland elsewhere within the province. This strict policy significantly limits the overall supply of land for urban development.

The supply of land for residential and commercial real estate development is even more limited. A significant part of the newly converted land for urban development is allocated for other purposes, including industry, public infrastructure, public administration, and affordable housing, which generate little or no revenue to municipal governments. There is just a small fraction (often about 15–20 percent) of new land left for commercial and private residential development. As the market expects less and less land for residential and commercial land development in the future, the prices for residential and commercial land soar. While this mechanism generates significant amounts of land revenues to the municipal governments, it also contributes to the high prices of commodity housing. It is widely believed today that the cost of land accounts for at least 30 percent of the final sales price of commodity housing in the major cities. The high prices attract housing investment from the high-income households and induce the sector to supply more housing units every year.

It is clear that China's unique land policy and municipal finance are important factors shaping the peculiar behavior of commodity housing supply. In November 2013 the central government initiated a new round of economic reform, which includes reform actions in land policy and municipal finance. Land policy reform has aimed at allowing the village collectives to sell the use rights of their rural construction land directly to developers for urban use, under the premise that the intended urban land use conforms to urban planning. The reform will likely break the monopoly supply of urban land by the municipal governments and reduce the heavy reliance of the municipal governments on land revenues. The introduction of a property tax on the ownership of private residential properties is part of the municipal finance reform. It is hoped that the property tax will become a sustainable source of municipal revenue and also play a role in reducing housing vacancies.

4. Factors influencing housing demand

The strong demand for commodity housing in China is driven by a number of factors, some of which are common, while others are unique. Among the common factors are per capita income growth, urbanization, and change of family size. According to the World Bank database, China's per capita GDP in current US dollars grew from US$959 in 2000 to US$8,827 in 2017. Even with a real GDP growth rate of 6 percent a year under the New Normal scenario, China's per capita GDP will reach US$15,000 in about 10 years. China's urban population increased steadily from 2005 to 2015 (Figure 13.7). Today, the urbanization rate has reached 58 percent (National Bureau of Statistics of China, 2017). It is expected to reach 70 percent in 2030, according to the National Population Development Plan of 2016–2030 issued by the State Council. Although most of the rural migrant households to cities are unable to afford homeownership in the first- and second-tier cities, some of them are able to purchase homes in the third- and fourth-tier cities. The average household size has decreased from 4.03 in 1988 to 3.11 in 2017 according to the data from *China Statistical Yearbooks* (National Bureau of Statistics of China, various years); there are more and more mini-households with one or two persons. The decline of household size not only creates more housing demand, but also changes the structure of housing demand.

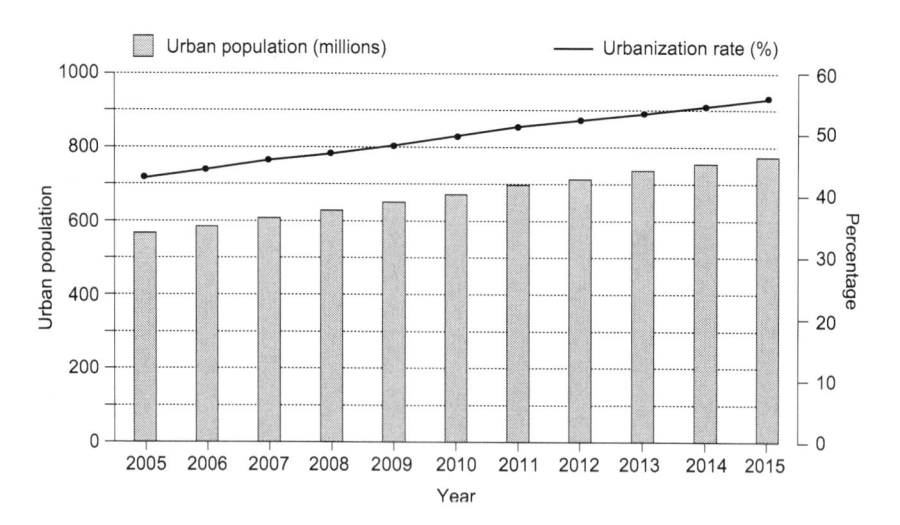

Figure 13.7 Population and urbanization rate in China, 2005–2015.
Source: China Statistical Yearbook, various years.

Chinese urban households display a very strong preference to invest in housing properties. Just as elsewhere, growing housing prices tend to attract demand for housing as a means of household investment. But in the context of mainland China, many households purchase housing units as the only viable means to household investment. The saving rates of urban households, at 25–30 percent of their incomes, are high (Xie and Mo, 2015); the high-income households save a lot more, over 50 percent (Gan, Yin, and Tan, 2014). The propensity for household savings is high in China, mainly because the social security system is far from strong, and households generally expect that the costs of education, medical care, and aged care will rise. At present, however, there are very limited viable investment options for households to keep the value of their savings. The Chinese currency, RMB, cannot be easily converted to international currencies; thus, investing savings overseas is not feasible for most households. Interest rates paid to domestic savings accounts with the commercial banks are low. The stock market is poorly regulated and not transparent, and its overall performance for the last decade has generally been money-losing for most investors, despite the continuing growth of the economy. Under this circumstance, and with few other choices, households tend to park their savings in housing properties. This behavior is further encouraged by the absence of a property tax on the ownership of private residential properties.

Finally, households expect that the housing prices will continue to grow in real terms in the future, due to expected factors such as continuing urbanization, limited land supply, high population density, and improvement of urban public services. The soaring housing prices reinforce this expectation. The situation creates a strong incentive for parents to save more and purchase houses for their children ahead of time, often long before they reach adulthood or the age of marriage. This is particularly true for the parents of boys (Wei and Zhang, 2011). As there are more boys than girls, boys having a ready housing unit would be more competitive in finding a bride.

5. Rental market development

The rental housing market should have played an increasing role amid the rising housing prices. However, China's rental housing markets are poorly regulated and managed, with inadequate and unsecured tenant rights. According to a national population sample survey carried out in 2015, 90 percent of the rental housing units are supplied by individual homeowners, and less than 10 percent by the specialized leasing agencies (National Bureau of Statistics of China, 2015). There is little incentive for real estate developers to supply rental housing due to its much longer payback period, compared to the highly profitable sales housing market.

According to the Sixth National Population Census (National Bureau of Statistics of China, 2010), the share of urban households in rental housing in China was 27 percent, significantly lower than the levels seen in European countries (often over 40 percent). The bigger the city, the higher the share. The

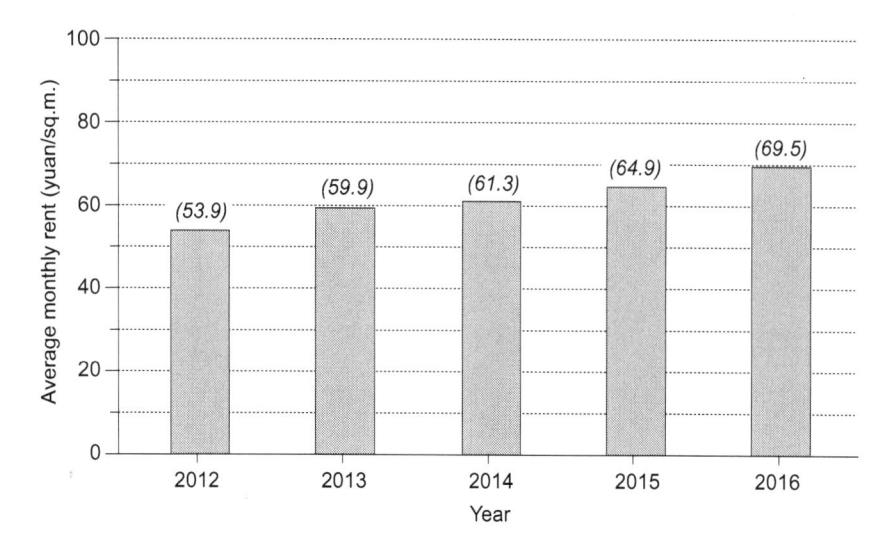

```
100 ─
 80 ─                                                        (69.5)
                                              (64.9)
                                   (61.3)
                       (59.9)
 60 ─    (53.9)
 40 ─
 20 ─
  0 ─
        2012     2013     2014     2015     2016
                         Year
```

Figure 13.8 Average monthly rent in Beijing, 2012–2016.
Source: HomeLink Research Institute (2017).

share is particularly high in cities that attract a large influx of migrant workers. For example, the share was 37 percent in Beijing, 38 percent in Shanghai, and 47 percent in the cities of Guangdong Province. In contrast, the shares in the cities of less prosperous provinces were lower, e.g., 24 percent in Sichuan Province, 19 percent in Hebei Province, and 13 percent in Liaoning Province. A population sample survey in 2015 found that the share of urban households in rental housing was 20 percent nationwide, 34 percent in Beijing, 33 percent in Shanghai, 15 percent in Tianjin, and 18 percent in Chongqing (National Bureau of Statistics of China, 2015). The two sets of survey data indicate that, between 2010 and 2015, the share of urban households in rental housing declined nationwide, and in Beijing and Shanghai.

Rents have increased steadily, but at much slower rates than the increases of housing prices. According to data from 35 major cities, the average monthly rent per square meter increased from 15.75 yuan in 1999 to 28.18 yuan in 2016, with an average growth rate of 3.5 percent per year (Zhang and Zhou, 2016). Figure 13.8 shows the monthly rents of Beijing between 2012 and 2016; the levels of monthly rents in Beijing are more than twice as much as the 35-city average, and the growth rate is 6.6 percent a year.

Within a major city, the rental housing market contains several segments. At the higher end is the market segment for the very high-income and expatriate communities. This market segment is generally well managed. In the middle range are a large number of rental housing units for the skilled labor and the young middle-income households. Many low-income individuals, including those who just graduated from college, choose to share a rental unit in this segment of the rental market. At the lower end are the urban villages—formally suburban rural settlements that are increasingly surrounded by urban spatial expansion. Urban villages have become the homes of migrant workers and households. As the settlements are not built according to urban standards, the housing and neighborhood conditions are generally poor; thus the low rents are relatively affordable for the low-income groups. In the worst cases, the very poor migrant workers reside in informal units using underground spaces (Kim, 2014).

Inadequate regulations and weak enforcement of rental contracts are the main problems of rental housing markets. According to a social survey carried out by the China Institute of Real Estate Appraisers and Agents (2016), up to 80 percent of the rental contracts either have a contract duration of less than a year, or no specified duration at all; nearly 60 percent of renters move from one rental unit to another within a year. Many renters complain that the owners discontinue the rental contracts at will. Another well-known problem is the lack of equal access to urban public services. Until very recently, tenants were not given the same rights to urban public services as the owners have. For example, the children of tenants were not given access to local public schools.

These problems are long-standing and well recognized. The governments are now taking measures to support the healthy development of rental housing markets, such as improving regulations, increasing land supply for rental housing construction, and encouraging rental housing business. In July 2017, the local government of Guangzhou issued a new regulation that gives tenant families the same rights to local education as homeowners have. Before this measure, the right to attend the local public schools was limited to the homeowners in most Chinese cities, and Guangzhou became the first city to grant equal rights to tenants who hold a Guangzhou *hukou* or a skilled worker certificate, allowing them to enroll their children in the elementary and middle schools.

In order to promote public rental housing development, in 2017 Shanghai tendered two new parcels of land designated only for rental housing development. Due to the restriction of land use to rental housing, the concession fee was expected to be lower than the land for commodity housing development. This measure is considered a pioneering way to promote the supply of rental housing.

In August 2017 the central government launched a pilot program in 13 major cities, including Beijing, Shanghai, and Guangzhou, to build public rental housing on rural land. The program encourages rural village collectives to build housing on their rural construction land and lease it to urban residents. It is expected that the pilot program will pave the way for the direct supply of public rental housing on rural land, without going through government expropriation of rural land.

At present, the rental housing available from the urban villages provides shelter for a large number of migrant workers and households in the major cities. However, with continuing urban spatial expansion, the locations of urban villages become increasingly valuable, thus creating incentives for the municipal governments to redevelop the urban villages into modern high-rise residential neighborhoods. Unfortunately, such large-scale urban village redevelopment often results in gentrification (i.e., a spatial process through which the rich move in after the redevelopment) and marginalizes the residential locations of migrant workers to more distant urban villages.

6. Affordable housing policy and implementation

Affordability has been a major policy consideration ever since the housing marketization reform of 1998. The 1998 Notice of the State Council on Further Deepening the Reform of Urban Housing System and Accelerating Housing Construction set broad guidelines for housing provision to different income groups: the housing needs of the lowest-income households (bottom 20 percent) were to be met by affordable rental housing provided by the government; the housing needs of low- to middle-income households (middle 60 percent) were to be met by "economical comfortable housing" provided by the market; and the housing needs of high-income households (top 20 percent) were to be met by commodity housing. "Economical comfortable housing" refers to housing built on land appropriated by the government without charging the land concession fee, with sales prices regulated by the government to ensure affordability. To some extent, this housing is similar to the affordable housing in Hong Kong and the Housing Development Board (HDB) housing in Singapore.

However, the real estate sector policy of 2003 that emphasized the role of commodity housing supply in economic growth overshadowed the affordable housing policy. In 2007, the State Council called for greater actions to meet the housing needs of the low-income groups through redevelopment of dilapidated urban neighborhoods (such as shantytowns). The 12th Five-Year National Social and Economic Development Plan (2010–2015) set an ambitious target of providing 36 million affordable housing units over the five-year period. As the data in Table 13.2 indicate, a total of 28 million affordable housing units were provided during 2011–2015. Although the actual number fell short of the planned target, it was equivalent to 12.3 percent of the total number of urban households in China

Table 13.2 The planned and actual provision of affordable housing in China, 2010–2015.

	Construction (1,000 units)	
	Plan	Actual
2010	—	3,700
2011	10,000	4,320
2012	7,000	5,050
2013	6,300	5,440
2014	4,800	5,515
2015	7,900	7,720

Source: National Bureau of Statistics of China, various years.

(227 million). In other words, 12.3 percent of the urban households should have enjoyed newly provided affordable housing units over the period of 2011–2015. This achievement is certainly highly notable.

Despite the impressive achievement, affordable housing remains a major challenge on China's urban housing agenda. In addition to the mistargeting and poor management that are typically seen in various government assistance programs, two other issues are significant and notable. First, the coverage of the affordable housing program is limited to urban *hukou* holders in most cities. In fact, the migrant workers and their households in cities are in most need of basic housing, but they lack access to the affordable housing program. While a few cities pioneered eligibility schemes (such as the years of working, living, and paying taxes in the city) to provide *hukou* to eligible migrant workers, it would take many years for most migrant workers to become eligible. This is a daunting task as there are 250 million rural-urban migrants in cities today.

The second issue is that many affordable housing units are located in distant locations far from employment centers and without adequate public transport, schools, and other urban services. In the widespread practice of land-based finance, most municipal governments keep land parcels in the premier locations for commodity housing development and select the distant land for affordable housing. To meet the planned target for affordable housing provision, they often quickly complete the construction and leave the public service provision to the future, when more public funds become available. Due to the inconvenient location and poor services, some tenants/owners do not stay in these housing units.

The Housing Provident Fund (HPF) is considered a policy tool to help improve the affordability of housing for urban employees. It is a mandatory saving scheme for housing, launched first in Shanghai in 1992 and later adopted in other prefecture-level municipalities. According to the HPF regulations, all urban employees are required to contribute a percentage (typically 12 percent) of their salaries to the HPF, and the employers must match the contribution. The HPF is managed locally by the prefecture-level municipalities; each municipality collects, manages, and uses its fund within its jurisdiction. Participating employees are not allowed to close their HPF accounts until their mandatory retirement. They are allowed to withdraw for housing down-payment or housing rehabilitation expenditures, or to apply for low-interest home loans available from the HPF. The interest rates of HPF loans are typically two percentage points lower than the interest rates of housing mortgage loans offered by the commercial banks. According to the Ministry of Housing and Urban-Rural Development (MOHURD), by the end of 2016, the number of HPF participants (or depositors) reached 130.6 million, and the outstanding amount reached RMB 4,563 billion; further, 3.3 million participants had applied for HPF home loans in 2016.

By design, the HPF would work well if the housing prices remain low and affordable. Unfortunately, the effectiveness of the HPF was seriously affected by the rapid rise of housing prices (or price-to-income ratios) and the availability of mortgage loans. The only viable function of the HPF for the participants is the possibility of getting low-interest housing loans, but that function does not work well either. The size of the HPF loans is relatively small compared to the total prices of most homes. Few borrowers can finance their homes with just one HPF loan; they have to seek another mortgage loan. Moreover, most of the participants who could afford the down payment (often 20–50 percent of the total price) are the better-off people who already own a housing unit. The low-income participants who cannot afford the down payment do not borrow from HPF. Their mandatory participation in the HPF actually helps expand the pool of the HPF that makes the low-interest loans possible, thus benefiting the higher-income people who borrow from the HPF. This is an undesirable phenomenon of the poor subsidizing the rich, a major problem caused by the rising housing prices. It is thus understandable that low-income participants complain about the HPF loudly, saying that it keeps their money dormant, which would have been better spent on more urgent needs such as paying medical bills.

Another major problem is the limited coverage of the HPF. The 130 million urban employees currently covered by the HPF account for 32 percent of the total urban employees. Expanding the coverage is increasingly difficult. The employers who are able to provide the HPF contribution are government agencies, state-owned enterprises, publicly listed private enterprises, and large private firms. Many small private firms are unable to provide HPF contributions for their employees. Some provide the HPF contribution only for skilled employees. A large number of unskilled workers, rural migrant workers, and self-employed individuals, who belong to the social groups most in need of housing, have not been covered in the HPF system. Reforming the HPF appears to be necessary and urgent, though a daunting task.

7. Small property rights housing

"Small property rights" (SPR) housing is a widespread phenomenon in urban China, especially in the urban fringe areas. It is built by villagers on their collectively owned land in peri-urban areas and urban villages, and sold to urban residents and migrant households. As the Land Administration Law prohibits the urban use of rural land without going through government expropriation, SPR housing is considered illegal. While no official statistics are available, the number of SPR units was estimated at 70 million—perhaps one-quarter of all housing units in urban China (Shen and Tu, 2014).

The genesis of SPR housing is discussed by Shen and Tu (2014) and Sun and Liu (2015). It arose from China's dual management system for urban and rural land. By the Land Administration Law, only the state has the legal power to expropriate rural land and convert it to urban use. Villages have no land development rights. Compensation to villages for expropriated rural land is based on the land's agricultural production value rather than its higher market value for urban use. As the land concession revenue to the government is often many times higher than the compensation to the village, the village desires a share of the land development benefits. Moreover, with the rapid growth of urban housing prices over the last 15 years, villages have realized that they could make a great profit from building and selling houses by themselves. Therefore, a large number of SPR houses have been built.

Sold primarily to individuals without a local *hukou*, SPR housing violates the Land Administration Law. Unlike buyers of legally built homes on urban land, buyers of SPR housing do not receive a property rights certificate from the housing administration agency of the municipal government; they sign only a property purchase contract with the village committee. Because Chinese laymen often see the state as the "big" institution, housing units purchased from village committees are popularly called "small" property rights housing (Sun and Liu, 2015).

The development of SPR housing provides affordable shelters for many lower- and middle-income groups that the government and the market have not been able to provide. The SPR housing units are much less expensive than comparable formal housing units in the same location. Their prices are typically 40–60 percent cheaper because villages do not pay land concession fees as the urban real estate developers do, and do not incur heavy administrative costs for obtaining the planning and construction permits (Sun and Liu, 2015).

As discussed earlier, the rental market in urban China is poorly regulated, and contract enforcement is weak. Tenants often face the risk of unexpected rent hikes and premature termination of leases. In addition, participation in the affordable housing programs run by the municipal governments is not an option for most migrant workers because they do not have local urban *hukou*. Thus, SPR units have become the rational housing choice for many migrant households, and even for some urban households with *hukou* in their city of residence.

Despite the recognition of the positive role of SPR housing, the government maintains that SPR housing is illegal and further development should be stopped. The bigger concern for the government is the impacts of SPR housing development on the housing market, municipal finance, and future urban forms. Additional provision of SPR housing units would weaken formal market demand and lower formal market prices, leading to lower prices of land concessions and lower land concession revenue. Moreover, China's city-planning efforts do not cover rural land outside designated planning areas. The spread of SPR housing in these areas would therefore lead to undesirable urban development patterns. Today, there is no clear way out of the SPR housing dilemma. The land policy reform that aims to give villages land development rights will be the key to solving the problem.

8. Conclusion

China's urban housing market experience is relatively short, if compared with the experiences of OECD (Organisation for Economic Co-operation and Development) countries and East Asian high-income economies. China's experience is also unique. The growth of commodity housing markets has been driven strongly not only by rapid income growth, urbanization, and demographic changes, but also by some unique factors such as real estate sector policy, dual land management systems, land-based finance, and lack of alternative investment opportunities for urban households. These unique factors have also distorted the housing market and contributed to the undesirable outcomes, such as the soaring housing prices due to limited land supply, high vacancies due to the absence of property tax and household investment alternatives, and excessive housing stock due to over-supply.

These problems are now well recognized. In response, President Xi Jinping recently stressed that housing is for living in, not for speculation.[8] A few major housing sector reform measures followed. The central government now calls for greater effort to develop rental housing markets and instructs local governments to develop a housing supply system with multiple supply modalities and measures (instead of only commodity housing and affordable housing). Recognizing the polarization of housing market performance among cities of different tiers, the central government has assigned the primary responsibilities for stabilizing local housing markets to the municipal governments.

More fundamental reforms will be needed. The deepening of reforms in municipal finance, land policy, and the financial sector is expected to help address the urban housing issues. Despite significant public resistance, the central government continues with the legislation process for levying a property tax on the ownership of private residential properties. The property tax is expected to generate own-source revenue to municipal governments and reduce the incentives to households of multi-homeownership and of keeping properties vacant. The tax would also eventually help municipal governments to reduce their heavy reliance on land concession revenues. Land policy reform is moving in the direction of eventually permitting villages, by law, to supply rural land directly for urban use. This reform is expected to help break the monopoly role of municipal governments in urban land supply and reduce the cost of residential land supply. The financial sector reform is moving in the direction of more openness, more liberalization, and more competition. Excessive household investment in multi-homeownerships would be tamed if other investment alternatives were available for households.

Notes

1 European Commission: Housing space per person: https://ec.europa.eu/energy/en/content/housing-space-person (accessed August 27, 2018).
2 When it comes to the housing market, Chinese cities are usually grouped into four or five tiers by the real estate sector and mass media. The criteria are loosely defined on the basis of the city's national ranking, economic significance, and population size. The first-tier cities include Beijing, Shanghai, Guangzhou, and Shenzhen. The second-tier cities include most of the provincial capitals and a few other major cities. The third-tier cities include a number of provincial capital cities and prefecture-level cities. The fourth-tier cities are mainly the county-level cities. The China National Bureau of Statistics monitors the housing prices for 70 cities, which include all first- and second-tier cities and some selected third-tier cities.
3 The data of housing floor area per capita are from the National Bureau of Statistics of China.
4 For more discussion, see Chapter 14 of this volume.

5 China Index Academy (www.industry.fang.com) reports the estimated price-to-income ratios for 35 major cities. In 2016, the ratio was 40.42 in Shenzhen, 24.29 in Shanghai, 24.21 in Xiamen, and 21.35 in Beijing.
6 The survey defines the housing vacancy rate as the ratio of vacant housing units over the total number of housing units. This survey excludes the finished but unsold housing units.
7 The *mu* is an area measurement of farmland in China. One *mu* is equivalent to 666.67 square meters.
8 President Xi Jinping made the statement in an economic working conference of the Central Party Committee in December 2016. He repeated it in his speech at the 19th National Congress of the Communist Party of China on October 18, 2017.

References

China Institute of Real Estate Appraisers and Agents. (2016). Current Issues in the Rental Housing Market in China. China Real Estate Appraisal and Agent Services, December 2016.
E-house Real Estate Institute. (2015). Ranking of Housing Price-Income Ratios among 35 Large and Medium-sized Cities. *Beijing Review*, (46) (in Chinese).
Gan, L., Yin, Z., and Tan J. (2014). *China Household Finance Survey Report 2014*. Chengdu, China: Southwestern University of Finance and Economy Press.
HomeLink Research Institute. (2017). *The Rental Market of Beijing*. The HomeLink Rental Market Research Series (in Chinese).
Kim, A. (2014). Hidden City: Beijing's Subterranean Housing Market. *Land Lines*, October 2014. Cambridge, MA: Lincoln Institute of Land Policy, pp. 2–7.
Liu, Z. and Wang, J. (2014). An Analysis of China's Urbanization, Land and Housing Problems. In L. Wei, S. Min, and S. Tiyan, (Eds.). *Annual Report on the Development of China's New Urbanization*. Beijing, China: Social Sciences Academic Press.
Man, J. Y. (2015). An Evaluation of China's Land Policy and Urban Housing Markets. In G. W. McCarthy, S. Moody, and G. K. Ingram (Eds.). *Land and the Cities*. Cambridge, MA: Lincoln Institute of Land Policy, pp. 277–291.
Man, J. Y. (Ed.) (2011). *China's Housing Reform and Outcome*. Cambridge, MA: Lincoln Institute of Land Policy.
National Statistics Bureau of China (2018). Frequently Asked Questions: Statistics on Prices, www.stats.gov.cn/tjzs/cjwtjd/201308/t20130829_74324.html. Accessed August 27, 2018.
National Bureau of Statistics of China (2015). 1% National Population Sample Survey, www.stats.gov.cn/ztjc/zdtjgz/cydc/. Accessed August 27, 2018.
National Bureau of Statistics of China (2010). Tabulation on the 2010 Population Census of the People's Republic of China (in Chinese), www.stats.gov.cn/tjsj/pcsj/rkpc/6rp/indexch.htm. Accessed August 27, 2018.
National Bureau of Statistics of China (2017). Total population. (in Chinese).(http://data.stats.gov.cn/easyquery.htm?cn=C01. Accessed December 20, 2018.
National Bureau of Statistics of China (Various Years). China Statistical Yearbook. China Statistics Press (in Chinese).
Shen, X. and Tu, F. (2014). *Dealing with "Small Property Rights" in China's Land Market Development: What Can China Learn from Its Past Reforms and the World Experience?* Working paper. Cambridge, MA: Lincoln Institute of Land Policy.

Sun, L. and Liu, Z. (2015). Illegal but Rational: Why Small Property Rights Housing is Big in China. *Land Lines*, July 2015. Cambridge, MA: Lincoln Institute of Land Policy, pp. 15–34.

Tencent Housing. (2015). 2015 Report on an Urban Housing Vacancy Survey in China. http://cangzhou.house.qq.com/a/20150604/043770.htm#p=1. Accessed September 14, 2018 (in Chinese).

Wang, Y. P. (2011). Recent housing reform practices in Chinese cities: Social and spatial implications. In J. Y. Man (Ed.). *China's Housing Reform and Outcomes*. Cambridge, MA: Lincoln Institute of Land Policy, pp. 19–44.

Wei, S. and Zhang, X. (2011). The Competitive Saving Motive: Evidence from Rising Sex Ratios and Savings Rates in China. *Journal of Political Economy*, Vol. 119, No. 3 (June 2011), pp. 511–564.

Xie, S. and Mo, T. (2015). Differences in Corporate Saving Rates in China: Ownership, Monopoly, and Financial Development. *China Economic Review*, (33), 25–34.

Yang, Z. and Chen, J. (2014). *Housing Affordability and Housing Policy in Urban China*. Springer. DOI: 10.1007/978-3-642-54044-8.

Zhang, Z. and Zhou, X. (2016). Rental Housing Market and Rent Bubbles in China: An Empirical Analysis of Panel Data from 35 Major Cities. *Price Theory and Pricing Practice*, (12), 77–81 (in Chinese).

14 Migration and urban housing market dynamics in China

Ya Ping Wang*

GLOBAL CITY FUTURES; GCRF CENTRE FOR SUSTAINABLE, HEALTHY AND LEARNING CITIES AND NEIGHBOURHOODS (SHLC) AND UNIVERSITY OF GLASGOW

1. Introduction

Housing market fluctuations and price inflation are common problems faced by many countries in the world. This has also been the case in China, where in particular the urban housing market shows some distinct features. The Chinese urban housing market was established through a series of housing reform policies that commercialized the socialist welfare housing system in the late 1980s and the 1990s. Ever since its introduction, the urban housing market has been volatile, with short waves of spikes and troughs. To help control this volatility, government policy makers have alternatively used "adjustment and control" (调控) and "market rescuing'" (救市) over the last 15 years. Where market support policies were introduced in one year, the property market would overheat, followed by price inflation. The inflation would lead to adjustment and control policies that often resulted in property market slowdowns and a decline of property development activities and investment. The slowdown of the housing market would normally affect the general performance of the economy in cities and the country. A decline in the economic growth rate would force policy makers to support and rescue the property market through various financial, taxation, and administrative stimuli. The consequence of these alternatives of control and support policies was continuous housing price increases and the fast accumulation of family assets by rich urban residents.

Recently, significant regional variations in the urban housing market have emerged. Housing prices in Beijing, Shanghai, Shenzhen, Guangzhou, and some second-tier cities have experienced hyperinflation, putting these cities among the most expensive places to live (relative to income) on Earth, while in other cities and smaller towns, market stagnation and slow sales of over-supplied housing stock have been prominent features.

Why is the Chinese urban housing market so volatile? What causes urban housing price inflation in some cities but not in others? Why does the property boom in some Chinese cities not bust, as many experts have expected? Answers to these questions are very important in designing the right policies to manage the urban housing market in the future. Some researchers have addressed these questions from the national perspectives of large cities (Li and Chand, 2013; Wu, 2015; Dreger and Zhang, 2013). In this chapter, I focus on the dynamics of local housing markets and the impacts from migration, and look at the county towns where the markets have a simpler structure and the migration patterns are more easily identifiable. Given this understanding of the local market in small towns, the general features of the housing market dynamics at higher levels will be discussed. The arguments presented here emphasize the importance of the linkage of housing markets at different levels in the country, and the relationship between the housing market and migration.

The first section of this chapter provides a short review of urban housing market development in China and highlights the short cycles of hyperinflation and government control efforts. This will be followed by a case study of the dynamics of commercial housing development and the housing market in a county town. Using the findings of this local case study, we will discuss the patterns of migration in the country and their impacts on the urban housing market in general. Finally, the conclusion provides some answers to the above questions. The work on the case study is based on fieldwork and continuous monitoring of the progress of housing development in the chosen town. Interviews with residents, officials, and developers have also been used.

2. Control vs. support policies

Economic and housing reform in China has brought significant changes to housing provision in cities and towns in the last 35 years (Huang and Li, 2014). Based on locally focused reform experiments in the 1980s, national housing reform policies aimed at commercializing urban housing were implemented in the early 1990s. These polices led to large-scale sales of public housing to sitting tenants and the re-establishment of an urban housing market (Huang, 2004; Li, 2000; Wang and Murie, 1999; Wang et al., 2005). In 1998 the most significant housing reform policies were announced; they were subsequently implemented in 2000. The policies banned direct housing distribution by public-sector employers and created a differentiated provision system with three types of housing in recognition of differences in incomes and affordability.[1] Government-supported affordable housing (*jingji shiyong fang*经济适用房) was to be the main source of housing in urban areas. It was to cater to low- to middle-income urban households (about 70 percent of urban households), including most public-sector employees. Commercial housing (*shangpin fang*商品房) would target the top 15 percent of rich households. On the lowest end of the income scale, some 15 percent of families with insufficient financial power to become homeowners would require subsidized rental housing (*lianzu fang*廉租房) (State Council, 1998; Wang and Murie, 2000).

The re-establishment of the urban housing market, however, was easier than maintaining the stability of this market. East Asia entered an unstable period after the financial crisis of 1997. China's exports fell dramatically, and the growth of GDP showed signs of decline for the first time since the early 1990s. The government sought ways to reduce the negative impacts from this international crisis by increasing domestic demand and consumption. The East Asian financial crisis effectively pushed housing provision toward the market, with commercial provision swiftly becoming a dominating sector. Conditions in the urban housing market had changed substantially by 2002. The rapid growth of personal income and increased demand for expensive commercial housing had not been anticipated by policy makers in 1998. Subsequent policies issued by the State Council in 2003 shifted emphasis away from government-supported affordable housing and toward the development of *ordinary commercial housing*.

By 2004, house price inflation had become a major concern in most cities. Because of the gap between commodity housing prices and prevailing low non-market labor wages, the average house price in urban areas was about 15 times the average annual salary. Affording housing was particularly difficult for the young and the poor. The sustained demand from richer households and the serious affordability problems experienced by low-income groups created what was referred to as "structural problems" in the housing market (the mismatch between housing available and the requirements of most urban residents). Faced with a property market bubble and complaints from middle- to low-income urban residents, the government introduced a macro-economic adjustment programme in 2005 (General Office of the State Council, 2005). Policies that aimed at cooling down the urban housing market relied essentially on old-style political, administrative, and land-use planning measures. Their effects, however, were limited, as they focused on price inflation symptoms and not on the underlying issue of the lack of supply. House prices in most cities continued to climb, and the "structural problems" of the market persisted.

In 2006, more economic and administrative measures were deployed to control the housing market. Among other measures, regulations significantly tightened the rules regarding mortgage down payments and housing transactions. Despite these efforts, house prices rose continuously in late 2006 and 2007. Regional variations also began to emerge, with coastal regions (which were attracting the highest share of migrants) having much higher prices than inland regions (Yi and Huang, 2014; Huang, 2012).

The much-desired slowing down of the housing market eventually commenced in the fourth quarter of 2007 and continued into 2008. This was mainly due to the Global Financial Crisis that lowered Chinese exports and temporarily disrupted industrial production chains across Asia (Figure 14.1). The decline of house prices was more dramatic in cities that had experienced the largest increases in previous years. This fall in house prices was initially seen as a positive result of macro-adjustment policies. After the Beijing Olympic Games, house price reductions spread more widely among large cities, and the effects of the global credit crunch became more evident.

In the following few months, house price deflation was accompanied by a sharp slowdown of the economy and exports, and a large increase in unemployment,

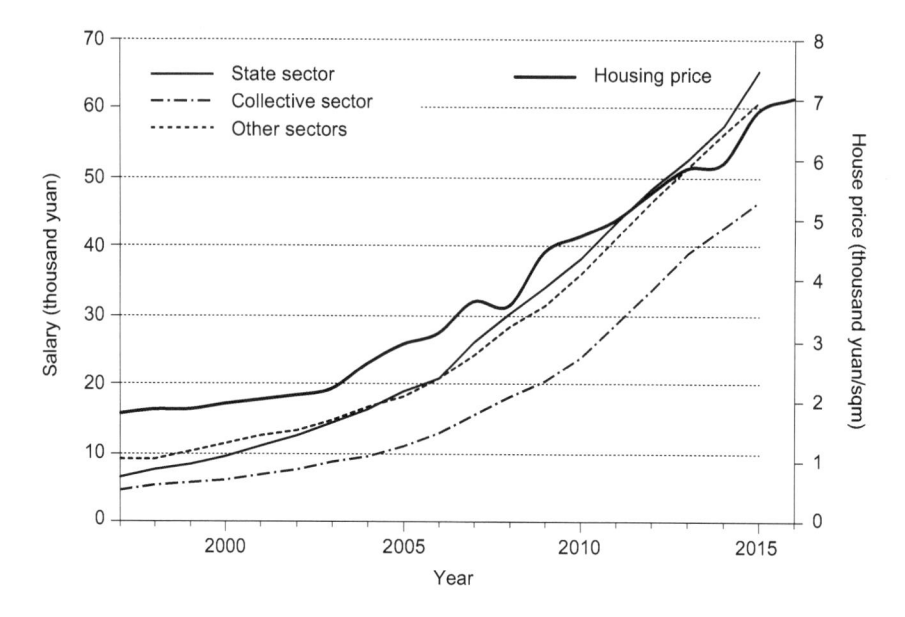

Figure 14.1 Average annual salary and housing price in urban areas.
Note: Housing price for 2015 and 2016 are based on estimates by various online reports.
Source: NBSC, 2009, 2016.

particularly among migrant workers. This was a similar economic situation to that caused by the 1997 Asian Financial Crisis. The government quickly shifted the direction of the national macro-adjustment policies and replaced controlling policies with more supportive ones to boost the housing market. In a short period of four months, from September to December of 2008, bank interest rates were cut five times from 7.74% to 5.94% in order to reduce the cost of borrowing and stimulate house purchases in cities.

By December 2008, it became obvious that the Global Financial Crisis could have a major impact on economic development in the country. While exports slowed down rapidly, the government had to stimulate internal consumption in order to maintain economic growth. Housing was again seen by central government policy makers as the biggest domestic consumption sector that could balance the loss in exports. A major housing policy document published by the General Office of the State Council (2008) focused on providing subsidized and affordable housing (a move back to the direction of the 1998 policies). Housing investment was increased as part of the major increase in capital investment to maintain economic growth over the next three years. Policies designed to stimulate the market included lowering deposits for house purchasers and giving preferable mortgage interest rates to people who lived in overcrowded housing to help them improve their living conditions. In 2009, the policy also cut the housing transaction tax to 2005 levels and reduced the transaction tax liability period from five years to two years, *reversing* the increases made in 2006.

From April 2009, house prices began to increase again (NBSC, 2009). This uptick marked the beginning of another cycle of house price inflation. The temporary market support policies introduced in December 2008 came to an end in December 2009, and the levels of deposit, mortgage interest, and transaction tax reverted to the 2007 levels. These policies, however, did not halt house price increases. In the first quarter of 2010, the average price for new commercial houses in the 70 largest cities increased by 15.9 percent compared with the first quarter of 2009. These alarming increases made another round of strict control policies inevitable. In April 2010, a 10-point policy instruction was issued with immediate effect. It included a more stringent housing finance regime. First-time buyers, if buying a house with floor space over 90 square meters, needed deposits of 30 percent or more. For purchasers of a second home, deposits had to be over 50 percent and the mortgage rate had to be 1.1 times the basic mortgage interest rate. For purchasers of a third home, deposits and mortgage rates needed to be "substantially higher." These rules were seen as tough, but they did not stop house price inflation. In January 2011, another series of policies was introduced, which increased the deposit for the purchase of a second family house to 60%, and banks were banned from issuing mortgages beyond second-home purchases.

Many cities began to limit all second-home purchases, and each household was allowed to own only one home. Discussions were held among policy makers to revise the national Real Estate Management Law to control the housing pre-sales, a dominant real estate development practice for the previous 15 years. Only first-time buyers of smaller properties were encouraged. At the beginning of 2011, the long-debated idea of property tax was introduced on a pilot basis in Shanghai and Chongqing (but still stalled nationally in 2017). At the same time, the central government demanded local governments to expand social housing construction and provision.

These control policies continued into 2013 with various reinforcements. From 2014 onward, the housing market began to show important variations between different types of cities. In most second-tier cities[2] and small towns, especially those located in inland areas, housing prices and sales of new housing stagnated. A serious problem of over-supply emerged. Municipal governments in these cities began to encourage house purchase through various kinds of taxation and regulation. At the same time, China entered another period of economic slowdown. Reduction of unsold housing stock (*qu ku cong* 去库存) became a policy priority from late 2014 to 2016. The control on bank loans and deposits, the limitation on second/third homes, and purchases made by people from outside of the city (including migrants) were relaxed.

In the first-tier cities (Beijing, Shanghai, Guangzhou, and Shenzhen), housing prices increased once again. The measures aimed at cooling down the housing market were tightened up. To release the pressure of demand for housing, municipal governments in these cities were encouraged to allocate more land for the construction of social housing, including affordable commercial housing. This

312 Ya Ping Wang

Table 14.1 Average price of commercial housing in selected cities, February 2016.

City	House price Yuan/m²	City	House price Yuan/m²
Shenzhen	47,248	Tianjin	11,029
Shanghai	37,144	Wuhan	9,113
Beijing	35,160	Chengdu	7,647
Guangzhou	17,268	Shenyang	7,066
Hangzhou	16,410	Chongqing	6,977
Nanjing	15,323	Xian	6,417

Note: The average annual personal income in China in 2015 was 21,966 yuan; for urban areas only, it was 31,195 yuan. A reasonably sized flat in a city is around 100 square meters of floor space (National Statistics Bureau, 2016).
Source: Fang.Com, http://newhouse.sy.fang.com/fangjia/. Accessed on March 28, 2016.

period is referred as "dual direction control and regulation" (*shang xiang tiao kun*双向调控); see Table 14.1.

Various mechanisms used by the Chinese government to regulate the housing market are summarized in Table 14.2 (based on Wang and Murie, 2011). In comparison with other economies, Chinese policy makers appear to have more options to manipulate the housing market. Conventional economic and financial measures, used by most economies, that control the levels of deposits, interest rates, and taxation, are relatively new to Chinese policy makers. At the beginning, Chinese housing market regulators relied on other social, economic, and political measures and the interactions between them. They also used the planning powers of the socialist state and the public ownership of land as policy alterations. In political and social terms, the Communist Party's control was, theoretically, based on the support of the ordinary people. The government had to consider the living conditions and affordability of low-income groups in cities, for example, by promoting the development of social housing to maintain social harmony. Though there are many regulation policy options available, not all of them, as demonstrated in the discussion above, are very effective, especially in stabilizing the housing prices in major cities. However, they have nonetheless prevented catastrophic property market crashes to date (Wang, 2011; Chen et al., 2014). The features and impacts of the various national policies affecting housing can be much better understood by examining the detailed characteristics of the local urban housing markets. The next section explores these characteristics through a case study of a county town.

3. Housing development and the housing market in a county town[3]

This section examines housing development and the housing market in a county town (towns located at the bottom of China's urban settlement hierarchy) in the inland province of Shaanxi. China has more than 2,000 counties, so this particular case may not be the best example. It does, however, reveal many common features in land and housing development in county towns located in inland

Table 14.2 Housing market regulation policy options.

Category of policies	Housing regulation and policy	Boost the market	Cool down the market	Support affordable housing
Taxation and fiscal	Basic bank interest rate (set by the People's Bank)	-	+	
	Property transaction tax	-	+	
	Transaction tax period (from purchase to sale)	-	+	
	Amount of deposit required for mortgages	-	+	
	Other property-related charges	-	+	
	Government instructions to banks to tighten up housing development loans for developers and mortgage lending to individuals		√	
Economic	Support economy growth during global crisis and promote domestic consumption	√		√
	Limit the profit of property developers (e.g., 3% for affordable housing)			√
	Promote the healthy development of the real estate industry	√	√	
Social	Increase social housing development and improve living conditions among low-income groups	√	√	√
	Give priority to developing low-price, ordinary commercial housing and affordable housing, targeting middle-income groups	√		√
	Relax population control: allow non-local residents to buy properties	√		
Land-use control	Release more land for the development of various types of housing, especially low-cost housing	√		√
	Control the development of low-density, high-standard luxury housing		√	
Planning and design	Control of density and size: construction floor space-to-land plot ratio (e.g., over 10); individual house unit (e.g., smaller than 90 m^2); proportion of social housing		√	√
	Municipal housing development plan; municipal social housing plans, etc.	√	√	√

(*continued*)

314 Ya Ping Wang

Table 14.2 (Cont.)

Category of policies	Housing regulation and policy	Boost the market	Cool down the market	Support affordable housing
Market administration	Differentiation between first-time buyers and second homes; ban on third/second homes		√	
	Strict rules on approving new property development companies and agencies		√	
	Data collection and monitoring market performance	√	√	
	Penalties on land waste, illegal advertising, and cheating in transactions		√	
Political	Housing market performance as evaluation criteria of local officials by the center	√	√	√

Source: Wang et al., 2011.

regions. I have visited the county almost every year and developed a better understanding of the development and functioning of the housing market there.

3.1 County background

The county has a total administrative area of 856.4 square kilometers. This territory is divided into nine towns, each with its own rural hinterland (there were 14 towns/townships in 2004; recent administrative boundary changes saw the merger of some smaller townships). Together they administrate over 180 village committees, which control 1,318 village settlements (The County Government, 2004; 2016). The total population in the county (including the rural areas under its administration) was 480,000 in 2015. The level of urbanization reported officially was 49 percent (The County Government, 2013), while the actual non-agricultural population was rather small, about 85,000 in 2009 (Table 14.3).

Of the nine towns, seven are township seats, which are slightly bigger than a normal village with some administrative offices and limited commercial activities. The two large towns host most urban residents in the county: one is the county's government seat; the other is an important local industrial center along the national railway line (several large, state-owned enterprises including textiles, machinery, lorry manufacturing, printing, and a fertilizer factory were placed here near to the railway station during the period of socialist planned economy). This urban and rural relation in the county reflects the overall picture in the country. The following discussion focuses on the town where the county's government seat is located.

The county seat has a land area of 127 square kilometers, including the town's own built-up area and around 30 village committees with more than 250 village settlements. The total locally registered population was 117,000 – of which

Table 14.3 Population changes in the county.

	2003	2007	2009	% of increase 2003/09	% of increase 2007/09
The whole county					
Population	461,156	468,143	468,833	1.7	0.15
Number of households	130,442	136,640	138,040	5.8	1.02
Non-agricultural population	81,653	83,868	85,145	4.3	1.52
Agricultural population	379,503	384,275	383,688	1.1	−0.15
The county seat					
Population	74,187	83,669	83,529	12.6	−0.17
Number of households	23,533	26,498	27,317	16.1	3.09
Non-agricultural population	24,784	27,967	30,705	23.9	9.79
Agricultural population	49,403	55,702	52,824	6.9	−5.17

Note: After 2010, major administrative reorganizations affected the population composition in the county seat, mainly by adding more rural hinterland and villages under its administration.
Source: a) The County Government, 2004; b) The County Statistics Bureau, 2008, 2010.

69,900 were agricultural population – in 2015 (The County Government, 2016). While the town covers a very large territory, the planned built-up area is only 6.5 square kilometers for a planned population of 65,000 people (Figure 14.2). In 2008, the total population in this central built-up area was 55,304, of which 25,108 were non-agricultural residents and 6,746 were agricultural residents; the rest (23,450) were temporary residents (The County Plan, 2010). No new data are available on how many people currently live in the central built-up area. Considering the trend of local population movement from villages to the town, total residents in the area must be more than the 2010 projection of 65,000.

The main function of this town is administration. In the past there were a few small industrial factories in the pre-reform period, which have since closed. The main economic activities in the town consist of agricultural production (in the surrounding areas), administration, education, health services, and shopping (servicing the surrounding rural areas). More recently, the county's government has promoted the development of the tourism industry, benefiting from local specialty foods and a historical temple located north of the town.

3.2 Housing development

Before 1980, aside from a few major government office buildings and retail stores constructed during the post-1949 period, most buildings along the single street were in the traditional style: single-story, simple structures constructed with timber, bricks/sun-baked bricks, and tiled roofs. Most residents lived in privately owned houses or dormitories provided by their employers, which were mainly government departments. From the 1980s onward, new buildings were constructed by government departments along the main street (the black line on the aerial image, Figure 14.2). These buildings normally occupied a small

Figure 14.2 Aerial image of the county seat.
Note: The historical town (the area along the main street marked by the black line in the south in the lower part of the map) was surrounded by few villages (low rise and irregular areas); the northern part (the top half) of the town (with a row of rectangular plots, partly showing rows of tenement buildings) is the new expansion.
Source: Based on Google Maps, accessed in April 2016.

piece of land, with an office building in the front and a housing block at the back, or housing units on top of the offices. From the mid-1990s, government departments began to acquire land outside the traditional built-up area and to build housing for their employees. From the late 1990s on, when national housing reform policies banned housing construction and allocation by public/work units, independent housing estates emerged in the town. These were initially set in the immediate surrounding areas of the old town (mainly between the old and new areas). The first large-scale housing estate in the town was constructed in 2003 to provide improved housing for officials employed by the county's government.

Similar to many other county towns, government-led real estate development (following the steps of major cities) began to take off around 2009, when the national government promoted "economic and comfortable housing" and other residential property development in response to the Global Financial Crisis. Large-scale housing development projects (each with over 1,000 units) emerged on the north side of the town. New streets were laid out on fresh agricultural land (Figure 14.3). Along these streets, large housing estates were

Figure 14.3 A new street paved for housing expansion in the town.
Note: The street ended in a field in 2010; it is now linked to the second ring road-the northern bypass of the town.

planned (Figure 14.4). In 2010, at least four large estates were under construction. Table 14.4 provides some details on the scale of these estates. By 2011, 13 large housing estates (including the government-supported affordable housing schemes) were under construction.

There were many reasons for this large-scale market-led housing development in the town.[4] Property development had been a profitable business, and competition among development companies in big cities had become intense. Some property developers extended their operations into smaller cities and county towns. Local county governments welcomed the investment and development; these governments made land available to attract property developers. Demand for better design and higher-quality construction of housing estates had increased substantially in cities, while requirements for design style and quality were more flexible and less competitive in smaller towns. Furthermore, land prices were much lower, resident relocation was relatively easier than in cities, and local county government officials were willing to help with difficult relocation issues. Most developers working in this town had businesses in the cities.

3.3 Market operation and customers of the new housing

Housing conditions outside of these new estates in the town have become very poor in general. Most traditional houses became dilapidated; residential buildings

Figure 14.4 Housing estates under construction in 2012.

Note: Typical style of development, mostly 6-story walk-up tenement buildings, with a very high building/plot ratio.

Table 14.4 Details of a housing estate under construction in 2010.

Estate design elements	Standard
Total land area of the estate	49,518 m^2 (about 75 *mu*)
Total construction floor space	173,313 m^2
Floor space for commercial use	13,500 m^2
Floor space for residential use	15,7713 m^2
Floor space for nursery	1,250 m^2
Floor space for clubs	850 m^2
Open space in the estate	36%
Building area-to-land ratio (construction density)	0.32
Total construction floor space-to-land ratio	3.5
Total number of housing units	1,272 (about 124 m^2 per unit)
Parking spaces	140 on ground
	600 underground
Parking spaces-to-housing unit ratio	0.58

Source: Taken from marketing material provided by the estate's marketing office.

constructed in the 1980s and 1990s were not carefully planned and often mixed with offices. By 2000, many of them were overcrowded and showed signs of age, with blocked drainage/sewage pipes, cracked walls, and broken doors and windows. Newly constructed housing estates, on the other hand, offered much better living conditions and were very attractive to government employees and officials. Two of the new housing estates were also constructed under the affordable housing policy, with government-subsidized prices targeting public-sector employees. Older and smaller houses acquired early by government employees through housing reform and privatization could be sold or rented to rural residents who worked or were studying in the town.[5]

New housing in the purpose-built estates had better facilities: gas supply, internet broadband, larger kitchens and bathrooms, more bedrooms with bay windows/balconies, and connections to the district heating system in the town; they also offered families more privacy. Gardens, open spaces, and parking between residential buildings were other attractive features for the potential purchasers. Professional real estate management systems were set up, as they had been in large cities, where residents pay a fee for maintenance, sanitation management, and security.

The county seat provides homes for most officials employed by the county government. It also accommodates many officials and other public-sector employees whose offices are located outside the county seat itself (e.g., in the administration of other towns in the county, or schools located in rural areas). Officials working in other towns during the pre-reform period normally had dormitories where they worked. With improved transport, most of these officials and their families now live in this county seat (or in the other towns on the railway line). There is no commercial housing estate or market in smaller towns in the county. Commuting is the way of life for those officials with offices in smaller towns. It is also common practice for the county's government to shift officials very frequently from one town to another; officials therefore do not tend to relocate to rural townships. Aside from officials, other professionals (such as school teachers) working in the rural hinterland also have their family homes in the county seat.

The total number of flats in these new estates would be more than that required to provide one new home for every non-agricultural household living in this town. These large-scale housing developments were clearly targeted at a much larger market. There were strong signs of rural village residents purchasing flats in the town. In one of the estates, 50 percent of the sales were to rural residents from surrounding villages. Through the 1980s and 1990s, housing construction in the county had happened mainly in villages through self-construction as large extended families were divided into smaller ones. Village expansion was a typical outcome. Old and run-down houses occupied by the elderly in the village center were surrounded by new and larger houses used by the young. After 2000, individual, family-organized house-building in villages gradually died out.

This trend also reflects the demographic changes in rural areas. Due to the family-planning policy, rural population growth slowed down substantially. In many villages, there were signs of serious depopulation. Many newly built houses were left empty as their owners migrated to cities and towns. In the 1980s and 1990s, many rural families had more than one son. A separate house had to be built for each son for his marriage prospects. Following the one-child policy and also the recognition of economic factors, few families now have more than one son. This single son tends to have a better education and prefer life in cities or towns. Many better-off families began to buy houses in the county town.

Rural depopulation is further compounded by the impacts on the rural education system. Due to large-scale depopulation, many schools located in rural areas have either been closed or are held to a lower standard. Young migrants working in the city now send money home for their children to be educated at better schools. As a result, schools located in the county's town seat expanded substantially and now have more resources. In fact the county's high school in the town was able to relocate to a new and larger site with entirely new buildings and facilities.

In the 1980s and 1990s, the first generation of rural migrants from the county brought back their savings and built new houses in the village. Now, most younger-generation migrants prefer an urban life (Sui et al., 2014; Huang et al., 2015). Many of them, however, cannot afford to buy a flat in the city, but most of them do not want to return to the village either. The cost of a flat in the county seat is not far beyond their financial reach, and many of them (with help from other family members) have bought flats in the town.

The house prices in 2010 in the town were around 2,000 yuan/m^2. The requirement for the deposit was low; banks were also happy to provide mortgage loans. Based on their experience in cities, young people began to understand the market values of houses in their hometown. Where potential price increases in properties bought in towns may bring some profit in the future, a house in the remote village is seen to bring no financial benefits of any kind. In some of the surrounding villages, more than 10 percent of residents had bought a flat in the town in 2015 (Figure 14.5). These include richer families who have people doing business or working in the cities and towns as migrant workers, and retired state-owned factory workers who receive a pension. There is no policy limitation or *hukou* restriction on rural families buying flats in the town, especially if they do not require a bank loan. Mortgage lending is available to potential buyers, as farmers' income including that from land can be used as assurance for loans. For ordinary village families relying on only agricultural production, though, it will be impossible to dream of a flat in the county seat, though the price seems only a fraction of that in large cities.

Figure 14.6 shows the housing market catchment area and the relationship between the county seat, villages, and other towns/townships located in the surrounding areas. It also indicates another important relationship between the town and the higher-level regional city. The county discussed in this chapter is under the administrative control of a prefecture-level city. In the centralized political and administrative system of China, there is important movement of officials between the county government and the city, especially at senior levels. In order to allow officials to gain more experience in more than one place (and also to avoid corruption), municipal governments often move officials under their control from one county to another, or from a county to one of the departments in the city at same level, or vice versa. Similar to the relation between the county seat and other towns, county government officials used to live in the county with their

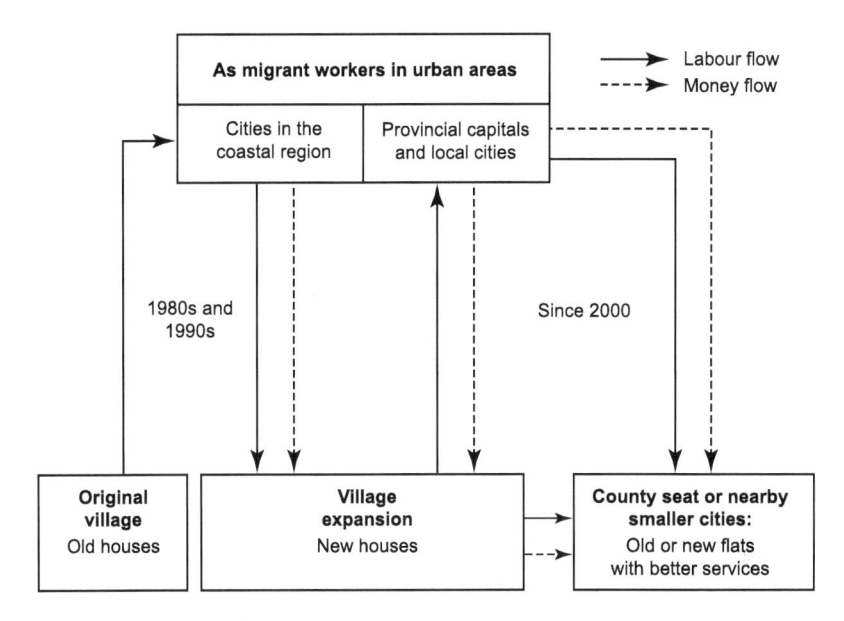

Figure 14.5 Rural labor migration and remittance flow.

Note: Figure shows two circles of flows of migrant laborers and remittance: the left-hand one represents the first generation rural migrants and the model of "going out to earn money and returning home to build a new house" (外出打工, 回村盖房); the right hand one represents the second- and third-generation migrants and the model of "going out to earn money and buying a flat in the town" (外出打工, 县城买房).

families during the pre-reform period. Since the 1990s, it is very common that top leaders in the county stay in the county seat only for weekdays and return to the city for weekends, as their families live there; some may even commute daily.

This type of movement and commuting has important consequences for housing. Senior county and city government officials could have houses in more than one location (in the county's towns and the city). With continuous housing price inflation, this multi-house occupancy and ownership also enables officials to accumulate family assets. During the pre-reform period, everyone lived in government-owned houses; once one moved away, the house was returned to the government for reallocation. In the marketized system, officials normally "bought" the house through their work (with subsidy); there is no need for them to return the house to the government when they leave.

Another aspect of the housing market relation between the county seat and the city is investment in second homes by better-off county residents. In China, the long-established philosophy is that "people move from low to high position like water flows from high to low ground" (人往高处走, 水往低处流). Ordinary farmers dream of a flat in the county seat, as the town is seen as having a higher position than the village. County seat residents, especially well-paid government officials, however, tend to invest in a property in the city, as the city occupies a

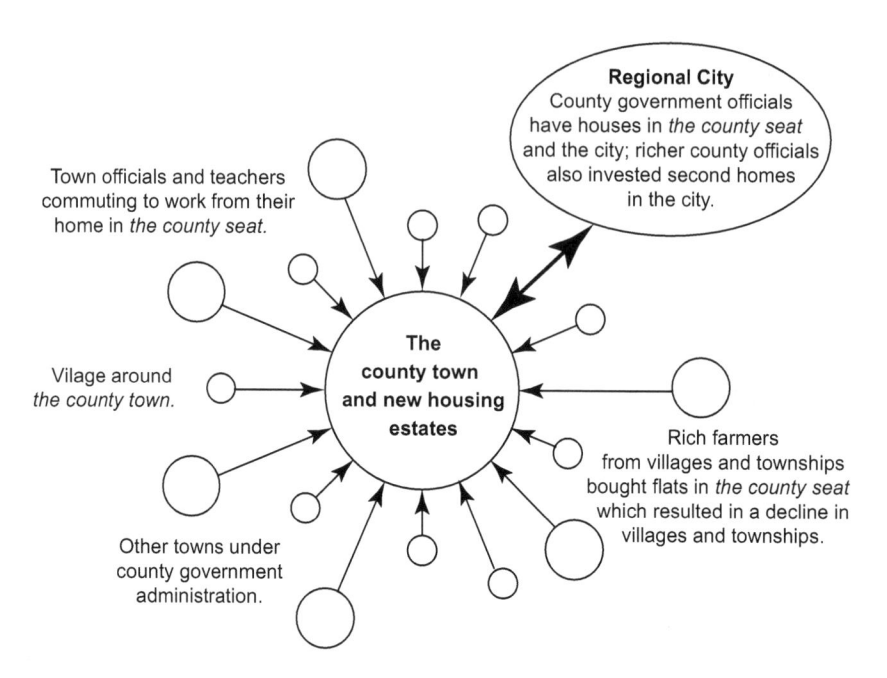

Figure 14.6 Housing market catchment areas in the county seat.

higher position in the national urban hierarchy than the county seat. For many reasons, a property purchase for the younger generation in a better or higher place is the most important investment for families. While most county officials are happy with their own lives in the town, not many would wish their offspring to stay there. Similar to farmers sending their children to the town for school, town residents would like to have their children educated in a city rather than mixed up with the farmers' kids in the town. For officials, children staying in the county seat would be seen as a failure, unless the official could secure a higher office position; without proper higher education in the cities, this is unlikely.

4. Migration and its Impacts on the urban housing market

The housing market relations in this small town, due to its simplicity, help us to understand housing development and market relations in cities and in the country as a whole. During the period of fast economic growth and urbanization, property development and the housing market fulfill two important functions. One, they satisfy the ever-growing demand and appetite for new and better housing by local residents. Two, they improve the urban living environment and increase the attractiveness of cities/towns for newcomers who can find employment (and a *hukou*) in these cities/towns (Wang et al., 2009; 2010). While the former is the concern of various urban housing reform and finance policies and

Migration and the housing market in China 323

mechanisms, the latter relates directly to the process of migration. The migration patterns between cities in China are determined by several important factors including *hukou* status, political and government association status, employment and economic power, educational achievement, and so on (Fan, 2001; Wu, 2004 and 2006). Migrants who arrive in a city with their *hukou* registration tend to include government and party officials through job relocation, similar to that discussed in the county seat. This sort of movement often relates to promotions. Each year, there are many officials who move from one city to another; it could be from a smaller city to a large city, or from a large city to a smaller city but at a higher rank. This trend also applies to officials employed by state-owned enterprises. These movements generate some housing demand in the host city and also help these officials to accumulate housing wealth. In general, this group of officially sanctioned migrants have a direct impact on the housing market. As they are in advantageous positions and have already accumulated some resources, their promotions will lead to better houses in better locations in the same city or in a higher-tier city. This movement is upward, or "buying up" (Figure 14.7).

Migrants who arrive in a city without transferring their *hukou* to the host city tend to include all rural-to-urban migrants, some new graduates, and residents making simultaneous moves from towns (like that in the case study) or smaller cities to larger cities. This group tends to work in the city as cheap labor, manual

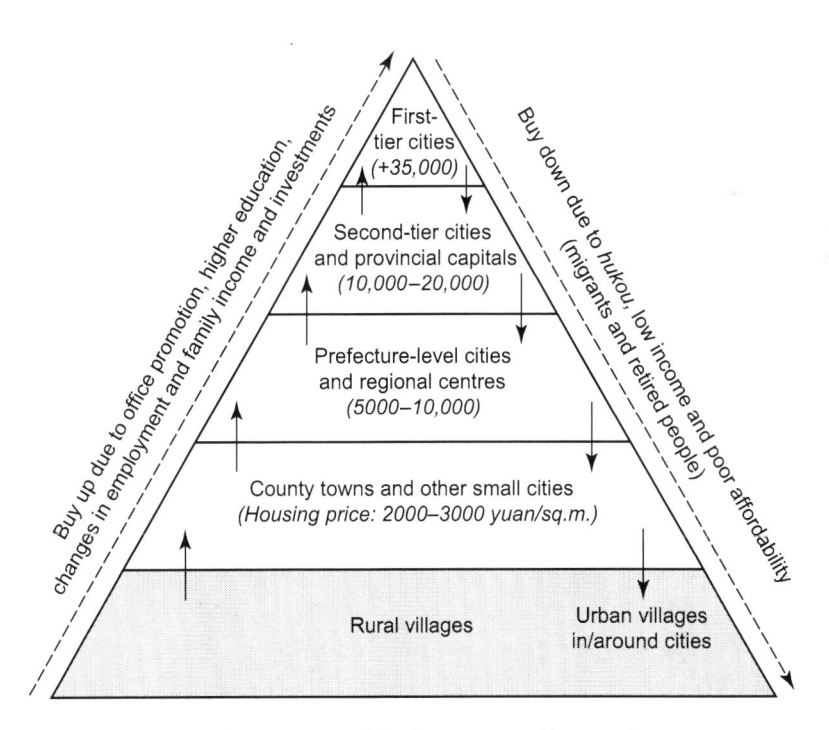

Figure 14.7 Trends of migration and the buying-up and buying-down processes.
Note: Figures in brackets show the average housing price in 2015.

workers, or junior office clerks. Most of them have relatively low pay and cannot afford to buy commercial housing in the city where they work and are not qualified for public housing allocation (Huang et al., 2014). They may, however, use their limited savings to buy houses in their hometowns or in a smaller city near their hometowns. As the case study shows, rural migrants have becoming a major force in the housing market in small cities and towns. To these individuals, their housing purchase still marks upward movement in the society. Although they cannot afford to buy houses where they work, the life in a town or a smaller city is still a step up from the village in the social hierarchy and the urban housing ladder. Within the national urban housing market system, this sort of housing purchase shows a trend of "buying down" (Figure 14.7).

Apart from these two main types of migration (the government arranged official movements and simultaneous movements between cities and from rural to urban areas), there are also other population movements that have significant impacts on the urban housing market. Education is a major reason for migration in China and is maybe the only mechanism for social upgrading. Every year, about 6.5 million high school graduates enter university; another 5 million enter various further education colleges; about half a million university graduates go on to postgraduate studies, many in a different city from where they did their undergraduate courses. All universities and most further education colleges are located in cities. Education provides rural youth a way to leave the village and the only chance to change their social and *hukou* status. For urban youth, to get into a good national university (located in few large cities) is a great achievement that parents would willingly support with a lot of their savings. It is not very common yet for ordinary students to buy houses in the city where they study, but most students tend to stay in a city to work after graduation. This is when they start to have a major impact on the homeownership market. At the start of their careers, and without much money saved, many graduates rely on their families for support to buy a flat in the city. For richer families, this is also seen as an opportunity for investment (indeed, some families buy houses in the city long before their children or grandchildren reach school or university age); for the many not-so-rich, supporting their child's housing purchase becomes a burden. This route of intercity and inter-region migration (along with the cash young people bring) is another main driving force for housing market expansion and urbanization. Housing purchase by graduates is normally a sort of buying up. Outside the tier-one cities, other cities with over-supply problems have introduced special policies to encourage students to move to and buy property in their city.

Cash and investment upward flow between urban areas is an important feature of the Chinese urban housing system. Better-off people from rural areas or small towns bought houses in regional cities, provincial capital cities, and even tier-one and national capital cities, which could have important impacts on local housing markets. Bosses of small coal-mining companies from Shanxi Province invested a lot in the Beijing housing market, and Shanxi people could be found in many upmarket new housing estates in Beijing. People from the north of Shaanxi Province (who also become rich from coal mining and gas exploration) bought houses in Xi'an City – the provincial capital. It was widely reported that housing purchase delegations from Wenzhou (where small businesses were very successful over the years and created many rich families) targeted cities with expected housing price inflation all over the country. Some of them even had their eyes on overseas property markets (e.g., the U.S.), as housing price inflation was so high that these investors could afford to leave their properties empty for the period of their ownership.

5. Conclusions

China has experienced an urban housing revolution over a short period of 40 years. The dominant urban landscape of residential poverty of the 1970s has been replaced by modern and spacious homes in well-designed housing estates. The socialist welfare urban housing system, with an average of about 3 square meters of housing floor space per person, has been replaced by a nation of homeowners in which a majority of urban families (mainly non-agricultural residents) live in properties of their own (on state-owned land). Apart from dramatic improvement in the living conditions of the existing urban residents, Chinese cities and towns have provided housing and shelter for many millions of migrants. Urban housing reform policies and the re-establishment of the urban housing market are the two most important contributing factors for this housing revolution.

Urban housing reform is also a very complex social, economic, and political process in a huge country like China. Having achieved such a huge success, China's urban housing market now faces many new challenges and problems. China has a centralized political and administrative system in which places and communities form a rank order, with rural villages and their residents at the very bottom (Wang, 2004; 2005). Recent urbanization has strengthened this spatial and social hierarchy in the urban settlement system, starting with county towns at the bottom, moving up to regional cities, provincial-level cities, second-tier national cities, and finally the first-tier cities and the national capital. Traditional cultural wisdom and current socio-economic and political ideologies encourage the upward migration of people and the vertical flow of money and other resources toward the top of this urban settlement pyramid. The urban housing market to some extent has reinforced this centralization. It encourages the concentration of people with talent (political, managerial, academic, and intellectual), money, and other resources in higher-level cities, where the housing market is dynamic and housing investment returns are high. Housing price inflation has become a major mechanism for asset accumulation by the rich and powerful. The integration and linkages of housing markets in different locations enable the exploitation of rural areas by urban areas, and lower-level cities by higher-level cities.

In the process of fast economic growth and urbanization, family wealth is created at every level, but many people would like to gain the advantage of higher-level markets where property is more expensive and value grows quickly. This desire made the house prices in larger cities grow much faster and become volatile. Because of the intercity linkages and the constant demand from newcomers, the housing bubble became bigger and bigger, but did not bust that easily, as the authorities, especially the municipal government, could control the migration flow. They can limit outsiders from buying properties in their city when prices increase too quickly, and open the gate to welcome outsiders when the market shows sign of slowing down. This control of the flow of outsiders is indeed one of the most effective housing policies so far.

Urbanization and vertical upward migration between cities effectively link together the urban housing market at different levels and form a very long housing ladder in the whole country, with county towns at the bottom and Beijing, Shanghai, Guangzhou, and Shenzhen at the top; other regional and local cities form the intermediate sections. In a huge country with such a large population, the dynamics of such a huge urban property market could lead to inequality in housing consumption and asset ownership between different cities and urban areas, between different socio-economic and political groups, and between the rich and the poor. The social, economic, and spatial motilities of people constitute a major step forward in comparison to the communist egalitarian society dominated by people's *hukou* and residence status in China, but large-scale and rapid concentration of people and resources into a few cities may lead to regional disparities and create problems for future generations, especially when property purchase in cities and towns is seen as investment and asset accumulation.

Future housing policies may need to pay more attention to the use values of housing and the effective use of houses in all locations, rather than housing market values in large cities. Municipal and town governments need to look at sustainable development ideas and the potential benefits of property ownership taxes to balance their books and manage urban growth, rather than continuous relying on real estate development, housing price inflation, and attracting more people into their cities and towns.

Notes

* The work presented in this chapter is based on fieldwork from two recent research projects: a Lincoln Institute of Land Policy-supported project on Village Lead Land and Property Development in China (CTW012413, 2013) and an Economic and Social Research Council of UK (ESRC) and China Academy of Social Sciences (CASS) jointly funded International Centre Partnership Project on Urban Development, Migration, Segregation and Inequality (ES/N007603/1, 2015). I would like to thank the editors of this book, especially Rebecca and Bertrand, for their advice and constructive comments.
1 Historical housing reforms were part of simultaneous major reforms in 1998 which aimed to do three things: (1) free enterprises from the burden of providing housing to their employees to enable them to focus on their primary business objectives and in the process to facilitate the emergence of a specialized real estate industry; (2) raise wages to market levels and improve nationwide labour markets by eliminating wages in kind that lock workers into their firms; and c) privatize non-strategic state-owned enterprises to develop new private enterprises and prepare China for its entry into the World Trade Organization. (For details, see in particular the collection of directives in *Zhu Ronji*

on the Record, vol. 1, in *The Road to Reforms 1998–2003*, Brookings Institution Press.) Many development economists see that the housing privatizations of 1998–2003, and the massive transfer of wealth to households that accompanied them, were one of the two most important moments in China's post-1978 development. The other was the development of the "household responsibility system" that transformed rural areas after the disasters of central planning.

2 There is no agreed classification of cities by tier. Every major institution analyzing housing markets has its own classification of cities by tier.

3 To comply with professional research ethics principles and avoid any potential negative publicity impact on the case study town, I have replaced the county name with "the county" throughout the chapter, including the references. Where information came from the county government's web pages, the websites are not identified for the same reason. If any reader would like to find out more about these sources, please contact me.

4 The imbalanced intergovernment fiscal reforms of 1994 drove local governments to speculate on real estate development projects in order to fill the gap between their expenditure obligations and their revenues by taking advantage of their conditional monopoly over urban land.

5 The wealth transfer associated with the privatization of 1998 accelerated housing trades, especially after 2003, when restrictions on the resale of privatized state housing ended. Prevailing housing inequalities, which were significant across different state officials before privatization, were amplified further after 1998–2003 through housing trades.

References

Chen, J., Yang Z., and Wang, Y. P. (2014). The New Chinese Model of Public Housing: A Step Forward or Backward? *Housing Studies*, 29(4), 534–550.

Dreger, C. and Zhang, Y. Q. (2013). Is there a Bubble in the Chinese Housing Market? *Urban Policy and Research*, 31(1), 27–39.

Fan, C. C. (2001). Migration and Labour Market Returns in Urban China: Results from a Recent Survey in Guangzhou. *Environment and Planning A*, 33, 479–508.

General Office of the State Council. (2005). Suggestions on Works of Stabilising Housing Price, Documents issued by the General Office of the State Council, no. 26, 2005.

———. (2008). Some Suggestions on Promotion of the Healthy Development of Housing and Property Market. Documents of the General Office of State Council, no. 131, 2008.

Huang, X., Dijst, M., van Weesep, J., and Zou, N. (2014). Residential Mobility in China: Home Ownership Among Rural–Urban Migrants after Reform of the *Hukou* Registration System. *Journal of Housing and the Built Environment*, 29(4), 615–636.

Huang, Y. (2004). The Road to Homeownership: A Longitudinal Analysis of Tenure Transition in Urban China (1949–1994). *International Journal of Urban and Regional Research*, 28(4), 774–95.

Huang, Y. (2012). Low-income Housing in Chinese Cities: Policies and Practices. *The China Quarterly*, (212) 941–964.

Huang, Y. and Li, S. (Eds.) (2014). *Housing Inequality in Chinese Cities*. London and New York: Routledge.

Huang, Y. and Ran, T. (2015). Housing Migrants in Chinese Cities: Current Status and Policy Design. *Environment and Planning C: Government and Policy*, (33), 640–660.

Hui, E. C. M., Yu, K. H., and Ye, Y. C. (2014). Housing Preferences of Temporary Migrants in Urban China in the Wake of Gradual *Hukou* Reform: A Case Study of Shenzhen. *International Journal of Urban and Regional Research*, 38(4), 1384–1398.

Li, Q. and Chand, S. (2013). House Prices and Market Fundamentals in Urban China. *Habitat International*, 40, 148–153.

Li, S. M. (2000b). The Housing Market and Tenure Decisions in Chinese Cities: A Multivariate Analysis of the Case of Guangzhou. *Housing Studies*, 15(2), 213–236.

National Statistics Bureau. (2009). *China Statistical Yearbook 2008*. www.stats.gov.cn/tjsj/ndsj/

———. (2016). *China Statistical Yearbook 2015*. www.stats.gov.cn/tjsj/ndsj/

State Council. (1998). The Notice on Further Reform of Urban Housing System and Speeding up Housing Development. Document No. 23, July 3, 1998.

The County Government (Yearbook Editorial Committee). (2004). *The County Nian Jian 2004 (The County Yearbook 2004)*, County Archive Bureau.

The County Government. (2013). County Economic Development and Planning Spread Sheet for 2013. Unpublished report (accessed online on August 8, 2016).

———. (2016). Enter The County – Urban Development. Online information.

The County Plan. (2010). Unpublished documents accessed at the County Construction Bureau.

The County Statistics Bureau. (2008). The County Economic and Social Development Statistics Report 2007. Unpublished report in booklet format (accessed online on August 8, 2016).

———. (2010). The County Economic and Social Development Statistics Report 2009. Unpublished report in booklet format (accessed online on August 8, 2016).

Wang, Y. P. and Murie, A. (1999). Commercial Housing Development in Urban China. *Urban Studies*, 36 (9), 1475–1494.

Wang, Y. P. and Murie, A. (2000). Social and Spatial Implications of Housing Reform in China. *International Journal of Urban and Regional Research*, 24(2), 397–417.

Wang, Y. P. (2004). *Urban Poverty, Housing and Social Change in China*. London and New York: Routledge.

Wang, Y. P. (2005). Low-Income Communities and Urban Poverty in China. *Urban Geography*, 26(3), 222–242.

Wang, Y. P. (2011). Recent Housing Reform Practice in Chinese Cities: Social and Spatial Implications. In J. Y. Man (Ed.). *China's Housing Reform and Outcomes*. Cambridge and Massachusetts: Lincoln Institute of Land Policy, pp. 21–46.

Wang, Y. P., Wang, Y. L., and Bramley, G. (2005). Chinese Housing Reform in State Owned Enterprises and its Impacts on Different Social Groups. *Urban Studies*, 42(10), 1859–1878.

Wang Y. P., Wang Y. L., and Wu, J. (2009). Urbanisation and Informal Development in China: Urban Villages in Shenzhen. *International Journal of Urban and Regional Research*, 33(4), 957–973.

Wang Y. P., Wang, Y. L., and Wu, J. (2010). Housing for Migrant Workers in China: A study of the Chinese Model in Shenzhen. *Housing Studies*, 25(1), 83–100.

Wang Y. P., Wang, Y. L., and Wu, J. S. (2010). Private Rental Housing in 'Urban Villages' in Shenzhen: Problems or Solutions? In F. L. Wu and C. Webster (Eds.). *Marginalization in Urban China: Comparative Perspectives*. Basingstoke: Palgrave Macmillan, pp. 152–175.

Wang, Y.P. and Murie, A. (2011). The New Affordable and Social Housing Provision System in China: Implications for Comparative Housing Studies. *International Journal of Housing Policy*, 11(3), 237–245.

Wu, F. (2015). Commodification and Housing Market Cycles in Chinese Cities. *International Journal of Housing Policy*, 15(1), 6–26.

Wu, W. (2004). Sources of Migrant Housing Disadvantage in Urban China. *Environment and Planning A*, 36, 1285–1304.

Wu, W. (2006). Migrant Intra-urban Residential Mobility in Urban China. *Housing Studies*, 21(5), 745–765.

Yi, C. D. and Huang, Y. (2014). Housing Consumption and Housing Inequality in Chinese Cities during the First Decade of the 21st Century. *Housing Studies*, 29(2), 1–22.

Zhu, R. (2015). *Zhu Ronji on the Record: The Road to Reforms 1998–2003*. Washington, D.C.: Brookings Institution Press.

15 Residential land supply and housing prices in China
An empirical analysis of large cities

Huina Gao

BEIJING INTERNATIONAL STUDIES UNIVERSITY

Zhi Liu

LINCOLN INSTITUTE OF LAND POLICY AND PEKING UNIVERSITY-LINCOLN INSTITUTE CENTER FOR URBAN DEVELOPMENT AND LAND POLICY

Yue Long

ABT ASSOCIATES INC.

1. Introduction

Concerns about the housing price hikes in many Chinese cities have been growing over the past 15 years. The driving force behind the price hikes has been a heated topic of public and academic debate. The question has been examined extensively from the demand side. Some researchers attribute the boom to the brisk demand for new housing by those who are preparing for marriage or demanding better and larger housing units.

However, few studies have examined housing prices through the lens of land supply. Indeed, the share of residential land supply out of total land supply in many large and medium-sized cities has been on the decline over the past few years. Figures 15.1–15.6 show the ratios of residential land supply in Beijing, Hangzhou, Taiyuan, Zhengzhou, Chengdu, and Huhehaote. The ratios of residential land supply fell while the urban population increased in these cities. We suspect that the local governments might have intentionally reduced the supply of residential land, which in turn might have contributed to the increase of housing prices in these cities.

Using the panel dataset of 31 major Chinese cities in 2000–2015, we constructed simultaneous-equation models to examine the impact of residential land supply on housing prices in the context of the government monopoly of the supply of urban land. The rest of the paper is structured as follows. Section 2 presents a literature review. Section 3 presents the evolution of urban residential land supply in China. Section 4 presents the analytical framework. Section

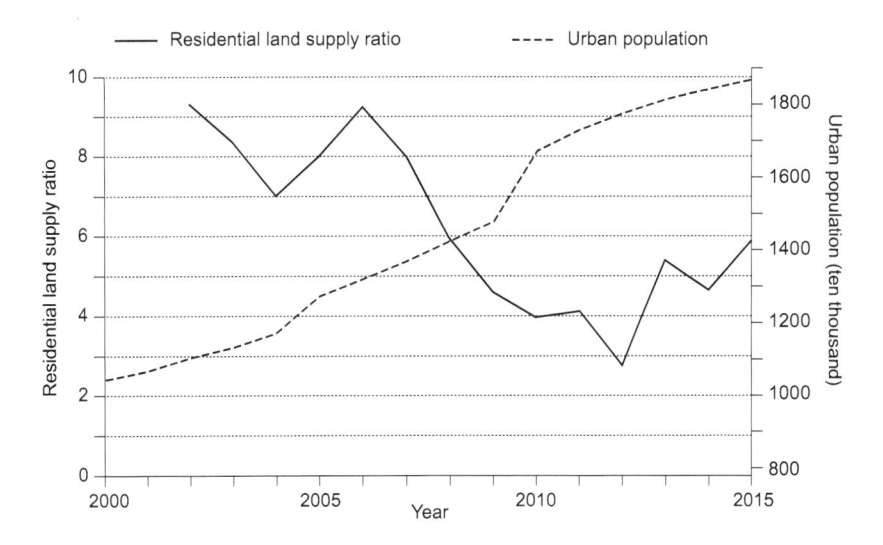

Figure 15.1 Residential land supply ratio in Beijing.
Source: Compiled from China Index Academy Database.

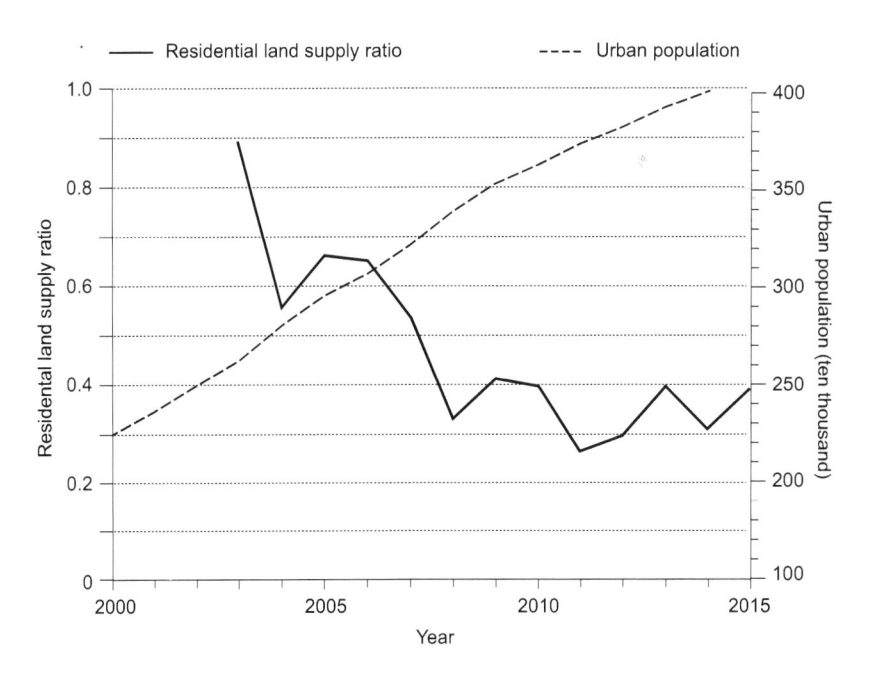

Figure 15.2 Residential land supply ratio in Hangzhou.
Source: Compiled from China Index Academy Database.

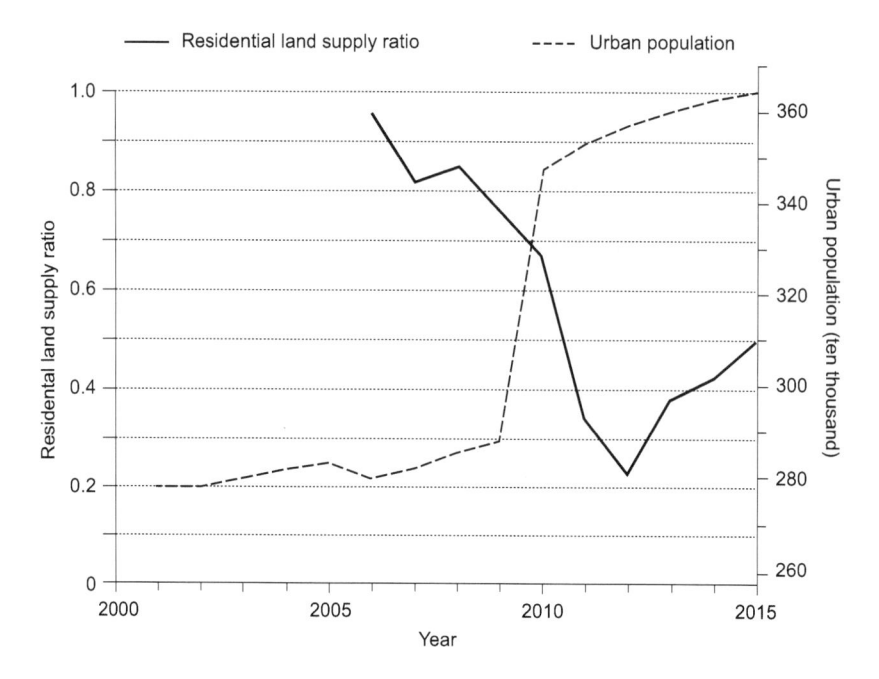

Figure 15.3 Residential land supply ratio in Taiyuan.
Source: Compiled from China Index Academy Database.

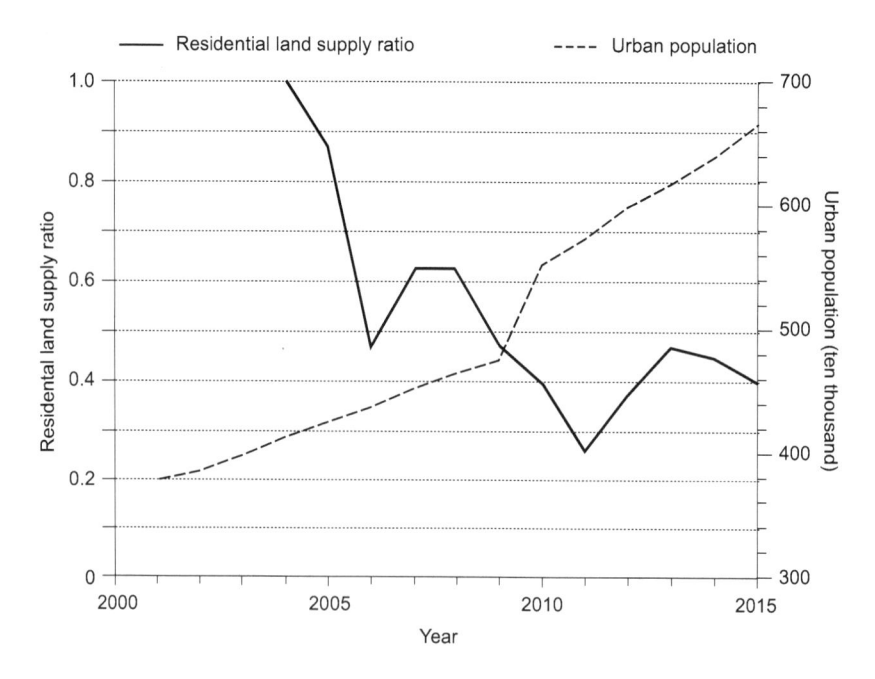

Figure 15.4 Residential land supply ratio in Zhengzhou.
Source: Compiled from China Index Academy Database.

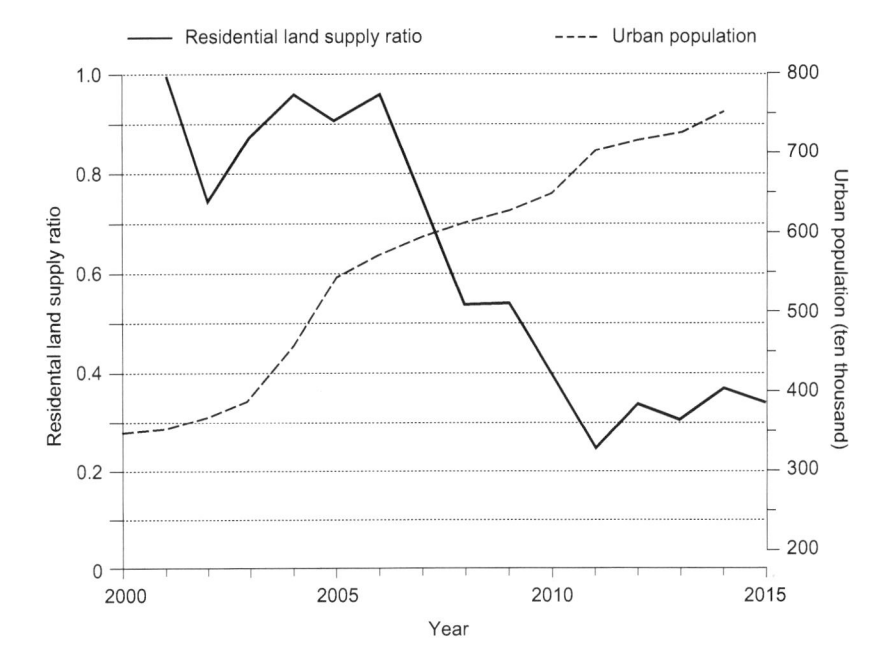

Figure 15.5 Residential land supply ratio in Chengdu.
Source: Compiled from China Index Academy Database.

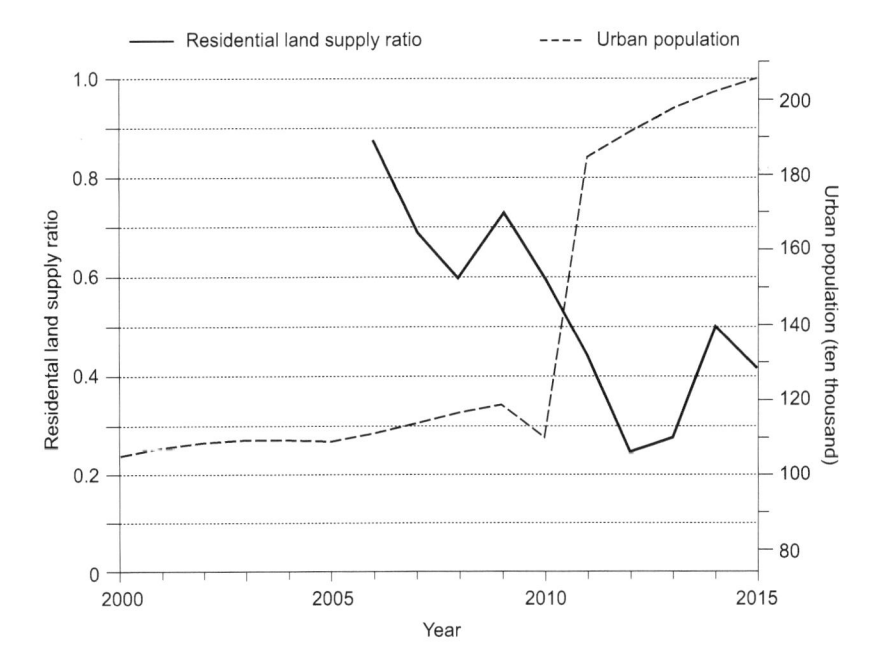

Figure 15.6 Residential land supply ratio in Huhehaote.
Source: Compiled from China Index Academy Database.

5 presents data and model specifications. Section 6 presents and discusses the empirical results. Section 7 summarizes the findings.

2. Literature review

There are three dominant views on the relationship between residential land supply and housing prices. The first is that land-use control limits the amount of land available for residential housing development, pulling up residential land prices, which in turn results in a decline in the supply of new housing (Quigley and Raphae, 2004; Glaeser and Ward, 2009) and appreciation in housing prices (Glaeser et al., 2005; Ihlanfeldt, 2007; Glaeser et al., 2008; Zabel and Dalton, 2011; Yan et al., 2014). This view has been challenged recently. There is also relevant comparative research in the UK and across Europe.

Using time series data, Huang et al. (2015) examined the effectiveness of the Hong Kong (HK) government policy to increase the supply of residential land in order to increase the housing supply. They found that the new housing supply in HK was independent of the land supply by the government. The structure of the HK real estate industry and the time horizon matter very much. However, there is significant previous research on HK showing that the large inventories held by the dominant real estate oligopoly have defeated the announced goals of government land release policies in the short and medium terms. For the purpose of the present paper, we can let Huang et al.'s claim stand as a useful conceptual statement, no matter what the truth is in HK.

The second view is that land-use control raises both residential land prices and housing prices. However, Ihlanfeldt (2007) has claimed that the appreciation in housing prices in the US resulted from the rise of construction costs instead of the decrease in residential land supply.

The third view based on the HK case is that land supply restrictions can cause higher housing prices from capitalization of higher expected rents, which encourage capital-land substitution in housing production (Peng and Wheaton, 1994). A few papers have also analyzed the heterogeneity of residential land supply. For example, Hui et al. (2014) have investigated the respective impacts of land sale and land exchange on housing supply in HK. The findings show that land exchange has a much larger long-term impact on housing supply than land sale does.

The relationship between residential land supply and real estate prices has been examined extensively in the context of private ownership of land. All HK studies are, by nature, based on a public land leasing system, which was the model for China's land reform after 1998. A difference in land supply between HK and mainland China is that land supply in HK is fixed, especially since land reclamation has almost stopped. However, few studies have examined the relationship in the context of public land leasing systems in a comparative way. In China, land-use rights and land ownership are institutionally separable. Municipal governments are able to lease residential land to individuals and social units with a use term of 70 years and have direct control of land supply.

Given the strong role of the government in land supply, it is inadequate to study China's housing markets merely from the perspective of housing demand. It is important to clearly identify the effect of the monopoly supply of residential land on housing prices.

Recently, Dong (2016) examined the influence of geographical and man-made land constraints on housing markets in urban China and found that cities with less naturally available land had experienced greater price appreciation, and the quantity response was less in those places. This paper examined the influence of land supply on housing markets in urban China from the perspective of geographical constraints. However, it did not clarify the mechanisms through which the land supply influences housing prices. There are also studies of the impact of physical constraints across the US and South Korean cities.

Our study makes two contributions. Firstly, it is among the first to attempt to empirically clarify the mechanism through which the control on residential land supply has on housing prices in Chinese major cities. Secondly, it adds to the limited literature on the interaction between residential land supply and housing prices in the context of public land ownership.

3. Evolution of urban residential land supply in China

China's economy was centrally planned from 1949 to 1978, when Deng Xiaoping initiated an economic reform that moved the economy to the market. Correspondingly, the supply of urban land was also centrally planned, with neither prices nor restrictions on the term of use during this period. Urban land was predominantly owned and allocated to users by the state. Transfers of land-use rights between users were prohibited. This land supply (or allocation) system produced enormous land-use deficiencies, which manifested in low and flat land density curves and disconnection of land use and transportation, among other factors (Bertaud and Renaud, 1997).

With the economic reform and marketization, the planned land supply system was no longer compatible with the new economic system. Since 1978, direct foreign investment and the number of joint ventures increased exponentially. Demand for access to land by joint-venture manufacturing plants skyrocketed. The previous land allocation system conflicted with the ultimate goal of economic reform, which aimed to introduce market mechanisms for resource allocation to improve economic efficiency, to correct government failures in land allocation, and to minimize negative consequences of the land tenure system (Ding and Knaap, 2002).

Adjustment of the previous land supply system became urgent. Land leasing was increasingly adopted following the example of the Land Leasing Tenure System in Hong Kong (Gu, Michael, and Cheng, 2015). In 1988, China's Constitution was amended to permit land leasing (Peng and Thibodeau, 2009). In 1990, the State Council formally affirmed land leasing as a public policy. By 1992, local governments in Shanghai and Beijing adopted land leasing as a local practice, which then spread to other cities (Peng and Thibodeau, 2009).

The establishment of the land-leasing system yielded both positive and negative consequences. On the positive side, land markets emerged and government revenue increased (Ding, 2002). On the negative side, problems such as decreases in available urban land for supply, abuse of land management power by local governments, and loss of state-owned assets appeared. After 1996, some municipal governments such as Shanghai, Hangzhou, Dalian, and Qingdao began to explore new ways to supply urban land. By 1999, local governments in Hangzhou and Qingdao adopted the use of urban land reserves, which was based on experiences in HK, and the practice soon spread to other cities.

"Land reserve" refers to the process by which local governments purchase land from previous users, convert it to productive use, or hold it until it is profitable to sell (Alexander, 2013). In China, the Land Reserve Center affiliated with the local Bureau of Land and Resources plays the role of agent of the local government to manage urban land. These agencies represent local governments in purchasing and servicing urban construction land (connecting electricity, water, gas, and communication lines, and building necessary transportation infrastructures). They can increase their land reserves by converting rural land into urban construction land and sell the land to the highest bidder through market means such as tender, auction, or public listing (Zhang et al., 2011). In 2007, the central government affirmed the local land reserve system as a new public land supply policy.

The new land supply system brings in considerable efficiency gains (Zhang et al., 2011). It enables local governments to enjoy the benefits of land value appreciation driven by rapid industrialization and urbanization processes, which greatly increases the local government's extra-budgetary revenues used for urban development, infrastructure projects, and land reclamation projects. More importantly, this system enables the local governments to monopolize the primary market of urban land (that is, the sales of new leasehold interest, also termed the granting of land-use rights, which is different from the trading of existing leasehold interests). The local governments have control over the timing, quantity, and structures in new urban land supply for commercial, industrial, residential, and public use (Tang, Zhu, and Xu, 2007).

Another change to China's urban land supply took place in September of 2010. On August 8, 2010, Order #117 from the Ministry of National Land Resources demanded that administrative departments of land resources at the levels of municipal and county governments had to make an annual Urban Land Supply Plan starting on September 1, 2010. Information including the quantity, spatial distribution, date for sale, and provided mode of land had to be included in the plan. The plan is specific to residential property, including the categories of private, affordable, and resettlement housing. Although the supply constraint is largely government imposed, market conditions must also be considered to optimize leasing revenue (Gu, Michael, and Cheng, 2015).

China's farmland protection policy also has a significant impact on urban land use. Farmland protection has been a national policy for many years. Amid rapid urbanization, the Chinese government opted to tighten urban land use as a way to preserve farmland. The annual land-use supply plan of every city has to be approved by the higher-level authorities. Municipal governments have to allocate limited land supply for various purposes, including industrial, residential, commercial, administrative, infrastructure, and so on. Under these circumstances, residential land supply could be more limited in the cities, where the municipal governments actively pursue industrial park development as a local incentive to attract firms, investment, and employment, as is the case in Shanghai, for instance.

In summary, while urban land-use policy in China has evolved significantly over the past six decades, the supply of urban land for housing development remains tightly controlled by the government. Local governments control the quantity as well as the timing of urban land supply for residential use. They can increase their land reserves by converting rural land into urban land through land releases, according to the local land-use plan. Faced with a monopolized land market system and strict land-use policy, residential land prices and housing prices are largely influenced by the residential land supply.

4. Analytical framework

4.1 Determinants of urban residential land supply

Residential land supply is affected by potential market demand, in which changes in per capita GDP and population play important roles. Since local governments need to consider industrial policy when determining residential land supply, per capita annual new supplies of industrial and commercial land were both included to measure the influence of industrial policy on residential land supply. Assuming a fixed total land area for public use, a larger supply of industrial or commercial land leads to less residential land supply. And since a decreasing capacity of residential land supply may increase residential land prices, we expect the expansion of industrial or commercial land supply to positively influence residential land prices. Finally, we included urban housing prices as one of the determinants of residential land supply.

We expect the influence of housing prices on residential land supply to be negative. At first sight, this negative expectation does not make economic sense: higher prices lead to a lower supply, everything else being equal. But this trend is true in mainland China's large cities, especially in Beijing, Shanghai, Guangzhou, and Shenzhen. It is well known that local public finance in Chinese cities is heavily dependent on land-leasing revenue, especially the housing land, since the tax sharing system reform launched in 1994, with more budgetary resources centralized in the hands of the national government, but with local government's remaining responsibility for many basic spending needs and the support of economic development. Thus, assuming the local government's pursuit of maximum benefit, local governments prefer higher land prices because higher land prices tend to equal higher local public finances. Thus, higher housing prices may lead to a lower land supply, because local governments have the economic motivation to raise land prices by lowering the supply.

4.2 Determinants of urban housing prices

We used a demand and supply framework to determine the major determinants of the levels of urban housing prices. On the supply side, housing unit supply and residential land supply are the main factors influencing housing price levels. When the supply falls short of the demand, supplying more residential units would lead to higher housing prices due to price competition. When the supply exceeds the demand, supplying more residential units would decrease housing prices. We measured housing supply using per capita annual supply of new housing units. We also expect residential land supply to have both positive and negative effects on housing prices. When supply lags behind demand, more residential land supply would lead to housing price increase due to the higher land prices bid up by developers. When housing supply exceeds demand, more residential land supply would lead to lower housing prices. We used per capita annual supply of new residential land to measure residential land supply.

On the demand side, household income was assumed to be a key variable affecting housing price levels. Conditional on housing supply, we expect a higher housing price with greater household income. Per capita disposable income was used to measure household income. The expected influence of household income on housing price is positive. The variations in urban populations can also explain the differences in urban housing prices. We used the non-agricultural population size to measure urban populations and expected it to have a positive effect.

5. Model specifications and data

5.1 Model specifications and variables

To estimate the interactions between residential land supply and housing prices, and also to avoid the measurement bias generated by the existence of endogenous variables, we formulated a simultaneous-equation model, as follows:

$$\begin{cases} HP_{it} = \alpha_0 + \alpha_1 LS_{it} + \alpha_2 X_{1it} + \mu_{1it} \\ LS_{it} = \beta_0 + \beta_1 HP_{it} + \beta_2 X_{2it} + \mu_{2it} \end{cases}$$

Here i refers to the cities; t stands for the year from 2000 to 2015. HP_{it} and LS_{it} stand for the endogenous variables Housing Price and Residential Land Supply, respectively; X_{1it} and X_{2it} are the predetermined variables; μ_{1it} and μ_{2it} are the stochastic disturbance term; α_0 and β_0 are constant coefficients; α_1, α_2, β_1, and β_2 are the coefficients of each variable separately.

There are five exogenous variables, namely: urban housing supply, disposable income, economic development level measured by city GDP, population, and industrial/commercial land supply. Each equation includes one endogenous explanatory variable. The equation on housing prices includes

Table 15.1 Descriptive statistics.

Variable	Unit	Obs	Mean	Std. Dev.	Min	Max
Housing price	yuan/m^2	496	5,406.805	4,154.125	1,077	33,661.39
Per capita annual new residential land	m^2	369	9,621.67	10,795.89	0	56,410.4
Non-agricultural population	10,000	369	453.58	374.78	8.82	1,877.7
Per capita annual newly built housing	m^2	369	1.59	1.08	0.18	13.13
Per capita disposable income	yuan	472	19,249.92	10,205.29	5,550	52,962
Per capita GDP	yuan	416	45,100.25	28,535.88	5,616	149,497
Per capita annual new industrial/commercial land	m^2	369	11,877.12	19,933.25	0	135,299.2

Source: Compiled from China Index Academy Database, which can be accessed from China National Knowledge Infrastructure.

two exogenous variables (economic development level and industrial/commercial land supply). It is over-identified. The second equation on land supply also includes two exogenous variables (urban housing supply and disposable income). The second equation is also over-identified. We use 3SLS to estimate the simultaneous-equation model.

Descriptive statistics of the main variables are detailed in Table 15.1.

5.2 Data

Our research samples are 31 cities in China, which include the four municipalities directly under the central government (Beijing, Shanghai, Tianjin, and Chongqing) and most of the provincial capital cities. The sample period is from 2000 to 2015. The data on residential land prices come from www.landvalue.com.cn. China's Ministry of Land and Resources collects this land data and posts it on the web. Data on other variables, including housing price, urban residential land supply, annual supply of new housing, income, population, GDP, and the annual supply of industrial/commercial land were compiled from the database of the China Index Academy. Data used in the research in the year 2015 are displayed in Appendix Table 15.7.

6. Empirical results

6.1 Benchmark regression

Table 15.2 reports the benchmark regression results. The results indicate that residential land supply has a significant negative impact on housing prices. The

Table 15.2 Estimation results of benchmark regression.

VARIABLES	(1) Urban housing prices	(2) Urban residential land supply
Per capita annual new residential land	−0.0868***	
	(0.0176)	
Per capita annual new housing	−194.5*	
	(100.1)	
Per capita disposable income	0.434***	
	(0.0241)	
Non-agricultural population size	1.033***	0.929
	(0.225)	(1.146)
Urban housing prices		−0.534**
		(0.256)
Per capita GDP		99.87
		(72.05)
Per capita annual new industrial/commercial land		0.336***
		(0.0298)
East	316.8	−1,189
	(248.4)	(1,062)
West	−1,144***	−1,279
	(209.1)	(923.5)
2001	−116.3	490.5
	(432.3)	(1,852)
2002	−270.0	1,183
	(425.7)	(1,821)
2003	−237.8	2,615
	(425.6)	(1,809)
2004	−387.6	3,512*
	(430.0)	(1,803)
2005	−149.2	6,127***
	(440.4)	(1,815)
2006	85.56	10,999***
	(480.3)	(1,844)
2007	397.1	12,157***
	(522.1)	(1,941)
2008	−572.4	7,259***
	(528.2)	(2,018)
2009	274.4	11,776***
	(600.7)	(2,262)
2010	996.0	12,199***
	(684.8)	(2,687)
2011	−293.8	8,407***
	(733.0)	(2,874)
2012	−1,376*	7,677***
	(745.6)	(2,784)
2013	−1,721**	9,493***
	(828.5)	(3,000)
2014	−2,662***	7,293**
	(822.6)	(2,919)

Table 15.2 (Cont.)

VARIABLES	(1)	(2)
	Urban housing prices	Urban residential land supply
2015	−3,759***	5,055*
	(862.2)	(2,952)
Constant	−1,438***	1,302
	(369.2)	(1,497)
Observations	361	361
R-squared	0.847	0.666

Note: Standard errors in parentheses; *** $p<0.01$; ** $p<0.05$; * $p<0.1$.
Source: The data of residential land price were web-scraped from www.landvalue.com.cn, and the remaining data were compiled from the China Index Academy Database, which can be accessed from China National Knowledge Infrastructure.

regression estimates indicate that, for each 100 m² decrease in residential land supply, housing prices would increase by nearly 8.68 RMB/m². Housing supply also has a significant negative impact on housing prices. For each 10,000 m² decrease in per capita annual new housing, housing prices would increase by nearly 194.5 RMB/m².

We also find that housing price appreciation in major Chinese metropolitan areas was accompanied by growth in per capita disposable income. Each 100 RMB increase in per capital disposable income would lead to a 43.4 RMB/m² increase in housing prices. Non-agricultural population size has a positive impact on housing prices. Each 10,000-person increase in urban population would lead to 1.03 RMB/m² increase in housing prices. Urban population is obviously one of the most important factors that determine housing prices.

Housing prices in the West region are 1,144 RMB/m² lower than other regions, after controlling for the variables, such as income, population, housing supply, and land supply. These lower prices are attributed to a lower degree of marketization in property and economic development in the West region. Since 2010, a number of first- and second-tier cities have implemented house-purchasing limitation orders to rein in market speculation and slow high housing prices. In 2011, 36 cities issued such orders, which seem to have an effect on housing prices. Estimation results show that housing prices in 2012–2015 were much lower.

6.2 Tests of alternative channels of influence between land and housing prices

As mentioned in the literature review, scholars have raised three possible mechanisms through which residential land supply influences housing prices. The first is through influencing the supply of housing units, which affects housing prices. Decrease in land supply limits housing development and thus raises housing prices. Some may argue that a decrease in land supply does not necessarily limit housing development because one can potentially adjust the density of housing units to maintain a similar housing supply. This is hardly the case in China, however. In China, the floor-area ratio is determined *a priori* during the urban planning process, and a land transaction can hardly change the original urban plan. The second channel is land supply influencing expectations of future housing and rental prices. In a rational market, a decrease in land supply leads to a higher expectation of future housing prices, which is capitalized into higher current housing prices. The third mechanism is through raising residential land prices. Decrease in land supply raises land prices, which in turn raises housing prices by increasing the cost of housing construction.

We applied a fixed effect model to test these three mechanisms by doing Hausman tests. We tested the first mechanism by examining the influence of the residential land supply on the annual supply of new housing. Estimation results in Table 15.3 show that residential land supply had no significant short-term influence on annual newly built housing. Note that we used a two-year-lagged supply of urban residential land to capture the influence of residential land supply on annual newly built housing. This is because, in practice, the median time from the start of land acquisition to completion of housing construction is two years. Thus, assuming that developers start construction right after they acquire the land, housing prices are most likely to be influenced by the residential land supply two years earlier. We also ran the same regression with a one-year lag of urban residential land supply, and the results from Table 15.6 (see Appendix) also show that residential land supply has no significant effect on annual newly built housing. This result still suggests that the first mechanism might not be valid among these major cities.

We then tested the influence of residential land supply on the expectation of future housing prices. In fact, it is difficult to measure expectations. Two methods are commonly used in academic research. One uses the actual variables in the future next period to measure rational expectation variables in this period (Blanchard and Kahn, 1980; Galí and Gertler, 1999; Kuang, 2010). The other uses substitution variables to measure rational expectations (Carlson and Parkin, 1975). We used the next year's housing prices to measure public expectation of future housing prices, and the estimation results showed that the residential land supply had a negative influence on rational expectations (see Table 15.4). This finding provides evidence in favor of the second mechanism, indicating that a reduction in residential land supply raises expectations of future housing prices, which in a rational market will push up current housing prices.

We then estimated the impact of residential land supply on residential land prices. The estimation results reported in Table 15.5 show that residential land supply has a negative influence on residential land prices, indicating that a reduction in residential land supply increases residential land prices, which may push up housing prices due to increased supply costs.

Table 15.3 Estimation results of mechanism test (1).

VARIABLES	Annual newly built housing
Urban residential land supply two years ago	4.28e-06
	(4.65e-06)
Annual real estate development investment ratio	−0.0414***
	(0.00980)
Per capita annual newly built housing	0.240***
	(0.0147)
Urban population	−0.000949**
	(0.000426)
2003	0.106
	(0.133)
2004	−0.0520
	(0.136)
2005	0.0774
	(0.141)
2006	−0.122
	(0.148)
2007	−0.341**
	(0.157)
2008	−0.547***
	(0.169)
2009	−0.407**
	(0.185)
2010	−0.642***
	(0.198)
2011	−0.757***
	(0.226)
2012	−0.936***
	(0.244)
2013	−1.080***
	(0.262)
2014	−0.798***
	(0.268)
2015	−0.774***
	(0.285)
Constant	1.152***
	(0.216)
Observations	295
Number of cities_n	30
R-squared	0.715

Note: Standard errors in parentheses; *** $p<0.01$; ** $p<0.05$; * $p<0.1$.

Source: The data of residential land price were web-scraped from www.landvalue.com.cn, and the remaining data were compiled from the China Index Academy Database, which can be accessed from China National Knowledge Infrastructure.

Table 15.4 Estimation results of mechanism test (2).

VARIABLES	Expectation of future housing price
Per capita annual newly residential land	−0.0556***
	(0.0126)
Per capita GDP	147.5***
	(39.37)
Urban population	9.812***
	(1.154)
2001	8.286
	(440.0)
2002	159.1
	(440.9)
2003	296.5
	(446.4)
2004	670.8
	(453.2)
2005	1,068**
	(466.5)
2006	2,052***
	(499.9)
2007	2,159***
	(532.9)
2008	2,612***
	(552.3)
2009	4,134***
	(615.3)
2010	4,306***
	(703.6)
2011	3,905***
	(750.8)
2012	4,292***
	(751.5)
2013	4,435***
	(825.8)
2014	4,646***
	(856.1)
2015	–
Constant	−1,892***
	(617.6)
Observations	357
Number of cities_n	30
R-squared	0.760

Note: Standard errors in parentheses; *** $p<0.01$; ** $p<0.05$; * $p<0.1$.

Source: The data of residential land price were web-scraped from www.landvalue.com.cn, and the remaining data were compiled from the China Index Academy Database, which can be accessed from China National Knowledge Infrastructure.

Table 15.5 Estimation results of mechanism test (3).

VARIABLES	Urban residential land price
Per capita annual new residential land	−0.0599***
	(0.0160)
Urban housing prices	0.988***
	(0.0886)
Per capita GDP	−46.54
	(31.85)
Urban population	5.133***
	(1.727)
Constant	−1,163*
	(652.8)
Observations	369
Number of cities_n	30
R-squared	0.600

Note: Standard errors in parentheses; *** p<0.01; ** p<0.05; * p<0.1.

Source: The data of residential land price were web-scraped from www.landvalue.com.cn, and the remaining data were compiled from the China Index Academy Database, which can be accessed from China National Knowledge Infrastructure.

7. Conclusions and policy implications

The question of what has driven the housing market boom in large Chinese cities over the past 15 years has been examined extensively by other studies from the demand side. However, some observers suspect that the monopoly supply of urban land by the local government may be a key factor in driving housing prices up in urban China. A major problem in China is that policy and institutional factors have had a negative impact on housing supply elasticity at the very moment that China was experiencing its peak rate of urbanization. The phenomenon that local public finance relies heavily on land-leasing revenue is one of the important reasons. Local governments in the first- and second-tier cities intended to reduce the supply of residential land as the supply of land for industrial and commercial development increased gradually. Using the panel dataset of 31 major Chinese cities from 2000 to 2015, we analyzed the influence of residential land supply on housing prices.

We found that the housing price increases in these major cities were a result not only of robustly rising demand, but also of the decreasing supply of residential urban land. We further examined the mechanisms through which constrained land supply raised housing prices in these major cities. We found that the reduction in land supply for housing might have resulted in the expectation of higher future housing prices, and this expectation might be capitalized into higher current housing prices in a rational market. Moreover, reduced land supply for housing might give rise to higher residential land prices, which might further push up housing prices.

The main findings of this study have important policy implications. First, these findings confirm that decreasing the supply of residential urban land was an important reason for rising housing prices. In this respect, we suggest that the local governments in those sample cities increase residential land supply as a key policy measure to slow down the increase of housing prices, which become unaffordable for the younger generations. Second, in the past decades, the central government of China carried out a series of macro-controls on the real estate market, hoping to effectively suppress the skyrocketing housing prices. However, the housing booms in some cities did not really cool down. The central government regulations seemed to fail. The lack of understanding of the ultimate reasons for rapidly growing housing prices may have contributed to the design flaws of the public policy, and hence its failure. This paper explains the reasons for the high housing prices in the major cities and clearly identifies the effect of residential land supply by local governments on housing prices. It can provide some guidelines when the central government draws up future relevant policies. Lastly, the main findings can also help investors to gain a better understanding of the housing market prices and make more prudent investment decisions.

Appendix

Table 15.6 Estimation results of mechanism test (1), with one-year lag.

VARIABLES	Annual newly built housing
Per capita new residential land in the last year	−1.08e-07
	(4.35e-06)
Annual real estate development investment ratio	−0.0391***
	(0.00849)
Per capita annual newly built housing	0.260***
	(0.0130)
Urban population	−0.000281
	(0.000361)
2001	−
2002	−0.0346
	(0.135)
2003	0.0385
	(0.137)
2004	−0.141
	(0.139)
2005	−0.0314
	(0.142)
2006	−0.261*
	(0.149)
2007	−0.471***
	(0.159)
2008	−0.682***
	(0.168)

Table 15.6 (Cont.)

VARIABLES	Annual newly built housing
2009	−0.550***
	(0.175)
2010	−0.844***
	(0.198)
2011	−1.045***
	(0.226)
2012	−1.137***
	(0.224)
2013	−1.355***
	(0.236)
2014	−1.111***
	(0.257)
2015	−0.925***
	(0.261)
Constant	0.832***
	(0.193)
Observations	327
Number of cities_n	30
R-squared	0.739

Note: Standard errors in parentheses; *** p<0.01; ** p<0.05; * p<0.1.

Source: The data of residential land price were web-scraped from www.landvalue.com.cn, and the remaining data were compiled from the China Index Academy Database, which can be accessed from China National Knowledge Infrastructure.

Table 15.7 2015 data used in the research.

City name	Per capita disposable income (yuan)	Investment in real estate development (billion yuan)	GDP (billion yuan)	Housing price (yuan/m^2)	Annual newly built housing (10,000 m^2)	Residential land price (yuan/m^2)	Annual new industrial land (m^2)	Annual new commercial land (m^2)	Annual new residential land (m^2)
Beijing	52,859	4,177.05	22,968.6	22,300.32	1,378.22	48,751	1.40E+06	1.40E+06	4.10E+06
Changchun	–	501.32	–	6,374.41	434.66	2,767	5.50E+06	616,874	4.40E+06
Changsha	39,961	1,006.84	8,510.13	5,543.73	949.47	4,614	7.80E+06	2.60E+06	3.60E+06
Chengdu	–	2,435.25	–	6,584.08	858.64	7,609	1.60E+07	3.70E+06	1.00E+07
Dalian	35,889	897.46	7,731.6	8,710.85	233.48	3,121	3.20E+06	862,162	2.10E+06
Fuzhou	–	1,381.12	–	11,332.75	745.09	11,102	5.90E+06	1.90E+06	4.40E+06
Guangzhou	46,735	2,137.59	18,100.41	14,083.32	981.3	28,196	3.40E+06	1.20E+06	3.60E+06
Guiyang	27,241	1,001.03	2,891.16	4,966.71	1,080	3,742	3.30E+06	1.80E+06	3.30E+06
Haerbin	30978	593.98	5,751.2	6,124.32	964.85	2,528	4.80E+06	1.60E+06	2.90E+06
Haikou	28535	456.39	1,161.28	7,635.91	172.05	4,086	198,929	696,236	1.40E+06
Hangzhou	48,316	2,472.6	5,660.27	14,747.94	1,070.27	17,061	5.30E+06	1.80E+06	4.80E+06
Hefei	31,989	1,259.14	3,090.5	7,511.89	709.5	4,192	1.20E+07	802,650	6.00E+06
Huhehaote	37,362	509.05	6,100.23	4,945.63	235.1	3,521	2.30E+06	630,425	2.10E+06
Jinan	39,889	1,014.4	3,970	7,526.88	370.55	3,871	7.00E+06	1.40E+06	5.90E+06
Kunming	33,955	1,451.31	2,095.99	7,177.86	482.05	7,047	8.60E+06	1.40E+06	3.00E+06
Lanzhou	27,088	320.56	4,000.01	6,088.92	164.8	2,820	4.10E+06	4.00E+06	3.80E+06
Nanchang	31,942	485.37	–	6,955.35	346.31	5,156	9.80E+06	1.50E+06	3.30E+06
Nanjing	–	1,429.02	–	11,260.37	1,063.87	9,764	7.00E+06	811,063	4.90E+06
Ningbo	–	1,228.84	9,300.07	11,021.82	613.39	12,669	5.40E+06	2.30E+06	3.30E+06
Qingdao	–	1,122.35	24,964.99	8,436.55	1,047.67	6,152	1.40E+07	2.30E+06	5.50E+06
Shanghai	52,962	3,468.94	–	21,501.07	1,588.95	35,767	2.00E+06	1.30E+06	7.00E+06
Shenzhen	–	1,331.03	–	33,661.39	202.37	42,251	750115	382,072	157,583
Shenyang	–	1,337.66	5,440.6	6,416.07	768.76	2,803	8.20E+06	616,483	3.30E+06
Shijiazhuang	28,097	965.13	2,735.34	7,797.7	259.91	3,976	6.40E+06	1.30E+06	6.00E+06
Taiyuan	27,727	597.83	–	7,303.07	322.01	2,976	2.40E+06	600,881	3.50E+06

Tianjin	34,101	1,871.55	16,538.19	9,869.62	2,182.99	6,291	1.40E+07	1.90E+06	5.30E+06
Wuhan	36,436	2,581.79	10,905.6	8,403.58	654.63	6,392	1.30E+07	1.60E+06	6.40E+06
Xi'an	33,188	1,820.85	5,810.03	6,220.53	747.84	5,476	5.40E+06	1.80E+06	5.20E+06
Xiamen	–	774.07	3,466	18,928.07	274.17	21,836	813,035	1.20E+06	988,660
Zhengzhou	–	2,000.2	7,315.2	7,222.57	670.5	4,056	1.20E+07	4.60E+06	1.10E+07
Chongqing	27,239	3,751.28	15,719.72	5,012.45	3,185.9	4,165	4.30E+07	9.40E+06	3.60E+07

Source: The data of residential land price were web-scraped from www.landvalue.com.cn, and the remaining data were compiled from the China Index Academy Database, which can be accessed from China National Knowledge Infrastructure.

References

Alexander, F. S. (2013). Land Banking as Metropolitan Policy. *Social Science Electronic Publishing*, (44),3–28.

Bertaud, A. and Renaud B. (1997). Socialist Cities without Land Markets. *Journal of Urban Economics*, 41(1), 137–151.

Blanchard, O. J. and Kahn, C. M. (1980). The Solution of Linear Difference Models under Rational Expectations. *Econometrica*, 48(5), 1305–11.

Jensen, C. (2002). Foreign Direct Investment, Industrial Restructuring and the Upgrading of Polish Exports. *Applied Economics*, 34(2), 207–217.

Carlson, J. A. and Parkin, M. (1975). Inflation Expectations. *Economica*, 42(23), 123–38.

Dong, Y. (2016). A Note on Geographical Constraints and Housing Markets in China. *Journal of Housing Economics*, 33, 15–21.

Ding, C. and Knaap, G.-J. (2002). Property Values in Inner-city Neighborhoods: The Effects of Homeownership, Housing Investment, and Economic Development. *Housing Policy Debate*, 13(4), 701–727.

Ding, C. R. (2002). Land Policies and Urban Housing Development. *Urban Studies* (in Chinese), 9(2), 61–66.

Galí, J. and Gertler, M. (1999). Inflation Dynamics: A Structural Econometric Analysis. *Journal of Monetary Economics*, 44(2), 195–222.

Gereffi, G. (1999). International Trade and Industrial Upgrading in the Apparel Commodity Chain. *Journal of International Economics*, 48(1), 37–70.

Glaeser, E. L., Gyourko, J., and Saiz, A. (2008). Housing Supply and Housing Bubbles. *Journal of Urban Economics*, 198–217.

Glaeser, E. L. and Saks, R. (2005). Why Is Manhattan So Expensive? Regulation and the Rise in Housing Prices. *Journal of Law & Economics*, 48(2), 331–69.

Glaeser, E. L. and Ward, B. A. (2009). The Causes and Consequences of Land Use Regulation: Evidence from Greater Boston. *Journal of Urban Economics*, 65(3), 265–278.

Gu, G., Michael, L., and Cheng, Y. (2015). Housing Supply and its Relationships with Land Supply. *International Journal of Housing Markets & Analysis*, 8(3), 375–395.

Huang, H. and Tang, Y. (2012). Residential Land Use Regulation and the US Housing Price Cycle Between 2000 and 2009. *Working Papers*, 71(1), 93–99.

Hanson, G. H. (2005). Market Potential, Increasing Returns, and Geographic Concentration. *Journal of International Economics*, 67(1), 1–24.

Ho, W. K. O. and Ganesan, S. (1998). On Land Supply and the Price of Residential Housing. *Journal of Housing & the Built Environment*, 13(4), 439–452.

Huang, J., Shen, G. Q., and Zheng, H. W. (2015). Is Insufficient Land Supply the Root Cause of Housing Shortage? Empirical Evidence from Hong Kong. *Habitat International*, 49, 538–546.

Hui, C. M., Leung, Y. P., and Yu, K. H. (2014). The Impact of Different Land-supplying Channels on the Supply of Housing. *Land Use Policy*, 39(3), 244–253.

Ihlanfeldt, K. R. (2007). The Effect of Land Use Regulation on Housing and Land Prices. *Journal of Urban Economics*, 61(3), 420–435.

Kuang, W. (2010). Expectation, Speculation and Urban Housing Price Volatility in China. *Economic Research Journal*, (9), 67–78 (in Chinese).

Meen, G. and Nygaard, A. (2010). Housing and Regional Economic Disparities (Government Report), www.communities.gov.uk.

Murphy, A, Muellbauer, J., and Cameron, G. (2006). *Housing Market Dynamics and Regional Migration in Britain*. Department of Economics.

Paciorek, A. (2012). Supply Constraints and Housing Market Dynamics. *Journal of Urban Economics*, 77(1), 11–26.

Quigley, J. M. and Raphael, S. (2004). Regulation and the High Cost of Housing in California. *American Economic Review*, 95(2), 323–328.

Peng, R, and Wheaton, W. C. (1994). Effects of Restrictive Land Supply on Housing in Hong Kong: An Econometric Analysis. *Journal of Housing Research*, 5(2), 263–291.

Peng, L. and Thibodeau, T. G. (2012). Government Interference and the Efficiency of the Land Market in China. *Journal of Real Estate Finance & Economics*, 45(4), 919–938.

Rabe, B. and Taylor, M. (2010). Differences in Opportunities? Wage, Unemployment and House-price Effects on Migration. ISER Working Paper. No.05.

Skiba, A. (2006). Immigration, Firm Relocation and Welfare of Domestic Workers. 6th Annual Missouri Economics Conference Selected Papers, Columbia.

Tang, T., Zhu, T., and Xu, H. (2007). Integrating Environment into Land-use Planning through Strategic Environmental Assessment in China: Towards Legal Frameworks and Operational Procedures. *Environmental Impact Assessment Review*, 27(3), 243–265.

Yan, S., Ge, X. J., and Wu, Q. (2014). Government Intervention in Land Market and its Impacts on Land Supply and New Housing Supply: Evidence from major Chinese Markets. *Habitat International*, 44, 517–527.

Zabel, J. and Dalton, M. (2011). The Impact of Minimum Lot Size Regulations on House Prices in Eastern Massachusetts. *Regional Science & Urban Economics*, 41(6), 571–583.

Zhang, T. T., Zeng, S. L., Gao, Y., Ouyang, Z. T., Li, B., Fang, C. M., and Zhao, B. (2011). Assessing Impact of Land Uses on Land Salinization in the Yellow River Delta, China Using an Integrated and Spatial Statistical Model. *Land Use Policy*, 28(4), 857–866.

Part 4
Implications for China

16 A roadmap for housing policy
Lessons of international experience

Edward Glaeser

HARVARD UNIVERSITY

Rebecca L. H. Chiu

THE UNIVERSITY OF HONG KONG

Zhi Liu

LINCOLN INSTITUTE OF LAND POLICY AND PEKING UNIVERSITY-LINCOLN INSTITUTE CENTER FOR URBAN DEVELOPMENT AND LAND POLICY

Bertrand Renaud

HOMER HOYT INSTITUTE, U.S.

1. Value of international experience for Chinese housing policies

China's housing policies face a significant transition. The country is rebalancing spatially with the emergence of massive clusters of cities, economically with labor-intensive manufacturing beyond its peak and the economic rebalancing away from investments and exports towards domestic demand, socially with the rise of a large urban middle-class with increasingly diverse preferences, and demographically with the ageing of the population. China's urban population has quintupled over the past 40 years and is expected to reach one billion by 2030. Pressures for change continue to grow on the housing system. In spite of its large-scale and sometimes unique features, is there anything of value that Chinese housing policies could draw from the earlier experiences of other high-income economies?

Since the 1980s, an increasing consensus has coalesced around desirable features of housing policies and around ways to improve the social and economic performance of this major sector. Broadly agreed principles and criteria to evaluate the performance of a housing system and to design appropriate housing policies have emerged to facilitate the making of policy choices, nationally and city by city. This closing chapter focuses on how such principles and criteria can be helpful to a middle-income economy like China, whose urban system has grown massively since the 1980s and still faces about 250 million more new urban residents in the next 15 years. The returns from making the right choices of housing policies should be very high. However, the principles and criteria presented in this chapter form only a broad roadmap: they are the starting points to review and rethink existing policies and practices. Considerable operational work lies ahead for policy makers.

Several factors taken together explain idiosyncratic housing policies in the past and the difficulty of defining what was a well-performing housing system. One is that the bundle of characteristics of housing as an economic good makes it difficult to manage the sector well spatially and over time. Housing is a complex economic good to analyze and manage properly for four reasons. Its durability is measured in decades. It can be quite heterogeneous in terms of its structural features and unit design. Its value is not simply determined by the characteristics of the unit itself, but is also linked to the characteristics of the neighborhood where the unit is located. And the way housing is financed, built and taxed is usually extensively regulated by national and local governments.

Housing is a data-intensive sector, and the absence or scarcity of relevant, timely and accurate information long encouraged improvisation or the substitution of ideology for missing data and evidence. Because housing and other types of real estate are quintessential non-traded local goods, for a long time there was little incentive to carry out comparative analyses of the performance of national housing systems and to agree on principles and criteria to evaluate performance. Things changed as the weight of housing and its impacts on the national economy kept growing bigger in middle- and high-income economies in terms of economic stability and social welfare. The development of modern analytical methods for housing and urban development really took off only four decades or so ago.[1]

The rest of this chapter has two parts. The first part is a synthesis of the principles and criteria for evaluating prevailing housing conditions and designing sound housing policies learned from international experience. The second part recommends policy actions in China. This chapter is addressed to policy makers and to the researchers and analysts concerned with the performance and improvement of housing policies. It is appropriate to stress that while the ideas discussed here reflect collective experience, not everyone may agree with every idea discussed here in every housing context. The responsibility for the views presented in this roadmap lies with its authors alone.

2. Principles and criteria for designing housing policies

This section opens with four principles that are significant for the diagnosis of a housing system and the choice of policies. These design principles are: prioritizing housing consumption over asset ownership; promoting widespread housing affordability; always jointly considering housing structure and location; giving priority to helping the young; and aiming at tenure neutrality and robust rental markets. Then three efficiency criteria must be kept in mind in order to reduce the social costs of providing decent housing for all: efficiency in the use of resources; efficiency in the design of demand-side vs. supply-side subsidies; and efficiency in land-use regulation. Finally, we highlight four institutional qualities shared by well-performing housing and real estate systems, nationally and locally: uniform property rights; free entry and competition; transparent rules; and socially inclusive but balanced urban planning.

2.1 Four core principles for housing policy choices

A housing policy is typically the outcome of a cumulative process of individual policy decisions made over time and is very rarely a single policy event. Yet, drawing on the growing comparative experience of recent decades, evaluating the performance of a middle- or high-income housing system best begins with four core principles. The extent to which these four principles appear to be embedded in the behavior of a housing market tells how well this system is likely to meet desired social, political, and economic housing objectives nationally or, better, locally, because the performance of a housing system is local in nature and differs across cities.

1. Prioritize housing consumption over asset ownership

Housing is an unusual commodity because it is both a consumption good and an asset. For the occupants of a home, their house is the stage on which they play out their private lives. The shape and location of that house helps to determine their comfort, health, and access to various services. For the owners of a home, that house is a store of value that can be sold to provide cash at some future date. For those who are both owners and residents, a house is both a living space and a store of value.

Housing policy debates alternate between the asset side and the consumption side of housing. Such debates are usually shaped by the local political economy, the income distribution, the fiscal capacity of the state, and the supply responsiveness of the local housing industry. An emphasis on housing assets and homeownership has often reflected the combined pressure of rapid urbanization and the demand for housing from a growing urban middle-income class, together with a limited fiscal capacity of the state to face urban infrastructure needs. The emphasis on the consumption side has been driven by the social goal that every citizen should have access to a minimum bundle of safe housing services, which typically increases over time as the income level of the country rises. For example, nineteenth-century housing regulation in Europe and the U.S. tended to focus on the health impacts of bad housing and regulated the structural safety and ventilation of the rented tenements.

After the 1980s, in the U.S., housing vouchers directed resources to lower-income renters to enable them to live in better private apartments and better neighborhoods. In Europe, there remain great variations from country to country in the relative size of the private and public social housing stock, its structure, and its investor composition (Taffin, 2007). While in the UK social housing policies have been subject to noticeable ideological swings over time, in continental Europe social housing policies have tended to be more stable and more pragmatic over time. A characteristic of East Asian countries has been a primary policy focus on homeownership throughout urbanization until its late stages (as discussed in Chapters 1 and 12).

Other housing policies aim at the expansion of the housing stock and asset accumulation, by increasing the level of homeownership and increasing the ease with which mortgage debt is repaid. In that context, the design of the housing finance system and its sources of funding play a critical role. Taxation can also be a very large factor in housing asset formation. Among Western countries, U.S. housing policies stand out for the dominance of indirect subsidies to ownership through the tax system, using mortgage interest deductions and capital gains tax exemptions which are larger in dollar terms than the value of the directly provided on-budget social subsidies, with other subsidies conveyed through the central role of government-sponsored mortgage finance institutions (Jaffee et al., 2007). Both common sense and the historical record suggest that public support for the provision of housing services consumption for lower-income groups should be prioritized. It is a better way of optimizing the use of scarce public resources than to subsidize homeownership, which favors owners of larger units and therefore the higher-income groups.

The opportunity cost of housing matters, and it is necessary to ask whether public and private resources going into new housing could have a better use elsewhere. Depending on national and local conditions, housing may not be the best investment for the nation's long-term economic future, and it may not be the right investment for the future of a particular family either. The multiplier effects of housing investment also matter. New housing doesn't necessarily generate the same employment multipliers as a new factory. Japan has also taught us that the myth of ever-rising land and housing prices is quite dangerous as it feeds very large bubbles that are extremely costly when they burst and have a lasting aftereffect.

Ownership of housing exposes households to the risk that housing prices may fall as well as rise. America's rush to create an "ownership society" through housing policy seems to have created instead a "mortgage foreclosure economy," since so many buyers were unable to pay for their mortgages once the 2008–2009 recession hit. Basic financial theory suggests that households should diversify their risks by holding a portfolio diversified among multiple assets. Focusing exclusively on homeownership as an asset leads to less diversification and ultimately more household risk. Therefore, the development of alternative forms of investment for household savings with sound, risk-adjusted returns can be quite beneficial to society. This is particularly important for China, where the majority of households have few, if any, alternative investment opportunities to diversify their assets.

There are a number of good reasons for policies that aim to improve the quality of housing consumed by ordinary people. We may want to ensure that children grow up in healthy housing and live in adequate neighborhoods. Taxpayers may want to redistribute resources to the poor, but they may prefer that those resources come in the form of housing, rather than habit-forming alcohol consumption or gambling by men. There is a coherent economic argument for opposing all good-specific policies and favoring cash transfers, but there are also pragmatic arguments for supporting the consumption of housing.

It is important to recognize that sound housing policies attempt to ensure that everyone has access to adequate housing, not that homeowners see healthy housing price appreciation. This stance was recently stated publicly by China's President Xi Jinping. The case of Singapore is an informative exception. Since independence in 1965, creating a citizenry of homeowners with a stake in the country's future has been a major goal. There were several unique factors in the success of this policy: a sustained housing policy governance of high quality and free of corruption for decades; the explicit integration of housing and urban development into day-to-day macroeconomic management, which is not the case for most other countries; and the widespread, pragmatic public understanding that Singapore, being a totally open, small economy, would quickly run into trouble if housing was mismanaged (see Chapter 11 on Singapore).

Another reason to prioritize consumption over asset accumulation is that ownership is much more rigid in terms of labor mobility than rental housing. Owning creates much higher moving costs, which means that families can be locked into the wrong neighborhood within a city, or into the wrong city.[2] Renting makes it easy for people to change homes, and since renters are typically poorer, housing support for the rental markets can be easily targeted towards the most disadvantaged (Peppercorn and Taffin, 2013). In fact, rental demand has been on the rise, either due to stagnant wages and lower incomes for young households in the U.S. and Europe after the Global Financial Crisis, and very significant housing price inflation in Asia, or due to new housing consumption behavior spurred by changing social trends such as ageing and changed lifestyles. Thus, expanding the private rental supply to match recent housing consumption trends has been the concern of many governments. In a newly announced housing policy reform, the central government of China emphasized the development of the rental housing market (see Chapter 13).

2. Promote widespread affordability

The choice to focus on housing consumption rather than on housing asset accumulation has one major implication for the public's outlook: higher housing prices become a negative rather than a positive factor. If the objective is to ensure that homeowners accumulate assets, then the objective of public policy is to ensure that housing prices go up. If the objective is to ensure decent housing consumption, then the objective of public policy is to ensure that prices stay stable or even go down in real terms. High prices hurt those who would come to cities, just as they benefit those who already own urban housing. Typically, buyers are poorer than owners, so that higher prices essentially transfer funds and wealth from the poor to the rich. This problem is potentially more serious in China as 250 million of the rural population are expected to move to cities in the next 15 years. The dilemma is also sharper in democratic, homeowning market economies where public policy is often shaped by the electorate who, in these cases, are dominated by homeowners.

Housing always combines a structure and land. When supply is not constrained by regulations, the price of housing directly reflects the costs of building a structure and the value of using the land. If building costs fall, then it will be cheaper to replenish and improve the nation's building stock. Consequently, the improved affordability that comes from technological change in the building industry is a social benefit. Public policy should not limit competition or other sources of technological improvements in the construction industry.

The value of land is more complex. When land supply is not constrained by policy, there are two reasons why land becomes more valuable. One is that local governments have improved the quality of local services, and we should certainly hope that local services improve. The other is that the growth of the urban population and the economy drives up demand for premier urban locations. Yet land can also become more valuable if government acts like a local monopolist and artificially restricts the supply of land, if it owns all or a significant amount of land, as Hong Kong shows. Land can also become more valuable if governments use land-use regulations to limit the supply of new homes. Such artificial limitations on the supply of land or homes mean that they produce places that attract too few workers, and the families that come pay too much for their housing.

The current Affordable Housing Movement in the U.S. has unfortunately chosen to prioritize the production of specifically designed affordable housing units, rather than ensuring that there is enough housing supply so that everyone has access to moderately priced housing space. By focusing on this special class of housing, which is accessed only through a lottery, the movement has distracted public attention from the larger goal of promoting better homes for all and has often implicitly taxed new developments by requiring builders to set aside a small number of apartments that will be given away at below market prices. In Hong Kong, the large public housing sector has ensured housing quality for a major percentage of the population, currently of the order of 46 percent. However, there remains a "sandwich class" consisting of those outside the subsidy net who are left to face market forces in the rapidly appreciating private sector. Singapore's successful housing policy has entered a new era where the small rental market is increasingly squeezed between sharpening land shortages on the supply side and increasing pressure from incoming high-income immigrant labor on the demand side.

3. Always jointly consider the housing structure and its location

When a family chooses a home, it also chooses the neighborhood where it will be living. One of the most common mistakes in social housing policy is to focus on structure while ignoring its location, often a deliberate choice to reserve better-located land for private housing development for higher fiscal or investment return. Both location and structure are important, and in richer societies, location becomes more important than structure.

There has long been a tendency of social housing policy advocates and of public agencies to focus on structure, with little attention—deliberately or out of ignorance—to the neighborhood and the part of the city where it is located. Public housing projects in post-war America tore down reasonably well-functioning, if somewhat dilapidated neighborhoods, and replaced them with towers that became places of crime and social dysfunction. Brazil replaced traditional favelas that were close to city centers with housing projects, such as the infamous City of God project, that were far more distant and isolated. In many Chinese cities, the choices of inconvenient locations—where land is cheap but far from jobs and urban services—for affordable housing development are often the main cause for failure. However, in Singapore and Hong Kong, the two developed Asian economies with government control of land supply and large public housing programs, the publicly built housing neighborhoods are comparable to or sometimes even better than those of private housing, and a substantial portion of them are in good locations because, though they were built in the urban fringe then, subsequent city development has rendered them rather centrally located.

The simplest illustration of the need to consider location occurs when we are considering affordability alone. Consider two units: one is cheap but far from the city center, and the other is expensive but close to the city center. The first unit may seem like it is promoting affordability, but when commuting costs are taken into account, the distant unit may end up being more expensive to the user, though cheaper to the provider when land costs are taken into consideration. It is always appropriate to consider both housing and transportation costs when considering the total cost of housing consumption.

The City of God trade-off in Brazil is somewhat more complex, especially since the favela residents who were relocated to the new housing project never had a proper title to their original homes. The essential decision lay between leaving the favela in place or moving the poorer residents to a newly built housing project. As the favela's current location was quite valuable, relocating the residents generated enough value to justify the costs of higher-quality buildings. It seems clear that the relocated residents were enjoying better buildings after relocation, but the quality of their lives may well have deteriorated because of the isolation and danger of their new location. The decision may still have been optimal for the city as a whole, because of the social value of intensifying the uses of the central land, but the relocation does not seem to have helped the favela residents themselves because structural improvements did not make up for locational degradation.

Urban housing projects in post-World War II America emphasized improved structure. Large towers were built without much attention to street life. Densities in some cases actually decreased despite increased height. As Jane Jacobs has most persuasively argued, the older streetscapes ensured that parents could watch their children from low-rise windows and that there was adequate pedestrian traffic to ensure safety. These older benefits were sacrificed in search of nicer apartments, and the results were largely disastrous. Similarly, massive urban redevelopment in many cities in China has replaced the lively old streetscapes with unfriendly, monotonic high-rise compounds.

In the U.S. today, structure type is rarely a problem. Even the poorest fifth of America's citizens consume over 80 square meters of space per household (Glaeser and Gyourko, 2008). Structural amenities, like electricity and plumbing, are universal. The larger problem for the American poor is that children grow up in neighborhoods that are rife with poverty and crime. There is little reason for housing policy to focus at all on better structures, but housing authorities still devote far too many resources to this objective, rather than to neighborhood locations.

The situations in developed Asian economies with large public housing programs are different. Again, in both Hong Kong and Singapore, the housing estates—comprised of residential blocks, a range of community facilities, and often landscaped public or social space—offer neighborhood environments which are often comparable, if not better than, those of average private housing clusters. In these two cities, professional housing management practices were introduced in either the 1960s or early 1970s to ensure a safe and clean living environment. Albeit due to different eligibility policies and housing policy objectives, the population mixes in the public housing estates in these two cities are much more diverse than in the U.S., especially in the subsidized owner-occupier housing estates. Thus, while there is still a certain degree of stigmatization attached to the public rental housing estates of Hong Kong, it is a non-issue in Singapore, which provides predominantly owner-occupied housing to the majority of its citizens.

In the U.S., Chetty, Hendren, and Katz (2016) have examined the lifetime impacts when poor Americans receive housing vouchers and move to lower-poverty areas. They find that children who moved when they were young earn materially higher wages at age 28 because of the move away from depressed neighborhoods. Children who move when they are older receive much less benefit. In housing, as in education and other investments, the biggest benefits often occur when we focus on the young. In the Netherlands, social housing projects are often mixed with market housing. The rationale behind this planning approach is to avoid the isolation of children living in social housing from the mainstream society.

There are three clear implications for focusing on the future. First, it again emphasizes the importance of prioritizing affordability over housing price increases. Typically, the old are owners who benefit from the price increases. The young are buyers who lose as they must pay more, unless, as in the Singapore case, the government provides subsidies to assist the young to buy. Certainly, some of the young may benefit from the real estate investments of their parents, but it hardly seems fair to have a housing policy that favors only children who are so lucky.

Second, rental assistance may be at least as sensible as help for homeowners. Rental apartments make particular sense for the young, who do not know where they want to permanently live and do not know how prosperous they will end up being. Rental apartments enable the young to move more readily in response to changing conditions. This does not imply the need to subsidize rental housing, but it does suggest that public policy should not prioritize owning at the expense of renting.

Third, if there are going to be housing subsidies, then it almost surely makes sense to target those subsidies towards parents with younger children. It may make sense to provide those families with rental assistance which can enable them to live in a better neighborhood during the child's critical formative years. In principle, the rental subsidy can be phased out if the family acquires more income or as the child enters adulthood.

4. Promote tenure neutrality and robust rental markets

Many nations and systems, including the U.S., Singapore, and China between 1998 and 2011, disproportionately favor homeownership for different reasons. Pro-homeownership policies include tax breaks for homeownership, government-guaranteed mortgages, and other pro-homeownership schemes. Singapore has publicly captured land capital gains and redistributed them to homeowners (see Chapter 11). Other government policies deter renting. Artificial controls on rent increases can often discourage people from allowing their units to be rented. America's homeownership rate, for example, shot up between 1940 and 1950 during a decade in which much of the country embraced rent control.

In many developing countries, the situation is particularly problematic when rents cannot be raised and tenants cannot be evicted. What private investor would rent an apartment under such conditions? Consequently, policies that are put in place allegedly to help renters end up destroying the rental market.

Our starting principle is that both renting and owning are valid choices, with their respective benefits and costs. Renting permits mobility and works well in large, multi-unit complexes, since the landlord bears responsibility for common infrastructure. Owning is sensible for older people with fixed locations and for single-family detached houses which require plenty of individualized care, which is often best provided by the resident. If ownership creates some public benefits by encouraging people to build local ties to their neighbors, then renting creates other benefits by enabling people to move in response to a labor demand shock.

We do not see a strong case for preferring one form of tenure to another and consequently the wisest course of policy action is neutrality. Public policies should neither nudge people towards ownership nor towards renting. If money is available to subsidize housing for the poor, then ideally the poor could use the aid either to help fund their mortgage payments or to pay their rent.

Alternatively, the government could use the money to build public housing directly or to subsidize developers or non-governmental organizations (NGOs) to build cheaper housing for sale or rent. Where the government owns land, as in the cases of Hong Kong and Singapore, the government could manipulate land resources for public housing provision. Or if the government owns the development right, as in the case of the UK, planning permission can be granted with conditions of affordable housing provision.

We do, however, see a need to ensure that stable institutional arrangements are in place to enable a robust rental market. Renting is a somewhat more complex legal contract than owning, and consequently there is plenty of scope for contractual disagreement where the rule of law is not observed, or where a robust legal system including enforcement is still in the making.[3] For example, a landlord and an owner may disagree about who is responsible for maintenance or for the damages created by a bathtub leaking. They may disagree about the owners' rights to evict a tenant. These disagreements ultimately require some form of legal adjudication. The simplest approach is to have a few contracts that will be adjudicated by a specifically targeted renting court, which is funded by a modest tax on landlords—but many other structures can work.

The larger point is that rental markets need appropriate legal institutions to function well. This principle is important for Chinese cities, as their rental markets are poorly regulated and the rights of renters are not well protected. Resolution of contract disputes is cumbersome and costly. Germany is a country noted for its stable policies of tenure neutrality, which also include the smoothing of individual rent increases reflecting wider changes in neighborhood conditions. In Finland, stable housing contract designs have facilitated the cohabitation of different types of renters and owners in the same building.

2.2 Three efficiency issues in housing policy

In the previous section, we discussed four core principles of housing policy. However, we did not discuss anything about the costs of achieving those objectives, and it is important that costs always be weighed against benefits. Therefore, in this section we discuss efficiency in the use of resources, which is one means of ensuring that waste does not overwhelm the housing system. Such waste can either come about because of misallocated government spending, like a publicly funded bridge to nowhere, or because of government regulations that induce work commuting trips that are too long and prices that are too high.

1. Efficiency of resource use

Our policy roadmap rests on the fact that housing is intrinsically important and worthy of considerable policy attention. It does not, however, accept that every investment in housing is a sensible use of scarce resources. It is just as easy to waste money by building housing structures as it is to waste money with any other projects. Consequently, it is important that housing policy rigorously ensures that benefits outweigh costs.

The use of cost-benefit analysis is sometimes challenged because the benefits of housing policy can be the relatively amorphous goal of improving the lives of poorer families. If that is the goal of housing spending, then the natural benchmark is to ask whether the same amount of money could achieve more benefit through an alternative use. For example, the benefits of housing vouchers for poorer families can be compared with the benefits of equivalent spending on early childhood education or pure cash transfers. A related question is whether the benefits of the spending accrue primarily to the intended beneficiary (the poor family) or to the landlord.

As we will discuss later, demand-side subsidies such as housing vouchers are less likely to result in extremely inefficient spending than actual public housing construction, provided that supply is elastic. When the public sector is actually directing resources to provide new structures, the possibility for inefficient resource usages is very high, though there may be social benefits if these structures are built with good neighborhood design, which is not common. Private-sector builders can make mistakes, but the need to cover costs with revenues creates at least some discipline.

Typically, we should expect that the benefits of such spending exceed the costs, which include both the construction costs and the opportunity cost of land. The benefits of purely public projects are not determined by revenues alone, since part of the goal is to benefit target populations. Yet, since those benefits are rarely directly measured, it is easy to make mistakes. Too often, the benefits of public housing projects have been overestimated and the costs have been underestimated.

One tool to reduce the potential for inefficient resource use from construction projects is to compare the cost of providing housing through public construction with the cost of leasing comparable privately provided space. This at least provides a check that the public sector will not spend more on building than the private alternative, although it may be hard to find truly comparable space. If the public housing programs are built on a large scale, there could be savings from economies of scale. This was a policy factor in Hong Kong and Singapore in the 1970s because the real estate industry was made up of small developers and was not organized. In Chinese cities where vacancy rates are high, there is a strong case for property tax on homeownership, as it would lead to more vacant housing units entering the rental market, thus lowering the rental prices. This would lower the need for public resources to build more affordable housing.

A particularly common mistake in the design of public projects is to underestimate the opportunity cost of land and to ignore the value of time in the journey to work. Using some currently empty space for a public project is not cost-free since the land can be used for alternative purposes. Taking currently used land through eminent domain can also lead to cost underestimates if the prices paid to existing residents by eminent domain leave those residents worse off than they were before the taking. Finally, large construction projects often cause traffic delays and other inconveniences across transport networks. These are also often unfortunately ignored by cost-benefit analysis.

2. Efficiency of subsidies (demand vs. supply)

A second efficiency consideration occurs when designing subsidy programs which are meant to improve housing outcomes. Demand-side subsidy programs are tied to the resident and help reduce the cost of purchasing or renting a home. Both housing voucher programs and tax credits for homeownership are examples of demand-side subsidies. Supply-side subsidies target the development of new construction, and typically provide some benefit to market builders based on the amount of space that they create, or the government directly builds on land that it owns and then sells or rents to target households at subsidized rates. In the U.S., the Low-Income Housing Tax Credit is an example of the former supply-side subsidy program. The public housing programs in Hong Kong and Singapore are examples of the latter.

In principle, either demand- or supply-side programs can be an efficient means of promoting housing affordability and improving housing quality, which conversely means that either program can be inefficient. While many factors help determine which policy is more efficient, *the elasticity of housing supply is a particularly critical factor*. At one extreme, consider a housing market with an absolutely fixed supply of housing, and consider the impact of a *demand-side* subsidy that enables all the residents of this market to spend more on housing. Basic economics predicts that this subsidy will only serve to increase the rents earned by landlords. Housing quality will not improve, since the housing stock is fixed. The prices paid by residents may not decline at all, since the same number of residents are still ultimately bidding for the same fixed stock of housing.

At the other extreme, consider a completely flexible market in which new housing can always be supplied at a price of $100,000. A *demand-side* policy that offered to pay $20,000 of the cost of a home would indeed reduce the effective cost of a home to subsidized buyers from $100,000 to $80,000. Subsidizing demand makes sense when supply is elastic but not when supply is inelastic, which is a locally determined fact.

This logic works in reverse for *supply-side* subsidies. In an elastic market in which builders can always supply housing at a price of $100,000, a supply-side subsidy will have no impact on price or quantity. Consider a subsidy that pays a subset of builders an extra fee of $20,000 per structure, but assume that this subsidy is not so widespread that it is received by every builder. As long as some of the demand needs to be satisfied by unsubsidized builders, then the market price will remain $100,000 and the $20,000 payments will only enrich the subsidized builders. In an inelastic market, however, there is some hope that supply-side subsidies may ultimately uncork a bit of supply and reduce prices. Direct provision of subsidized housing by government to the target group is the most effective in alleviating the affordability problem, but if the overall housing land supply is too small, this subsidy approach also does not work, unless the government has previously built up a land bank for public housing construction. It also has to be noted that a direct provision approach runs the risk of over-subsidizing the longer-standing public tenants as their income rises over time, unless a stringent subsidy policy can be strictly enforced (see Chapter 10 on Hong Kong).

This discussion highlights that there is no universally correct form of housing subsidy across every city and in every time period. Consequently, an attempt to have a national housing policy, as the U.S. does with both Section 8 Housing Vouchers and the Low-Income Housing Tax Credit, will inevitably generate some substantial (in)efficiency differences across the national system of cities. Ideally, subsidies should be targeted to local conditions. If not, then they should reflect at least the dominant average conditions in the country as whole. And in a country with a rapidly growing housing supply associated with increasing urbanization, the search for efficiency will push public policy towards demand-side subsidies.

A final point about the two types of subsidies is that we have so far ignored the possibility of corruption and inefficiency in the use of the subsidies. Demand-side subsidies are somewhat more cumbersome to allocate, since they are going to a large number of dispersed households. However, they are also less subject to corruption, since they go to residents, not to developers or government agencies, who may be politically favored. If there are substantial fears that the real estate industry is subverting local governments, then demand-side subsidies may be the safer course of action.

3. Efficiency of land-use regulation

The third efficiency relates to city planning and the regulation of urban space. A distorting urban planning rule can be just as inefficient as a bridge to nowhere, and consequently the need for cost-benefit analysis of urban regulations is just as great as the need for cost-benefit analysis of infrastructure. In the case of land-use regulation, relevant controls concern the use of property (residential, industrial, agricultural, or public) and rules relating to the structures that can be built on the land, including height restrictions.

Perhaps the most basic (and usually the largest) land-use regulation is the allocation of space for transportation: the road grid which is marked as off-limits to private uses. Few sensible people dispute the need to set aside public land for thoroughfares, although there is a potential to debate how much land should be allocated to different modes of transportation. Cost-benefit analysis applied in this setting means comparing the value of more space for transportation, which presumably reduces travel times or permits more travel, against the value of space for other competing uses. The most basic tool for estimating the social value of the land in a non-transportation use is just the market price of the land.

The allocation of land for public parks and open space would also seem to be an appropriate target for cost-benefit analysis, although there have been relatively few such exercises in wealthy cities. A sensible exercise is to compare the market value of the land with the social benefits produced by using the land as shared public recreational space. Valuing the usage of the space requires some means of estimating the social benefit from the recreation, and then adding the benefit of the open space to surrounding land users (which can be estimated using hedonic housing price regressions), and then of open space to the city as a whole (which requires more heroic calculations). The value of open space is highest in central locations, and the provision of open spaces improves as the income of a city rises and its density gradient rotates and flattens.

Parks and transportation yield primarily public benefits and, consequently, estimating their social benefits is challenging. When considering the private uses of land—agricultural, commercial, industrial, and residential—the benefits primarily accrue to the owner of the land. Consequently, if land-use regulations have efficiently allocated land across these different uses, then land should have roughly the same value if it is zoned for residential or if it is zoned for industrial purposes. If land is worth substantially more when it is zoned for residential than for industrial purposes, then there is a serious possibility that there is too little land zoned for residential purposes and too much land zoned for industrial purposes.

It is conceivable that there is a public purpose in having land zoned for particular private purposes. For example, there might be some health benefit associated with having agricultural land that is kept undeveloped, although locating agricultural land near cities can also have serious health consequences, such as endangering pollution effects. It is less clear why there are social benefits to having excess industrial or commercial land, but if such social benefits exist, then sound regulations require a serious attempt to quantify those benefits and then compare them with the price differential between land zoned for residences and land zoned for businesses. There is a plausible argument that zoning for specific uses (sometimes called "Euclidean" zoning in U.S. literature) is largely a mistake in developed cities. The best case for Euclidean zoning is that there are sizable externalities associated with some types of production, and it is reasonable to eliminate hazardous or especially unpleasant activities from residential districts. Yet many other productive urban activities have few negative effects, and there are significant accessibility advantages associated with intermixing residential and commercial spaces, since this can lower commutes and traffic congestion. Indeed, mixed-use zoning has been favored now for decades among many U.S. urban planners, since Jane Jacobs.

Land-use regulations don't just control the uses of a plot of land; they may also control the size of structures, or the number of housing units that can be placed on the plot of land. In Hong Kong, however, structure size and number of housing units are not specified in land-use zoning, which only specifies the total gross floor area permitted. Height restrictions, of different forms, are justified as tools for ensuring sunlight or just limiting the overall size of a new town. Almost never have these restrictions been justified with rigorous cost-benefit analysis, and one estimate finds that the costs of these restrictions are likely to significantly exceed their benefits in New York City (Glaeser, Gyourko, and Saks, 2005). Hopefully, future regulations of this form will only be imposed after more rigorous analysis showing that benefits outweigh costs.

2.3 Four institutional qualities underlying a well-performing housing system

Moving from choice of policy goals to operations, we now highlight procedural principles rather than outcome-related objectives. These procedural points are valued not only because they are thought to improve outcomes, but also because they are seen as having intrinsic economic and social value in themselves. There are certainly costs to some of these procedural principles, and they should not overwhelm all other concerns. These are ideals to work towards, not commandments that supersede all other objectives.

1. Uniform property rights

A foundational principle of housing processes is that property rights and rules should be uniform, not specific. In other words, there should be one law that rules everywhere, not a law for place A or person A and another law for place B or person B. The ideal of uniformity helps to create predictability and transparency. Uniformity also makes it easier for land to be used efficiently, since it will not be constrained by place-specific legal rules.

What are some examples of non-uniform property rights? Some individuals might be free to develop their land while others might not have such rights. In rural areas, land might be governed by totally different laws than in urban administrative areas. While it may be desirable to preserve some rural areas, this can be done within a single property system by creating subsidies for maintaining agricultural land or placing limits on development. China's property markets are distorted by the duality of urban and rural land property rights embedded in the 1988 Amendments to the Constitution. The current reform that aims to unify the urban and rural construction land markets is obviously a positive step towards uniform property rights.

2. Transparent rules

Transparency has become a popular objective and it can take on many different meanings. Transparent procedures can ensure that everyone understands what decisions are being made and who is making those decisions. Such transparency surely has advantages, but an even more basic transparency is to ensure that the rules themselves are transparent, meaning that all of the participants understand—and follow—the rules that structure real estate markets. These might be rules that concern land use and that largely impact private developers. Rules also shape the relationship between landlord and tenant. In every case, there is little social value to be had from confusion and uncertainty.

For example, transparency in the case of land-use restrictions might mean a predominance of "as of right" zoning, which means that as long as the project adheres to the land-use regulations, the project need not be subjected to any further review or debate, as in the Hong Kong land-use planning system. In other words, if the land-use map enables a floor-area ratio of 16 and permits residential uses, then a 32-story residential tower that occupies one-half of the area can be put up with minimal review and little uncertainty about the result from that review. To allow flexibility, it is also possible to specify uses that may be permissible subject to approval from the planning authority, such as a hotel in a residential area. There are "positive" lists of uses and "negative" ones. A positive listing of permissible and possibly permissible uses infers that any uses not listed will not be permissible. In contrast, in a negative list whatever is not forbidden is permitted. If the zoning maps and their rules are available easily online, this will increase the effective transparency level.

Transparency is also important for private real estate transactions. The responsibilities of landlords and tenants need to be clear from the start. That type of transparency will be easiest if there are one or two standardized rental or mortgage contracts, and a third institutional party has worked to ensure that the elements of that contract are clear and obvious, even to less well-educated tenants or borrowers. Similarly, the obligations of a seller in a standard real estate sale need to be as transparent as possible.

It is a somewhat higher bar to expect full transparency in political debates and regulatory reforms. We agree that transparency is desirable in those realms as well, but in some cases the paradox is that too much transparency can make it impossible to negotiate honestly and reach agreement among parties. While there is no advantage to a system that confuses borrowers or tenants, there are times in which full transparency during the process of making rules is just impractical in politics, and some cloudiness is necessary.

3. Inclusive but balanced planning

In the first two-thirds of the twentieth century, planning was largely seen as a technocratic exercise that did not involve popular participation. This process led to a number of both construction and highway projects that generated significant popular opposition. A landmark in the growth of popular opposition in the U.S. was the work by Shirley Hayes and Jane Jacobs in the 1950s that stopped the master builder Robert Moses from running a road through Washington Square Park in New York. Since the 1960s, urban planners in the U.S. and elsewhere have learned to engage with the community and to seek their input in designing urban change.

The more inclusive approach has great value. If we agree that the point of planning is to improve people's lives, it makes sense to hear what the people who will be most impacted by a project actually think about the project. In some cases, the views of the public may help planners determine whether or not to go ahead with a project. In other cases, inclusive planning may lead the project to be improved. Crowd-sourcing ideas is often useful: more inclusiveness can be generated both with in-person meetings and online discussions.

Yet, local residents should be given a voice but typically not a veto, at least when a project impacts people beyond the local area.[4] Residents of a neighborhood may strongly dislike a road being built through their neighborhood, but if the road benefits a large enough group of others, then it may still make sense to build. Residents will often dislike new construction projects that increase neighborhood density. Yet those new housing units will provide living space and opportunity for people who do not currently live in their neighborhood. Political leadership must balance the welfare of the neighborhood's current residents against the welfare of other city residents.

4. Free entry and competition

A final procedural objective is free entry and competition, especially among developers. In many markets, such as the U.S., there are thousands of potential builders of smaller structures. There are even sizable numbers of builders prepared to erect much larger structures. This limits the returns that accrue to any individual builder, and ensures that market structure doesn't delay the supply response to any increase in demand. However, oligopolistic local real estate markets are common in Asian cities, with their high densities (Shilling and Sing, 2006).

There are two natural reasons why competition may be limited in real estate development: returns to scale and regulation. In some cases, the scale and expertise needed for a project can be considerable. This may reduce the ability of smaller firms to compete. Yet it seems unlikely that scale economies will be a barrier to competition in larger markets in the absence of regulatory barriers to entry.

The more common reason why free competition is limited in real estate development is that a maze of regulations reduces competition and only a handful of well-connected contractors have the political connections to work their way through. Consequently, those builders are able to earn significant returns, some of which may be shared with politicians. If a market has been captured by a small number of builders, this means that they will limit the supply of homes and extract a significant share of the excess welfare generated by the city.

The natural antidote to limited competition is to loosen the regulations that restrict building. If the rules are modest and transparent then anyone can compete, which brings benefits to downstream consumers. In a longer time-horizon, free entry may increase the rate of technological improvements in the industry, which may yield even larger benefits for consumers.

3. Policy actions in China

The Chinese housing sector has moved mountains, literally and figuratively, in an extremely short time period. The scale, speed, and range of China's achievements in housing are awe-inspiring and historically unprecedented. Inevitably, there are places for improvement in today's housing system given the scale, speed, and geographic spread of change. There is a real need for further reforms to a new level. Compared to other middle-income and high-income countries, China's housing system is unique in terms of scale, the development path of its institutions, and the remaining reservoir of further urban population growth. Nonetheless, international experience is valuable because it is always wiser to learn from the mistakes of others rather than to make one's own. And by now, social scientists and housing policy analysts have reached a two-way street between the experiences of foreign countries and those across China.

What do the policy design principles of the roadmap imply for further policy reforms in China? This section proposes specific policy actions that appear relevant to current conditions across China's unusually large national and diverse housing system. These actions are of three kinds: institutional reforms, urban planning reforms, and financial reforms. Among these, institutional reforms represent the lion's share of the policy actions that we think should be considered. We discuss these policy actions in order of the least to the most difficult.

Less difficult reforms include technical improvements that can take advantage of China's increasingly strong capabilities in information, communications, and technology, but that do not challenge existing power relationships. One example is to bring the permitting and zoning process online. Another example is to improve the quality of property registration. More difficult reforms are those that require major institutional changes, such as reforming the financing structure of local government or altering the eminent domain process.

We recognize that some of these more important reforms may be politically unfeasible or unwise under present conditions. We write here as social scientists and policy analysts with limited understanding of the political intricacies involved in such major reforms in China's context. Yet in the spirit of completeness, we believe that it is important for us to include both the more and the less difficult paths towards a better housing and urban development process.

Housing cannot be fully understood without also considering the larger process of urban development within China. For further reforms, we highlight the importance of planning flexibly to allow for very rapidly changing local circumstances, and of considering new tools for infrastructure provision and finance.

This section closes with two proposals relating to the financial system. Larger financial issues are outside the scope of our discussions, but two items come up repeatedly in the discussions of country-specific housing systems. First, it is vital to ensure that the financial system is protected against the adverse consequences of a housing bust. Including dynamic real estate value assessment in the bank oversight process can help reduce boom-bust risks. Second, it is even more important that households have good alternatives to housing to save for their retirement and other future expenditures.

3.1 Less difficult institutional reforms

We first focus on those institutional reforms that are challenging but still feasible within a relatively short horizon because they face few entrenched interests who would stand to lose significantly from such reforms. In contrast, the most difficult reforms may affect the interest of powerful actors and interest groups, and consequently engender significant opposition. The less difficult reforms still need effort, but may require less political capital to implement successfully.

1. Establish a universal property registry and cadaster

The most basic ingredient for any real estate system is a clear delineation of land parcels and a reliable and up-to-date record of their ownership or usage rights. Yet, many localities in China lack a unified, clear registry of titles and ownership. One reason is that private land ownership was abolished in 1949 and since then public registries have separated building records from the land on which they are

Lessons of international experience 373

built. Rural land is typically owned collectively. Urban land is held in leaseholds. Yet neither of these dual ownership structures presents an information barrier to creating a universal system of recording ownership.

An appropriate recording structure is one in which all land should be registered and include both a registered owner and a registered user, who may be one and the same individual or entity. There should be only a small number of specific types of user and the nature of the lease should be recorded in the registry of deeds along with the end date of the lease. Comparisons with the legal and technical features of property registries of Northern Europe that are considered the best in Europe might be helpful. Similarly, there is merit in comparing with the registry systems of Japan, South Korea, and Taiwan (China).

In the case of collectively owned or leased land, the clearest means of registering the land is to create non-profit entities or trusts that represent the collective. These trusts should have a clear legal status and also a clear managerial structure, such as a board of trustees that is elected by a designated set of voters, or a board that is appointed by a particular governmental entity. The trusts would then have either ownership rights or leasehold rights. The existing village committees in rural areas which in fact deal with land-use matters could be the starting point for establishing such entities and trusts.

The ultimate objective is clear: an online registry containing complete rights to land throughout all of China. The process of moving towards that objective is more difficult, especially where usage rights over land are in dispute. One means of proceeding is to produce a map of plots and then invite claims to lease or usage rights. Once claims are submitted, there must be some means of adjudicating the claims. In some cases, the rights to a particular lot may be quite clear. In other cases, some means of sharing usage may be appropriate.

The long-term advantage of a clear titling regime is that future claims to land should be far easier to solve and rights to compensation for takings should be more straightforward. The process might be initially thorny, but the ultimate goal of a reliable, universal acknowledgement of rights seems invaluable.

While this procedure is meant to handle rights to land, there is a secondary question about *tenants* who live in apartments rented from individuals who own the structures built on the land and are either joint lessees or owners of the land. Ideally, the system would also include the tenant information, perhaps in a secondary phase. Going forward, it would be relatively straightforward that any time a new rental relationship is formed, that relationship is recorded within an appropriate section of the registry of ownership and use rights.

2. *Clarify land lease renewal rules*

When an individual owns the land, then planning for the future is straightforward. This individual can invest with confidence since he or she knows that any structures on the land will be owned in perpetuity. A temporary leasehold, however, carries the danger that the land's ultimate owner may retake the land and capture the investments made on that land. Many of the current leaseholds on China's land stand to end before 2050, which creates a potential final date for extracting returns from any investment. Despite this leasehold end date, many investors seem to believe that they will still be able to use their property and reap returns beyond that point. They may be influenced by the land leasehold policy of Hong Kong, which stipulates automatic renewal while charging an affordable land rent (Gurran, Gallent, and Chiu, 2016).

There are advantages and disadvantages to extending the leasehold to another end date, but there are only advantages to clarifying the conditions that will be used to determine whether and why the leasehold is extended. The government may wish to take over the land in 2047 and reap the revenues that would be gained from reselling property. Alternatively, the government may want to allow the current holder to retain control, perhaps for a fixed fee such as a share of the property's value.

Uncertainty is not good for investment, and in the case of land leases this uncertainty is entirely artificial. It could be eliminated if only the government could decide on its plan of action, even if that plan of action is initially contingent upon possible fine-tuning. One plausible route is to extend leases in exchange for ongoing property taxes. In other words, an owner must pay an annual fee (such as 1 percent of the value of the land and current property) to have the right to extend the lease automatically until a certain year. If that owner ceases to pay the fees, then he or she loses the right to extend the land-use period.

The government may also want to set conditions about continuation of the leasehold, which may depend either on the property in question or on the economy as a whole. For example, a lease for industrial property might be extended for 20 years if and only if there are 500 people employed on the land. The lease might be extended for 20 years if and only if the local unemployment rate is less than 6 percent. Alternatively, the site may be up-zoned or down-zoned for alternative use, and the lessee, if interested in staying on, will pay a new premium at lease renewal. The exact terms should depend on a strategic plan for using the land differently in the event of ending the lease. The larger point is that governmental clarity does not require tying the government's hands.

Similar to many countries, China exercises land-use control through city planning and its implementation. There may be possibilities that the local government wants to change the use of land when a current lease expires. In such a case, the government should announce its decision of not extending the lease well ahead of its expiration date, say 10 years before. This would allow the users sufficient time to adjust their investment decisions. Moreover, the government should have a policy in place to compensate for the replacement costs of the private properties on the land.

3. *Adopt internet-accessible zoning and permitting processes*

A nationwide register of land ownership goes together with a nationwide map of land-use restrictions. Besides competitive bureaucratic claims among administrative units, there is no reason why regulating authorities cannot produce online maps that illustrate usage and structural restrictions throughout China, especially at a time when China has already initiated the effort to integrate various spatial plans formulated by different localities and line agencies. When this map is connected with the ownership or usage rights maps, then it will be easy for owners to know what can be done with their land. It will also be easy for neighbors to know if someone else is abusing the zoning rules by building too much or by using the land for non-permitted uses.

Many jurisdictions already have such maps and there already exist good models to use. Typically, areas are subdivided into zones and the rules for building differ across these zones. A map makes those differences easy to see. Moreover, the structure of online links makes it simpler for readers to click through and learn about any idiosyncratic, area-specific rules.

Ideally, users of the website should also be able to click through and start the permitting process, which can also begin online. If enough of the area is under 'as of-right' zoning, then the permitting process should be straightforward. The builder submits plans which are then approved before construction begins. A secondary approval might be needed before occupancy to be sure that the builder has actually followed the rules. In both cases, the application for a permit can be easily done online.

One of the advantages of online permitting is that it becomes easier to track the course of the permitting process. The builder can check how the regulator is doing. The regulator's supervisors can evaluate the average time it takes to get a clear permit. Ideally, a single permitting entity is responsible for allowing the construction and occupancy. With online permitting, the speed and approval rate of that entity will be easier to monitor.

4. Draw on international best practice for joint ownership structures

Much of the new urban China consists of high-density development, yet joint ownership of a high-rise building can be challenging. Cooperative buildings in American cities are famous for the constant infighting between residents. In some cases, management companies are accused of skimming funds that should be owned by the tenants. Avoiding all such problems under cooperative ownership is impossible, which is one reason why a rental structure can work well for high-rise buildings. Yet, ownership is also appropriate for many urbanites and consequently it makes sense to adopt the best legal structures for jointly owning large structures.

The primary alternative to the cooperative model is the condominium model, in which case the owner does not own shares in the entire building, but rather owns just his or her unit. The land is co-owned but as undivided shares, meaning that the shares cannot be individually sold, such as in the case of Hong Kong. Consequently, the owner has independent control over the unit, which has tended to make this ownership structure more popular. Yet there are conditions under which the joint collective ownership model of the cooperative seems more functional than the separate ownership model of the condominium. For example, if the whole building was to be sold, the cooperative model should be somewhat easier to organize since only a predominant proportion of the owners (e.g. 80 percent, in the case of Hong Kong) would have to agree.

We do not mean to recommend any one particular model for China, but rather suggest that the right approach is to study the existing international models and then focus on one or two ownership structures that appear particularly well suited to the Chinese setting and its regional diversity. As already stressed, there are great advantages to having one or two well-understood contract types. There are also advantages in drawing the knowledge gained from over a century of experience with joint ownership of apartment buildings.

5. Improve information systems for demand-side subsidies

As discussed in Section 3.2, it may be desirable to use demand-side rather than supply-side subsidies for housing, such as housing vouchers. Yet such demand-side subsidies cannot be universal. To be efficient, they must be targeted towards particularly needy populations such as poorer parents with younger children.

To have a national policy that targets such subsidies, there must be a national information system that keeps track of incomes and family status throughout China. In the U.S., portable housing vouchers are administered on a first-come, first-serve basis to eligible households. Consequently, the public housing authorities do not need universal information on potential voucher recipients. When people apply, they must document that they are eligible. However, this system seems quite unfair to many, because aid is not targeted towards the neediest, but to those who apply first among a particular group. A better system would be to ensure that all poorer Americans have access to housing assistance.

China's national identity card system is the natural first step towards an information system that can be used to target demand-side subsidies. The cards already carry information about work history and place of residence. If the cards also included information on earnings and family status, then there should be enough information to target demand-side subsidies.

For example, if a national system contained information on family status and income, then a provisional group of needy persons could be targeted for preliminary aid messages. They could be informed that they might be eligible for housing aid, but that they would have to present documents to the relevant aid-granting entity. The entity would then check documents and administer the aid. The system would be considerably more efficient than the U.S. system because it would be more tightly targeted towards poorer persons.

We recognize that China's policy implementation and public service provision are presently strongly based at the provincial level. Establishing a national system will require integrating or dismantling the province-based systems. Past experiences show that it is very difficult to break the tradition of province-based service provision. However, this hurdle will have to be overcome if China wants to introduce a more efficient system in the future.

3.2 More difficult institutional reforms

We now turn to the more difficult building institutional reforms. In all cases, these reforms would involve both winners and losers, and consequently reform will assuredly be a fight. Despite the political costs of such a fight, these are areas in which reform could produce considerable benefits. Consequently, it seems sensible to at least seriously consider a reform process in each of these areas.

1. Integrate rural and urban land ownership systems

One major factor negatively affecting the quality of Chinese urbanization is the duality in property rights between rural and urban land embedded in the 1988 Amendments to the Chinese Constitution. The primary goal of these amendments was the revival of the trading of land-use rights within Chinese cities, which proved very successful. The perverse dynamic consequences of the duality of property rights between rural and urban land were probably not appreciated at the time because there was no evidence to rely on. Three decades later there is widespread evidence across China of the unintended perverse consequences of this duality. We will revisit some of these issues when discussing rural-urban land conversions in Section 4.3 on urban planning reforms.

Ending the current strong distinction between urban and rural land ownership in China would have major benefits. The primary focus of reforms should be to ensure *the continuity of property rights across all forms of land uses for the same land parcel*, whether in rural or urban use. Such continuity is taken for granted in market economies where real estate property rights are the oldest of all forms of property rights. This continuity is a major positive factor in rural-urban land conversion and in the spatial efficiency of cities where the use of each land parcel tends to go to the most productive among competing uses.

Presently in China, urban land is traded and associated with a single use, and opaque land-use changes are known to attract corruption, especially rural-urban conversions in rapidly growing cities. Rural land is collectively owned and this can make it difficult for individual farmers to benefit from the value embedded in their property, or for farmers to improve their land and trade the benefits as needed. Current reform experiments to unify the urban and rural construction land markets are an encouraging step in the right direction.

Just as with urban apartments, it may be appropriate to have joint ownership of rural land. One possibility is for the land still to be owned collectively, but for ownership to be formalized through a non-profit ownership entity. The farmers would have shares in this entity and use those shares to elect a leadership team. In principle, it would even be possible for the land of a village to be owned and managed by a trust that is appointed by government officials, but that would seem to be far less desirable.

This non-profit collective entity would have formal rights over the land, and the elected leadership would be able to manage the land if it remains rural. More importantly, if the land is to be improved, this formal ownership structure will enable the farmers to collectively bargain with would-be developers and local governments. This will essentially create a single national ownership system that can be regularized in the national online land registry.

Even if collective ownership systems dominate within rural areas, this will not create a separate system as long as urban land can also be collectively owned. In the West, there are plenty of trusts and non-profit institutions that own land. Well-defined property ownership doesn't preclude joint ownership and ownership for the good of a particular group of people. The notion of joint responsibility for land can be preserved, at the same time that the system moves towards regularizing control rights over land.

2. Reform of the eminent domain process

Complaints about the use of eminent domain principles to buy a private property occur around the world. Very often, people who are forced to surrender their property in exchange for a fee often determined by the (public) taker feel ill-treated. If the required compensation is too small, then the taking decision can underestimate the true social cost of the project, which will to lead to inefficiency as well as injustice. On the other hand, eschewing eminent domain altogether will make land assembly far more difficult and substantially reduce the ability to create projects which benefit the entire city.

Reformers have suggested three ways in which to improve eminent domain. Most commonly, eminent domain can be restricted to particular public uses, such as providing common infrastructure, in which benefits are likely to outweigh costs. The problem with this approach is that agreeing on appropriate uses is difficult and hard to monitor. The implicit definition of the public good varies from country to country, and from time to time. In the U.S., the Supreme Court essentially ruled in 1984 that it was unable to meaningfully determine whether a local government taking was in the public interest or not, and that consequently the court would not disallow a taking that meant to take land from one party and give it to another. We suspect that in other settings restrictions based on use will be similarly difficult to enforce.

A second approach is to focus on better assessments, as discussed in Chapter 2. More accurate assessment can ensure that the local government pays at least the fair market value of the property. Moreover, a reasonable view is that owners should be compensated more than fair market value, since they presumably valued the land more than the market did since they hadn't sold out before the taking. For example, in the U.S., the 2013 Indian Land Reform Law requires that farmers receive compensation equal to four times the value of their land and that rural dwellers receive compensation equal to twice the value of their land.

One significant problem with compensation is when ownership is not clearly defined and renters may not be receiving some of the surplus from the land. The national land registry is meant to address the problem of determining ownership or usage rights. If compensation exceeds fair market value (as it probably should), then some of that compensation could be shared (as a legal rule) with evicted tenants to compensate them for the inconvenience of moving. Urban village redevelopment in many Chinese cities encounters this problem. While the original villagers stand to benefit from the redevelopment through benefit sharing and compensation, the large number of renters, mostly migrant workers, have to move elsewhere without any compensation.

A third approach is to grant stronger property rights to owners or users, and allow them to reject proposed takings. The 2013 Indian Land Reform in the U.S. requires at least 70 percent of the population to approve a land deal for it to go through. Choosing the appropriate supermajority required to approve public takings addresses the desire to protect minorities against losses from expropriation with the larger need to implement a project of public interest. In the city of Shenzhen, urban village redevelopment requires at least 80 percent of the original villagers to approve the redevelopment project, but renters are not included in the voting process.

3. Reform of public property development institutions

China's urban property development institutions include Urban Development Investment Corporations (UDICs), state-owned real estate development enterprises, private developers, and land banks. Among the most difficult tasks ahead is to reform entities such as state-owned enterprises and UDICs that have been created by China's municipalities. These entities have often built engineering marvels, but they also seem responsible for some of the most egregious examples of over-building within China. Forests of skyscrapers have emerged in the wrong places at the wrong times.

At the heart of the over-building problem is a lack of transparency in the property development process. It is difficult to figure out specifically how these entities are finding their financing and what implicit or explicit subsidies flow to them from local governments. Stories certainly occur of cozy connections between UDICs and local governments, and of local governments that pressure commercial banks to lend to connected developers. As long as local government officials focus on GDP growth, they will have incentives to push resources towards real estate developers.

A reform of the public property development sector is needed. If there are going to be public building companies, they should be non-profit and clearly regulated. Their sources of funds need to be transparent. Their books need to be open to regulators from the central government, to the local community, to the media, and to ordinary taxpayers. Rules about conflict of interest need to be made clear. If a state-owned enterprise produces goods for export, then that enterprise is exposed to the discipline of foreign competition. If a state-owned enterprise produces only local real estate, then there is little discipline to prevent waste from ill-advised construction.

Real estate is a regulated industry that often has close relations with local governments. That is probably unavoidable, but with more transparency and less discretion, corruption-related risks can be reduced. The goal is to have a system whereby builders face the full social costs of land, credit, and construction, and then make the decision to build if and only if they believe that future market returns justify those costs. Any factor that undermines that decision, such as favors from local officials, creates the risk of over-building and instability in the real estate sector.

4. Reduce local regulatory barriers and lower the volatility of local housing supply

The prevailing view is that China has built too much rather than too little housing in many cities, especially since 2009. Still, housing costs have remained high and China runs the risk of underbuilding if reforms of land institutions also lower the elasticity of supply. When reforms occur, it will be important to ensure that regulatory barriers do not artificially restrict new construction and reduce competition in the building markets. There are legitimate reasons to regulate new construction, but sensible regulations need to be fully compatible with a transparent and accessible regulatory regime. Under present conditions, local leadership may be an important factor in avoiding costly boom-and-bust building cycles.

Ideally, the government would begin with an assessment of the costs of allowing different building types in particular locations. These costs might include the stresses on available public infrastructure and negative side effects associated with locating industry in a particular locale. With these cost assessments in hand, it would become much easier to design a specific schedule of fees for building in a given location. For example, a 10-story manufacturing facility may end up paying more than a five-story residential building. The assessment schedule should be subject to regular review to reflect major changes in the built environment such as transport infrastructure development. If the fee schedule is clear, every developer can build in every location as long as he or she pays for the social costs of the development. The approval process could be much shorter. As long as the costs have been estimated accurately, the city will also be better off after a new development.

A transparent fee process should also encourage competition within the development industry. Since outsiders and insiders will have to pay the same fees, insiders enjoy no comparative advantage over outsiders and there is no home field advantage. The openness of this system ensures uniformity of treatment for all and a level playing field.

5. Change the financing structure of local governments away from land-based finance

Various research projects find that Chinese local governments who have a local monopoly over land use have encouraged too much building as a means of raising revenues (see, for example, Man and Hong, 2011). After the 1994 fiscal reform, Chinese local governments faced an imbalance between their revenues and their obligations. Consequently, they turned to urban development as a means of raising funds. There is much that is good in Chinese urbanization. Nonetheless, history and pragmatism suggest that Chinese local governments would benefit from a predictable system of local revenues, including property taxes.

Property tax financing has several advantages for local financing. Most importantly, property cannot run away. If the incomes of rich people are taxed more in some localities than others, these people have incentives to move. But property is immobile, which means that taxing property is relatively less likely to distort locational choices. The advantages of property tax improve further if that tax is based more on land values than on building values, as a land tax will deter new construction less than a property tax.

Property tax-based financing will encourage local governments to invest in local amenities that improve property values. Such taxation has the virtue of being relatively stable, which may reduce the tendency to view real estate as an attractive means of investment speculation. Since owners have to pay taxes annually, a property tax reduces the appeal of a buy-and-hold strategy and socially undesirable high vacancy rates. Finally, property taxes can also be fair if the amount of real estate value reflects the availability and usage of local services, such as police and infrastructure services.

Policies should distinguish clearly between an ongoing property tax which is independent of ownership and a transaction tax which occurs only when the property is sold. A transaction tax is a far less desirable form of revenue since it distorts turnover behavior and produces a far more irregular—and pro-cyclical—stream of revenues. Local governments have little reason to specifically try to reduce the resale of homes.

Implementing property taxes requires an effective and impartial assessment of property values. At this point, there are numerous examples of how to carry out mass appraisal. This can easily be done at the provincial or prefectural level within China. The assessment system should be combined with the system of national land registry, and individuals who hold use rights over land must risk losing those rights if they fail to pay their local property taxes.

3.3 Urban planning reforms

Real estate is not independent of the larger environment of urban change. When discussing changes in housing policies, many people also see room for reform of urban planning. Here we briefly discuss three modest changes for urban planning policy.

1. Plan for long-term urban change

London's Green Belt was planned in the 1930s (see Chapter 5 in this volume), when the city was far smaller and it was less dynamic and less critical to the national dynamics of Britain than it is today. Consequently, the Green Belt was seen as a modest check on slow urban expansion that offered pleasant greenery to those on the edge of the city. Moreover, ahead of all other countries, the UK had essentially completed its long-term urbanization. Today, London is much larger and the demand for expansion is enormous, but the supply response is notoriously inelastic. Unsurprisingly, there is now a significant debate about whether the Green Belt unfortunately chokes urban growth rates, reduces affordability, and stymies positive change. By contrast, London's internal parks (e.g., Hyde Park, the Green Park, and St. James'), which are even older, are seen as gems that bring quality of life to the heart of the city.

We do not mean to take a stance on the Green Belt or any piece of English land-use planning, but rather to make a larger point. Cities change, and in China a very large population increase in cities is still expected. Land-use planning must plan for the future and accept the possibility of an urban China that is enormously different in 50 years' time. In some cases, this may require more regulation, such as setting aside spaces in the city center that will provide attractive greenery to an increasing wealthy urban population. In other cases, there needs to be more flexibility to accommodate an ever-changing urban environment.

The demand for flexibility pushes towards a different balance between fee-based permitting and long-term physical planning. For example, if builders get used to the idea that the price of building is to pay higher development fees and property taxes, then their investment choices can be changed as the city matures. If builders are faced with an unbreakable set of rules that do not adjust, such as the Green Belt, then the city will end up being less flexible.

Flexibility also argues in favor of some forms of infrastructure rather than others. Infrastructure networks shape the spatial dynamics of cities in a lasting way. Buses are far more flexible than trains. In the case of obviously successful cities, like Shenzhen, trains may be appropriate, but when long-term development is less certain, bus rapid transit is a far more flexible alternative. Bus routes can be rerouted and even turned over to private vehicles. Train routes, with their fixed right-of-way, are far more permanent, and many cities in the West are still coping with the legacy of train routes that were built in what is now the wrong place. Going underground, of course, avoids much of this problem.

2. Consider alternative infrastructure provision mechanisms

Infrastructure always needs public management. International experience shows that the public role in the initial stages of development of infrastructure is irreplaceable (Bhattacharya et al., 2012; Brown et al., 2015). Yet the development process itself can be by public, private, or even non-profit entities. China has moved to embrace alternative infrastructure provision mechanisms, such as public-private partnerships (PPPs). This is a sensible step. PPPs are not a panacea, but under the proper terms they have the potential to mitigate two of the largest problems in infrastructure provision: unwise investment and undermaintenance.

The unwise investment problem occurs when a public agency overestimates the benefits and underestimates the costs of a large piece of infrastructure such as an urban rail system. The benefits are particularly prone to overestimation, if there is no attempt to fund the system with user fees, such as tolls or rider charges. If the construction is done by a commercial/private entity, then the presumption is that this entity is supposed to recoup the costs largely or entirely with user fees. The entity therefore has stronger incentives to seriously estimate the expected number of users of the infrastructure.

The entity also has the right incentives to maintain the infrastructure, because otherwise the system would lose users. The undermaintenance problem occurs when public officials enjoy the splashy headlines generated by new infrastructure, but then avoid the hard work in maintaining that infrastructure. A PPP that is financed by users must pay for maintenance or go bankrupt. Consequently, user fee financing helps to ensure that roads will be maintained.

As just said, PPPs are not a panacea. In some countries, PPPs have corrupted local officials. Private provision can also lead to underinvestment in safety or prices that are too high. We recommend that PPPs be subject to serious oversight, including safety inspections and reasonable price regulation.

3. Manage rural-urban conversions

The land conversion multiplier (i.e., the ratio of the value of land in urban use to its value in current rural use) has been extraordinarily high in China due to the duality between rural and urban property rights created in the constitution and in addition to the monopsony granted to urban governments to purchase rural land for urban use. It is this very high land conversion multiplier that has driven local governments to over-invest in housing and other forms of real estate. It is one of the greatest sources of local corruption as well.

A major urban planning issue in any urban system is the conversion of rural to urban land. This is especially the case in a country like China, where the tempo of urbanization has been very high. Currently, only urban governments can undertake this step, and this limits both local innovations and the ability of farmers to benefit from widespread urbanization. In order to achieve greater social stability, reforms need to make it easier for rural groups to independently plan for rural-to-urban conversion and to profit from such conversions.

There are two particularly important steps needed for this reform, both of which we have already discussed. First, rural land ownership needs to be regularized and brought into a unique property rights legal framework. Greater clarity in collective ownership, such as a land cooperative, is needed. As long as a significant majority of the collective supports conversion, then conversion should be able to occur.

The second major step is to reduce the implicit zoning that makes it difficult to convert agricultural-use land to residential- or commercial-use land. Beyond food security issues and environmental issues, the rationales for privileging agricultural land use relative to urban land should be reviewed again. In some cases, it may be beneficial to move agricultural activities further away from large population centers. When conversion of land use is easy, and land ownership is securely lodged in a collective, then it should become straightforward for rural groups to plan for and profit from the conversion to urban uses.

From a purely technical perspective of urban development, this conversion process cannot be entirely unplanned. Certainly, the conversion would need to fit within the larger urban land-use, transportation, and infrastructure plan. Surely, there will need to be some constraints. But it seems desirable that China's farmers move from being passive witnesses to urbanization to being major actors shaping the urbanization of the country.

3.4 Housing finance reforms

This roadmap is meant to address reforms in real estate, not reforms within the financial sector. Nonetheless, two real estate-related recommendations emerge that are finance related. The first one is to insulate the financial system from the damage that might be caused by a collapse in real estate prices. The second is to lower the propensity of real estate to generate speculative bubbles.

1. Adopt dynamic real estate price evaluation for bank (lender) oversight

During price cycles, real estate prices typically display mean reversion. Empirically, in the U.S., if a metropolitan area's housing market rises above trend by one dollar extra over a five-year period, then it typically declines by 32 cents over the next five-year period (see Chapter 2). This tendency to display mean reversion can be particularly visible in cities where prices rise substantially above construction costs, where there are few regulatory or physical barriers to new building and housing supply is elastic, such as Las Vegas and Phoenix in the U.S. (Glaeser et al., 2008).

The propensity of metropolitan area prices to mean reversion has strong implications for banking sector oversight. Real estate is frequently used as collateral in loans, either to private borrowers who use mortgage financing or real estate development firms. This collateral is, in a statistical sense, likely to be worth less in the future when prices have recently soared to a peak value. They are likely to be worth more in the future when prices have recently plummeted. Consequently, when banks are highly exposed to real estate risk, it seems prudent to have capital requirements to reflect the stage of the real estate cycle.

Capital requirements are a primary tool to prevent banking firms from taking on too much risk. Bank lending tends to be pro-cyclical: more lending takes place when the economy is booming (and property values are rising). Higher capital is needed when the bank has taken on more risk. Capital requirements should reflect the risk in the financial entity's portfolio. More adequate capital requirement rules are part of the Basel III bank regulation rules following the financial crisis of 2007–2008, when consumers were forced to rescue the banks. Comparable capital rules are part of the 2010 Dodd-Frank Law in the U.S.[5]

The riskiness of a real estate portfolio is predictable. While more statistical work would need to be done in China, it seems that in China, as in the U.S., risks are magnified when prices have risen far above construction costs. Consequently, capital requirements for lenders should also be higher when prices are high above construction costs, especially in areas with abundant supply relative to demand. Lenders with exposure to real estate in Tier-3 and Tier-4 cities might be particularly subject to extra capital requirements

Germany has long had a bank regulatory requirement that the collateral value of a real estate loan not be based on the market value of the property at the time when the loan is made, but on the "stabilized value" over the property market cycle for that type of property. German regulators refer to the "mortgage lending value" compared to the "market value" (Roach et al., 2012; Kälberer and Lux, 2012). As a result, Germany has experienced low volatility in real estate prices for decades. The only exception was in East Germany after the 1989 reunification, where these rules could not be and were not applied, and a boom-bust occurred. Denmark has similar mortgage lending rules.

2. Provide better alternatives to real estate for household retirement savings

Owning one house can be a prudent investment. We all need to consume housing services, and owning a home is essentially a hedge against changes in the future costs of those housing services. But owning multiple homes introduces significant risk into a retirement portfolio, especially if those homes are all in the same metropolitan area and there is little geographical diversification. Moreover, the price of homes in an area tends to be tied to labor market earnings, so the saver has invested in an asset that magnifies basic labor market risk. So, using homes to invest for retirement seems like a potentially dangerous strategy.

Until now, lacking investment alternatives with reasonable, risk-adjusted rates of return, ordinary Chinese have been investing heavily in housing as a form of retirement savings. They are motivated by the very high real rates of return during the Great Housing Boom (Zheng, Sun, and Kahn, 2016). How can policies take into account high household expectations? Naturally, multiple housing ownership should be legal, but can steps be taken to mitigate against this practice? Ideally, investors should have a greater range of retirement options than real estate. Individual household investors should also be warned about the risks of non-diversified portfolios. The issue is the possibility of diversification and the existence of alternative forms of financial investments. Property taxes will then reduce the tendency to see property as the only sensible investment.

Perhaps a study group could be formed to consider the role of real estate in retirement savings in China. There is a need for a clear statistical grasp of the size of the problem, nationally and by region. There is also a need for a better understanding of the forces that push savers towards owning multiple homes. Among those, what are the factors behind the exceptionally high housing vacancy rates in major cities? Hopefully with more studies a clearer path of reform will emerge.

4. Conclusion

China's real estate markets have grown exceptionally quickly since the housing marketization reform of 1998, but they may have also put the nation's economy at risk. Individual savers and the banking system, especially its shadow banking system and its regional banks, may both be subject to significant downside risks if land and housing prices turn down sharply, which would affect, in addition, the position of the large number of financially leveraged local governments. In this roadmap, we have recommended a series of housing reforms meant to mitigate against downside risks both today and in the future.

Perhaps both the various levels of government and individual investors should focus more on housing as consumption and less on housing as investment. Buyers are less likely to be led into speculative frenzies when they focus on the flow of services created by a home. Government policy is more likely to achieve beneficial results when it focuses on ensuring that its population is well housed and that ordinary buyers have access to affordable homes. For many lower-income and young people, renting will be the better option than buying, save cultural factors stemming from, for instance, a society with an agrarian origin.

The roadmap proposes a series of reforms. The simpler reforms focus on regularizing the land ownership system, and providing online data about ownership rights, usage rights, and current land-use regulations. The more difficult reforms focus on transforming entities that regularly impact land and housing markets, including local governments and their UDICs. Not all of these reforms may be immediately feasible nationwide, but every reform we proposed is at least worth considering, if only as a means of better understanding the performance of China's housing sector after two decades of intense growth.

As in too many other countries, housing and real estate is becoming an increasingly important factor in rising income and wealth inequality, and China is experiencing the rise of many forms of housing divide. One critical strategic issue yet to be addressed is the excessive role played by housing in household savings because most Chinese households have no access to sound financial investments with a reasonable, risk-adjusted rate of return. Presently, the majority of households face a choice between two unattractive financial assets: either receive very low interest rates on their bank deposits (thereby subsidizing banks and state-owned enterprises), or invest in poorly regulated and volatile stock markets dominated by short-term insider trading and face the probability of being the loser. There is a danger that Chinese households may begin to adopt their own version of the Japanese fairy tale of ever-rising land and real estate prices that eventually led to the twin real estate and banking crash of 1990, followed by two "lost decades."

Notes

1 Much credit goes the epochal U.S. Housing Allowance Experiment at the start of the 1980s. It generated a large volume of information and high-quality research, which in turn stimulated the growth of international networks of researchers strongly motivated to compare their data collection methods, research tools, and findings.
2 A factor contributing to deepening income inequality and worsening social problems in the U.S. after the housing crash of 2007–2008 was that some homeowners not only lost wealth when their houses lost value, but, being unable to sell their houses, they

could not move to jobs elsewhere (Brown and Matsa, 2016). The negative impact of housing immobility on labor markets was a long-standing outcome of the shortage of state-provided housing in Eastern Europe.
3 The range of relationships between private landlord and tenant can vary very significantly in a mature, high-income rental market like that of the UK (Allen and McDowell, 1989).
4 In a city with a large and growing middle-class, personal values, household preferences, and policy opinions can vary greatly. German sociological research has shown that there might typically be a dozen broad groups holding different values and incomes in an urban society. Therefore, public committees have to address pragmatically Arrow's mathematical impossibility theorem, which proves that there is no possible voting scheme that can consistently and sensibly reflect the preferences of individuals with diverse views (Summers, 2017).
5 See BIS Paper 21 on *Real state indicators and financial stability*, Bank for International Settlements, Basel, 2003, www.bis.org/publ/bppdf/bispap21.htm and *Property Markets and Financial Stability* joint conference between the Reserve Bank of Australia and the Bank for International Settlements on "Property Markets and Financial Stability," Sydney, Australia, August 20–21, 2012.

References

Allen, J. and McDowell, L. (1989). *Landlords and Property: Social Relations in the Private Rented Sector*. Cambridge: Cambridge University Press.

Bhattacharya, A., Romani, M., and Stern, N. (2012). *Infrastructure for Development: Meeting the Challenge*. London: Grantham Research Institute in collaboration with the Group of Twenty-Four (G-24).

Brown, M., Kim, Y., and Romani, M. (2015). *Infrastructure Finance in the Developing World*. Seoul, Korea: G-24 Global Green Growth Institute.

Brown, J. and Matsa, D. (2016). Locked in by Leverage: Job Search during the Housing Crisis. NBER Working Paper 22929, December 2016.

Chetty, R., Hendren, N., and Katz, L. F. (2016). The Effects of Exposure to Better Neighborhoods on Children: New Evidence from the Moving to Opportunity Experiment. *American Economic Review*, 106(4): 855–902, http://dx.doi.org/10.1257/aer.20150572

Glaeser, E., Gyourko, J., and Saks, R. (2005). Why is Manhattan So Expensive? Regulation and the Rise in Housing Prices. *The Journal of Law and Economics*, Vol. 48, October 2005, pp. 331–369.

Glaeser, E. and Gyourko, J. (2008). *Rethinking Federal Housing Policy: How to Make Housing Plentiful and Affordable*. Washington D.C.: AEI Press.

Glaeser, E., Gyourko, J., and Saiz, A. (2008). Housing Supply and Housing Bubbles. *Journal of Urban Economics*, 64(2), 198–217.

Glaeser, E., Gyourko, J., Morales, E., and Nathanson, C. G. (2014). Housing Dynamics: An Urban Approach. *Journal of Urban Economics*, Vol. 81, 45–56.

Gurran, N., Gallent, N., and Chiu, R. L. H. (2016). *Politics, Planning and Housing Supply in Australia, England and Hong Kong*. London: Routledge. Chapter 7.

Jaffee, D. M. and Quigley, J. M. (2007). Housing Subsidies and Homeowners: What Role for Government-Sponsored Enterprises? [with Comment]. *Brookings-Wharton Papers on Urban Affairs*, pp. 103–149.

Kälberer W. and Lux. R. (2012). *Definition of Value: Market Value and Stabilized Values.* Washington D.C.: Joint Collateral Risk Network (CRN) and American Enterprise Institute, International Conference on Housing Risk.

Man, J. Y. and Hong, Y. (Eds.) (2011). China Local Public Finance in Transition. *Lincoln Institute of Land Policy*, Cambridge, MA.

Peppercorn, I. G. and Taffin, C. (2013). Rental Housing: Lessons from International Experience and Policies for Emerging Markets, the World Bank. http://dx.doi.org/10.1596/978-0-8213-9655-1.

Roach, S. D., Kälberer, W., Lux, R., and Nygard, W. A. (2012). The Quiet Giant: Explaining the Stability of Europe's Largest Real Estate Market, Appraisal Institute Meeting, August 2012, San Diego, CA.

Shilling, J. D. and Sing, T. F. (2006). Why is the Real Estate Market an Oligopoly? Paper presented at the 2006 Annual ASSA-AREUEA Conference, Boston, MA.

Summers, L. (2017). Farewell to Kenneth Arrow, a Gentle Genius of Economics. *Wall Street Journal*, February 25.

Taffin, C. (2017). Private and Social Rented Housing: Basic Principles and overview of Practices in Europe. Workshop on Middle and Low Income Housing in China, World Bank and State Council Development Research Center, Beijing, July 19.

Zheng, S., Sun, W., and Kahn, M. E. (2016). Investor Confidence as a Determinant of China's Urban Housing Market Dynamics. *Real Estate Economics*, Vol. 44, pp. 814–845.

Index

Note: Page numbers in *italics* and **bold** denote figures and tables, respectively.

Abbott, Andrew 44, 56, 58
Abercrombie, Sir Patrick 93
Abe, Shinzō 7
adjustment and control policies 23, 307
administrative centers, cities as 262
affordability 185, 189–193, 231, 291, 300–302, 309, 312; in Taiwan 178–180; and construction technologies 360; in Hong Kong 202–207, 212–214, 215, 222; and hukou (rural household registration) 5, 301; in Japan 119–122; and land-based finance 301; and land supply 360; and location 301; and migration 289, 301; policy 289, **300**, 300–302, 303, 308, 309, 363; policy design, international experience 359–360; promotion of 359–360, 362; in Singapore 243; in United States 360; urban housing challenges **300**, 300–302
aged society 6, 194
agglomeration 13, 63, 66–67; networks of 67; social value of 69–75
aging society 6–7
agricultural land 62, 91, 95, 99, 260; pre-industrial 262; *see also* Green Belts; rural land
Ahlfeldt, G. M. 66, 82
Aichi 108
Almere 82
Amati, M. 92, 94
Amsterdam 63, 68, 70–71, 75, 76, 77, 80, 81
ancestors' plot 56
ARAVA system 51–52, 53
Artificial Neural Networks 34

Asian Financial Crisis 5, 15, 16, 23, 145, 151, 155–157, 202, 206, 208, 212
"as of right" zoning 369, 375
asset ownership 94, 326; prioritizing housing consumption over 357–359
Atkinson, A. B. 230
automobile industry, sector policy 289

banking system: central banks 4, 10, 177, 181, 191–192, 274; financial oversight 40–41, 42, 384–385; mono-banking system 274
Bank of Japan 107
bargaining 35, 37
Barker Review 92
Basel III, bank regulation rules 384
Basic Farmland Protection Regulations 95
Bastagli, F. 231
Baum-Snow, N. 230
Beesley, T. 9
Behrens, K. 230
Beijing 95–96, 97; Urban Master Plan for 2004–2020 96
Berlin 66
Bernanke, Ben 4
bidding 37, 366
birth and death rates variation 6
Bischoff, K. 236, 237
Bishop, B. 231
boom-and-bust building cycles 380
Bourassa, S. C. 182
Brazil 361
Brexit 11
brownfield 88
Burns-Grebler hypothesis 21, 258, 265–267, 266, 281
Burtless, G. 74

cadaster, universal 372–373
Canada 4, 8, 10
capital-land substitution 334
capital requirements 384
Capozza, D. R. 78
Case, K. E. 32, 40
Cathy Life Insurance 198n2
causal ordering 259, 268
central banks 4, 10, 177, 181, 191–192, 274
central business district (CBD) 63–64, 64, 65, 78; and agglomeration 66; growth of the attractiveness of 83
central planning 5, 272, 273, 274
Central Urbanization Work Conference 98
Chang, C.-C. 199n15
Chetty, R. 231, 237, 362
China: and East Asian economies 270–271; financing 276; GDP 257, 280; Great Housing Boom (1998–2013) 280–281; Green Belts 95–98, 99–100; growth objectives and housing strategies 272–273; growth transition (economic development) of 20–24, 268–270, 269, 281–283; implications from Hong Kong study 222–225; investment-led growth policies 274–276; labor markets 269, 280; long-term dynamics of housing *see* long-term dynamics of housing, in China; macro-economic policies 275; mortgage loans 275–276, 278; National Bureau of Statistics 269, 279; peri-urban fringe 95–98; policy actions *see* policy actions, in China; post-WW2 East Asian development policies 271–272; property booms in 32–33; real estate overinvestment 276–277; resource transfer policies 275; service sector 269; socialist market economy 273, 284n8; structural changes in policies *see* structural changes, in Chinese housing policy; structural distortions 273–274; take-off stage (economic development) 274–281
China Construction Bank 275
China Index Academy 305n5, 339
China National Bureau of Statistics 304n2
Chinese currency conversion 275, 297
Chiu, R. L. H. 205–206, 215
Chivakul, M. 32, 33
cities: city planning 367, 374; definition of 76; environmentally friendly city 96; size, and wages 65, 66–67

city extension 75–79; city planners' right legal instruments to extract gains for private developers 83–84; land price at the optimal moment of construction 79; private developers and incentive for building at too-low densities 83; value and cost of construction 79–82
City of God project 361
climate change 11
collective ownership: collectively owned/leased land, registration of 373; rural land 377–378; and rural-urban conversions 383
commercial housing 23, 308, 309; average price of 311, **312**
commercial land supply, and residential land supply 337
commodity housing markets 288, 289, 292–293, 293, 300; demand, factors influencing 296–297; and land policy/municipal finance 295
Communist Party 312
commutation *see* transportation
comparable government interventions 23
comparable sales prices 33–34
compensation 30, 36–37, 167, 373; eminent domain 378–379; fairness in 35, 153; to villages for farmland 293, 302, 378–379
competitive tendering 243, 293
computer-vision method 30, 33, 34–35
condominium model 111, 115, 117–118, 243, 244, 375
construction: in Finland 51–53, 55; in Korea (*see* Korea); land price at the optimal moment of 79; and land rents 84–86; in Netherlands, the 67–69; and policy development 47, 48–49, 51–53; technologies, and affordability 360; in United States 334; value and cost of 67–69, 79–82, 334
contracts: mortgage 370; property purchase 303; rental 299, 363–364, 370
cooperative housing/ownership 58n1, 375–376
Copenhagen 74–75
corruption: and land conversion multiplier 383; and land-use changes 377; and subsidies 367
cost assessments 380; *see also* property valuation/value
cost-benefit analysis 364, 365, 367, 368
county town (case study) 312, 314–315, **315**; aerial image 316; government

officials 320–321; housing development 315–317, *316*, *317*, **318**, *318*; housing estates 317–319, **318**, *318*; housing market catchment areas 320, *322*; investment in second homes by better-off county residents 321–322; market operation and customers of new housing 317–322; rural depopulation 319–320; rural labor migration and remittance flow *321*
crowd-sourcing 370
Cutler, D. M. 40–41

Dalian 96
debts/loans 9–10; business debt 33; household debt 9–10; Housing Provident Fund 301–302; *see also* mortgage
De Groot, H. L. F. 63, 74, 75
De Koninck, R. 250n20
demand-side subsidies 365, 366, 367; improving information systems for 376
demographic aging: and GDP growth rate 10–11
demographic change: structural change 6–7
Deng Xiaoping 20, 257, 271, 273, 335
Deng, Y. 32
Denmark 45, 56, 385
Denton, N. A. 241
depopulation 62, 108, 127, 319–320
Devan, J. 233, 235
development economics 20, 257, 259
discriminatory residential registration system 5
diversification, and homeownership 8, 358
Dong, Y. 335
Dordrecht 71
dual direction control and regulation 312
durability 5, 356

East Asia: financial crisis (1997) 309, 310; housing experiences 14–20
East Asian perspective, China's development from: distinguishing features that China shares with other East Asian economies 270–271; growth objectives and housing strategies 272–273; post-WW2 East Asian development policies 271–272; structural distortions specific to Chinese housing 273–274
economical comfortable housing 300, 316

economic development, long-term: actual development record since WW2 267–270; Krugman model 263–265, *264*, 270, 271–272; Lewis model 258, 259–262, *261*; three main stages of 259, *261*; urban dimensions of 265, *266*; von Thünen model 262–263
economic geography 263
economic good, housing as 356
economic inequality 7
economic models 259; Krugman model 263–265, *264*, 270, 271–272; Lewis dual-sector model of development 258, 259–262, *261*; von Thünen model 262–263
edgelands 90, 91, 99
education: and migrants 324; rural 320; system, Singapore 246
Eeckhout, J. 230
eminent domain system 35–37, 41; reform of 378–379
enhanced value assessment 36–37
Enschede 71, 74
environmental constraints 11
Estonia 7
ethnic segregation 239–241, **240**
Euclidean zoning 368
Europe: housing regulation in 357–358; rental housing in 359; sustainable city model 114
European Union (EU): GDP growth rate 11

fair market value 35, 378; *see also* compensation
family-planning policy 319–320
Fang, Hanming 29, 31, 32
Farley, P. 90
farmland preservation policy: and residential land supply 336; and urban development 295; *see also* agricultural land; rural land
favelas, Brazil 361
finance: and asset ownership 358; East Asian systems 274; Hong Kong 217–221; housing, Chinese 276; Japan 123, 274; Korea 157, 168–169, 171, 173n16, 274; land-based 380–381 *see* land-based finance; land-based finance 293–296, 301, 380–381; land policy/municipal finance, and commodity housing markets 295; local public, and land-leasing revenue 337; oversight 40–41, 42, 384–385; policies 311;

property tax-based financing 381; repression policies 274, 275; Singapore 235, 247, 274; structure of local governments, changing 380–381; Taiwan 196–197; United States 40, 112, 117

finance reforms 304, 327n4, 372, 384; alternatives to real estate for household retirement savings 385; dynamic real estate price evaluation for bank (lender) oversight 384–385; and municipal governments 276

Finland 55, 58; ARAVA system 51–52, 53; construction stage 51–53, 55; housing question 50; housing tenure 50; introduction stage 50–51; law on housing companies 50–51; management stage 53; owner-occupation 50–51; public retrenchment stage 53–54;rental housing in 364

Five-Year National Social and Economic Development Plan (2010–2015) 300

fixed effect model 32, 342

flexibility, and urban planning 92, 382

Florida, R. 231

Ford Foundation 232

foreign direct investment, and land supply 335

free entry 24, 357, 371

free markets 18, 65, 134, 146

Fukuoka 108

Fukuyama, Francis 270–271

Gallent, N. 215
garden city 96–97
Gautier, P. A. 74
Geleen/Sittard 71, 74
gentrification 300
George, Henry 30, 38, 39, 66, 232, 249n3
Germany 127; mortgage lending value 385; policies of tenure neutrality 364; population aging in 6, 7
Gibrat's law 75, 77
Gini coefficient index 234–235, 235
Glaeser, E. L. 31, 32, 34, 40, 41, 67, 74
Global Financial Crisis (2008/2009) 3–4, 16, 134, 143, 145, 275, 309–310, 316; aftermath, in Korea 157–159; and rental demand 359
globalization: of real estate markets, impacts of 8
Google Street View 30
Gottlieb, J. D. 31
government economic policy 22

government failures 13
Great Depression 3, 265
Great Housing Boom (1998–2013) 8, 258, 275–276, 278–280, 385; spatial dimensions across the Chinese system of cities 280–281
Great Recession 3, 258; impacts of 3–4; structural changes and housing policy 4–11
Green Belts: China 95–98, 99–100; Seoul 154; United Kingdom 89–93, 99, 381; worldwide 93–94
green growth challenging cities 11
gross domestic product (GDP) 296, 309; China 257, 265–267, 266, 280; of China 257, 280; and demographic aging 10–11; European Union (EU) 11; evolution of ratio of housing investment to 265–267, 266; Japan 105–106, 108; Korea 161; Singapore 266; slower growth rate as normal 10–11; and sustainable growth 266–267; take-off stage 266, 267; United States 10; of United States 258
growth transition (economic development) 258, 259, 260, 261, 265; of China 268–270, 269; of East Asian countries 267–268; implications, of China 20–24; and political adjustments 268; urban and housing issues of China during 281–283
Guangzhou 96, 299
Gurran, N. 215
Gyourko, Joseph 31, 32

Habitat III conference (United Nations) 135, 232
Haskins, R. 250n15
Hawaii Housing Authority v. Midkiff 35
Hayes, Shirley 370
hedonic regressions 34
Heerlen 71, 76
Heller, M. 37
Helsley, R. W. 78
Hendren, N. 362
Hills, J. 231
Hills, R. 37
Himmelberg, C. 31
Holmes, Oliver Wendell, Jr. 241, 250n11
Hong Kong 17–19, 201, 335; affordability issues 202–205, 212–214, 215, 222; affordable rental housing shortage 205–207; British Crown Colony of 274; co-financing 218;

constrained land supply 215–217; foreign investments in 219–220; government's dual role 216–217; Green Belts 93–94; Harbour Protection Ordinance 215; homeownership rate 202–205; Home Ownership Scheme 212; Hong Kong Housing Authority 218–219; housing class concept 209; housing estates in 362; housing ladder concept 208–209, 209, 211; housing mobility and dualistic system 208–215; housing quality policies in 360; housing subsidies 209–211, **210**, 214–215; impacts of land sale and land exchange on housing supply 334; implications for Chinese cities 222–225; investment demand management 219–222; land management and land policy 216; land shortage 215; land supply 215–216; land-use zoning in 368; latent homeowners 211–212; limited financial tools for homeowners' risks mitigation 217–219; loan-to-value ratios 218; Long Term Housing Strategy (2014) 221, 222; low housing space standards 207–208; Mandatory Provident Fund 221; Minimum Wage Policy 206; mortgage insurance policy 218; mortgage interest rates 217–218; private housing 212; public housing in 272–273, 361, 363, 365, 366; public land leasing system 334; public–private duality of housing system 214–215, 222; refinancing 220–221; residential land supply in 334; Special Administration Region (SAR) 18; "Starter Homes" scheme 225n3; structural challenges 222; top-up financing 218; vacancy rates 221
household income: distribution of households in Singapore by 237, **238**; and housing prices 279, 338, 341; *see also* income inequality; income segregation
household responsibility system 277, 327n1
households, urban: in rental housing 297–298; savings 297; size, average 296
house-purchasing limitation orders, and housing prices 341
housing consumption: prioritizing over asset ownership 357–359
housing cycles 270, 279

housing demand regulation, in Singapore 244
housing estates 317–319, **318**, 318, 362; integration of house types, Singapore 245
Housing Movement, Taiwan 180
housing poverty 48–49, 54, 55
housing price indices (HPI) 279
housing prices 279; diffusion, pattern 281; and household income 279, 338, 341; and house-purchasing limitation orders 341; and housing supply 338, 341–342; and land valuation 279; and non-agricultural population 341; and per capita disposable income 338, 341; and residential land supply (*see* residential land supply, and housing prices); and rise of land values 279; Singapore 243; stability 171; in Taiwan 181–189, 190, 197; in United States 334; urban housing 309, 310 *see* residential land supply, and housing prices
Housing Provident Fund (HPF) 301–302; limited coverage of 302; loans 301–302
housing question 48, 50, 55
housing shortage 8, 54, 57; in Finland 50, 52, 54; in Hong Kong 205–207; in Japan 123; in Korea 142, 143, 146, 151, 157, 158, 170, 172n1
housing supply: elasticity of 366; in Hong Kong 334; and housing prices 338, 341–342; new housing supply, in Korea 159–162; private, in Singapore 243; volatility, lowering 380
housing tenure 45, 47, 50, 58n1
housing vouchers 357, 362, 364, 365, 376
housing wealth 230–231; distribution, in Singapore 235–236, 249n6
Howard, Ebenezer 96
Ho, W. K. 94
Huang, D. 99–100
Huang, J. 334
Hui, C. M. 334
hukou (rural household registration) 5, 21, 273, 288, 323; and affordable housing programs 301; and small property rights housing 303
human assessor 34
human capital, of East Asian societies 270
hyperinflation 307

Iceland 45
Ihlanfeldt, K. R. 334

impossibility theorem (Arrow) 387n4
Inclusive Growth in Cities Campaign 232
income inequality 7, 230; Gini coefficient index 234–235, 235; and intergenerational mobility 231; in Singapore 234–236
income segregation 231; in Singapore 236–239
index of dissimilarity 241
India: Land Law 2013 36, 37; premium over market value 36
individual housing investors 33
industrialization 278; Krugman model 263–265, 264, 270, 271–272; Lewis model 258, 259–262, 261
industrial land subsidization 295
industrial land supply, and residential land supply 337
industrial park development, by municipal governments 336
Industrial Revolution 262
information systems for demand-side subsidies 376
infrastructure: alternative infrastructure provision mechanisms 382–383; undermaintenance 383; unwise investment problem 382–383
institutional framework, of housing delivery system 170–171, 173n18
institutional reforms 372; changing financing structure of local governments away from land-based finance 380–381; eminent domain process 378–379; improving information systems for demand-side subsidies 376; integrating rural and urban land ownership systems 377–378; international best practice for joint ownership structures 375–376; internet-accessible zoning and permitting processes 374–375; land lease renewal rules, clarifying 373–374; lowering volatility of local housing supply 380; public property development institutions 379–380; reducing local regulatory barriers 380; universal property registry and cadaster 372–373
interest rates 4, 9–10, 18, 31
International Commission on Growth and Development 267
International Monetary Fund (IMF) 9, 143, 242, 257
internet-accessible zoning and permitting processes 374–375

inventory, excess 32
investment-led growth policies, of China 274–276

Jacobs, Jane 67, 74, 361, 368, 370
Japan 15, 108, 358; affordability problems, homeowners and tenants 119–122; ageing population 107–108; agency restructurings 123; Asian currency crisis 105; Basic Law of Housing Life of 2006 123; Big Issue Japan Foundation 125; Building Standards Law 122; capital investment 106; City Planning Acts 110, 112; City Planning Law of 1997 117; compact city policy 113–114, 127; condominiums 111, 115, 117–118; consumption tax 105, 107; corporate housing 125; deflationary cycle in 105–107; deregulation policies 110–114; economical changes 105–107; employment, regular vs. non-regular 106–107, 110; financial repression policies 274; GDP 105–106, 108; Government Housing Loan Corporation (GHLC) 122–123; government's housing policy, restructuring of 122–124; Green Belts 96; growth transition and 50% level of urbanization in 269; housing allowance and social housing provision 126–127; Housing Assistance Council 125; Housing Construction Plan Act of 1966 123; housing demand 126; housing experience, lessons from 127–128; Housing Finance Agency (HFA) 123; Housing Life Master Plan 123–124; housing market, polarization and high volatility 116–119; housing policy, and private rental housing market 124–125; housing poor 122, 126, 127; housing problems and housing policy 122–127; housing security 126; housing standards 116; housing stock 114–122, 127; illegal rental rooms 122; income distribution Inequalities 108–110; Japan Housing Corporation (JHC) 122–123; land fairy tale 270; land policy and urban planning in cities 110–112; life expectancy 7; Livelihood Protection System 125; local cities 113–114; manufacturers, elimination of 107; metropolitan cities 112–113; Minato Ward 115, 125; Ministry of Construction 116; Ministry of Land, Infrastructure, Transportation,

and Tourism (MLIT) 115, 116, 122, 123; National Consumption-state Survey 122; National Tax Administration Agency 110; overabundance and differentials in quality of housing stock 114–116; owner-occupied houses vs. rental houses 116–117; owner-occupied housing 126–127; planning permissions 110; polarized land and housing markets 127; population aging in 6–7; population decrement 107–108; post-bubble economy of 105–110; post-war reconstruction 272; poverty 108–110; private rental housing 118–119, 125; Public Housing System 122–123; public rental housing 125–127; rental housing 118–119, 122, 126, 127; rent subsidy schemes 126–127; reverse sprawl 114, 127; reversion to recession 107; singleton households 108; social housing system 127; Special Measures Law of Urban Regeneration 112; studio apartments 119; subsidy reduction policy 124; underinvestment in housing 276; Urban Development Corporation 123; urbanization control area (UCA) 110; urbanization promotion area (UPA) 110; Urban Regeneration 112; Urban Renaissance Agency (UR) 123; Urgent Redevelopment Zones (URZs) 113, 117; vacancy rates 114–115
Jinxing, Z. 97
joint ownership: of rural land 377; structures, international best practice for 375–376
joint ventures, and land supply 335
Jorgenson, Dale 260

Kahn, M. E. 67
Kaika, M. 90
Kanagawa 108
Katz, L. F. 362
Kelo v. City of New London 35
Kim, C.-H. 136
Kim, K.-H. 136
Kim, S. 264
Kinkaid, Michael 34
Kitakyushu 96
Korea 16, 108, 150–151, 170; AFC recovery phase 155–157, 175; *Bojeongbuwolse* 131; build-transfer-lease (BTL) scheme 168; *Chonse* Deposit Loan Program 141; *Chonsei* houses 131, 139, 158; construction cost linkage system, apartment prices 143, 147n2; dealing with squatter settlements 153–154; debt-to-income ratio (DTI) 139, 156; demand-side policy measures 153; developable land supply 162–164; financing 168–169, 171, 173n16; and Global Financial Crisis 136–137; governmental role and intervention 134–142; Happiness Housing 141; Homebuilders' Cooperative 171; homeownership 131; house prices in 137; housing allocation, inequality of 146; housing ceiling price system 147; Housing Construction Promotion Act (1972) 152; Housing Corporation 140; housing cycles 164–166; housing divides 146; housing investment as a share of GDP and of TFCF 161; housing market boosting measures 137–139; housing polarization 145–146; housing situation and housing market dynamics in 130–134; Housing Voucher Program 141; HUG 168, 171; *Ilse* 132; income polarization 146; infrastructure provision 167–168; institutional framework of the housing delivery system 170–171, 173n18; institution building phase 152–154, 173–174; integration between land development and infrastructure provision 171; interventionist approach to housing challenges 150; Korea Homebuilders' Cooperative 155; land and housing delivery system, evolution of 151–159; land development 167; Land Readjustment Act (1966) 152; land readjustment scheme 152, 162; loan-to-value (LTV) ratio 139, 156; local government 163; low-income urban housing, policy measures 139–140; massive supply phase 154–155, 174; milestone events and housing policy interventions in 173–175; Monthly Rent Assistance Program 141; National Housing and Urban Fund (NHUF) 168; National Housing Fund 140, 141, 143; National Housing Fund (NHF) 152, 159, 171; National Public Housing Fund 141; new housing supply 159–162, 169–170, 171; New Stay program 141–142, 158; on-site infrastructure 167–168; owner-occupied housing 137;

policy issues and implications 142–146; post-Global Financial Crisis market reshaping 157–159, **175**; price ceiling 135–136, 169–170; price control disputes 142–143; price control on the new apartment submarket 135–137; price control system 142–143; private market, housing production and allocation 134; project finance (PF) loans 157, 169, 172; public housing development 139–140; public housing for rent 140; public housing for sale 140, 143–144; public housing provision 139–142; Public Land Development Act (1982) 152; public land development scheme 152–153; public–private partnership (PPP) method 168, 171; public sector and the allocation process 143–144; public-sector housing institutions 151; public sector, housing production and allocation 134–135; real estate investment trusts (REITs) 158, 171; real estate lending restrictions, and relaxations by government 137–139; redevelopment of built-up areas 153, 156–157, 161–162, 172; rental housing 137, 162; residential development system 166–170; *Sakwolse* 131; Seoul Capital Region (SCR) 143, 150, 154; Seoul Metropolitan Government (SMG) 153–154, 157; social cohesion 145; social housing and non-social housing 145; social mix 144–145; social polarization 144; socio-tenurial marginalization 144; supply system and housing outcomes 159–166; Two Million Houses Construction Project 135; Two Million Housing Drive (TMHD) 154–155; Urban Redevelopment Act (1976) 152; *Wolse* 132; *see also* South Korea
Korea Housing Bank (KHB) 152, 171
Korea Housing Finance Corporation (KHFC) 156
Korea Housing Guarantee Corporation (KHGC) 155, 171
Korea Land and Housing Corporation (LH) 140, 143–144, 146, 152, 170
Korea Land and Housing Corporation Act 141
Korea Land Corporation (KLC) 141
Korea Land Development Corporation (KLDC) 152, 159, 163, 170

Korea National Housing Corporation (KNHC) 141, 152, 153, 159, 163, 170
Korean Housing and Urban Guarantee Corporation (HUG) 159
Krueger, A. 231
Krugman model 258, 263–265, 264, 270, 271–272, 283n3

labor markets, Chinese 269, 280
Lai, L. W. C. 94
Lam. W. R. 33
Land Administration Law 293, 302, 303
Land Bank 17n2
land-based finance: and affordable housing units 301; changing financing structure of local governments away from 380–381; impacts on housing markets 293–296
land concession fee, for rental housing 299
land concession revenue, and urban expansion 294, 294, 302
land consumption per person, near CBD 65
land conversion multiplier 383
land institutions and land operations across urban and rural areas, integration of 21
land leasing 334, 335–336; renewal rules, clarifying 373–374
land ownership: and eminent domain 378; and rural-urban conversions 383; systems, integrating rural and urban 377–378
land rents 63, 64; construction of 84–86; in the Netherlands 67–75; total land rent surplus and social value of local public goods 65
Land Reserve Center (Ministry of Land and Resources) 336
land reserves, urban 336, 337
land speculation/hoarding in Singapore, measures to prevent 243
land supply: and affordability 360; Hong Kong 215–217, 334; Korea 162–164; residential *see* residential land supply
land tax 39–40, 66–67
land use 13, 65; control, and leases 374; and fixed wages 65; Green Belt *see* Green Belts; map of land-use restrictions 374–375; planning 382; regulation, efficiency of 367–368
land-use rights 294, 335; of municipal governments 276; trading of 377

Index 397

land valuation 39; and housing prices 279; and market price 367; *see also* property valuation/value
Leamer, E. E. 270
lease/leasing: land lease renewal rules, clarification of 373–374
Lehman Brothers 147n3
Lehman shock 105, 107, 112, 137
Lewis dual-sector model of development 258, 259–262, *261*
Lewis turning point 260, 269, 280
life expectancy, increase in 6
Liu, Z. 302
loans *see* debts/loans
local governments 97–98; *see also* municipal governments
local regulatory barriers, reduction of 380
location: affordable housing units 301; and property valuation 34; and structure, jointly considering 360–363
long-term dynamics of housing, in China 257; East Asian perspective (*see* East Asian perspective, China's development from); economic models (*see* economic models); evolution of ratio of housing investment to GDP 265–267, *266*; four decades of profound change 257–259; Great Housing Boom (1998–2013), China 278–280; growth transition of China 268–270, *269*; investment-led growth compared with earlier East Asian experiences 274–276; overinvestment of China in housing and real estate after 1994 276–277; population 257; privatizations, housing (1998–2003) 277–278; spatial dimensions of housing boom 280–281; three main stages of long-term economic development 259, *261*; urban and housing issues during growth transition 281–283
long-term urban change, planning for 381–382
Lucas, R. E. 63, 67, 74

macro-adjustment measures/policies 290–291, 309, 310
macro-economic policies, of China 275
macro-prudential policies 15
maps: of land-use restrictions 374–375; zoning 369–370
market rescuing 307
Marlet, G. 63, 74, 76
Massey, D. S. 241

Mayer, C. 13
mean reversion 280, 384
Meiji reforms (Japan) 268
Mian, A. 31
middle-income trap 20, 258, 262, 265, 267, 268
migrants/migration 296; and affordable housing programs 289, 301; buying down process 323, *323*, 323–324; buying up process 323, *323*, 324; and county towns 320; and rental housing 298; and urban housing market 322–325, *323*, 326; and urban villages 299–300
Ministry of Housing and Urban-Rural Development (MOHURD) 301
Ministry of Land and Resources 336, 339
mixed-use zoning 368
mobility rates 37
mono-banking system 274
Moon Jae-in 142, 148n4
Moore, R. 209
mortgage: contracts 370; in county towns 320; foreclosure economy 358; in Hong Kong 217–218; lending value, in Germany 385; loans, in China 275–276, 278; loans, in Singapore 242; in Taiwan 180–181, 191, 192, 194, 197; in United States 358; and urban housing market 310, 311
Moses, Robert 370
municipal governments 299, 311; commercial borrowing with land as collateral 294; and industrial land subsidization 295; land concession revenue and local tax revenue 294, *294*; leasing of residential land 334; and local economic development 294, 295; and overinvestment in housing/real estate 276–277; property tax 295–296; and residential land supply 336, 337; sale of land-use rights 294–295; and urban expansion 293–294
municipal investment companies (*Chengtou*) 276, 284n10

Naik, Nikhil 34
Nanjing 96
Nathanson, C. 32, 42n1
National Bureau of Statistics (China) 269; housing price indices 279
national identity card system 376
National Population Development Plan of 2016–2030 296
natural histories 44, 54, 58

Nenov, P. T. 184
Netherlands, the 13, 62–63, 127; city extension 75–84; construction value 67–69; land rent differentials and social value of agglomeration 69–75; land rents 67–69; policy evaluations framework 63–67; social housing projects in 362
New Normal period 289, 296
New Taipei 188–189, 192
Ng, I. 247
Ng, K. H. 250n15
Noh Tae-Woo 154
non-agricultural population, and housing prices 341
non-profit entities/trusts 373, 377, 378
Nordic countries: Finnish housing policy *see* Finland; housing policies in 44–46, 56–57; housing tenure in 45–46; public housing policy 58; scrutinization of stage model of housing policy 54–57
Norway 45

Occupy movement (2011–2012) 230
online permitting processes 374–375
online zoning and permitting processes 374–375
open spaces 80, 96, 367
opportunity costs 358, 365
Organisation for Economic Cooperation and Development (OECD) 29, 108, 145–146, 232, 234, 235
Osaka 111–112, 113, 116, 125
Osborn, F. J. 93
Ostry, J. 9
over-building 38, 42, 379, 380
overinvestment in housing/real estate 276–277
over-supply, in urban housing market 311, 324
owner-occupation 46, 50–51, 58n1, 244, 273

Papadimitrou, N. 90
Park Geun-hae 141, 148n4
path dependence, of housing system 5, 46–47
Pavan, R. 230
Peng, C.-W. 182, 183, 184–185
per capita disposable income, and housing prices 338, 341
peri-urban fringe 88–89, 97, 99; Chinese Greene Belts 95–98; recommendations for 99; UK Green Belts 89–93; worldwide Green Belts 93–94
Persson, T. 9
Phang, Sock-Yong 231, 236
Piketty, Thomas 7, 108, 230, 235, 236, 249n6
planning: inclusive/balanced 370; urban, reforms *see* urban planning reforms
policy actions in China 371–385; finance reforms 384–385; less difficult institutional reforms 372–376; more difficult institutional reforms 377–381; urban planning reforms 381–384
policy design, international experience: core principles for 357–364; efficiency of land-use regulation 367–368; efficiency of resource use 364–365; efficiency of subsidies 365–367; free entry and competition 371; inclusive, balanced planning 370; institutional qualities 368–371; jointly considering housing structure and location 360–363; prioritizing housing consumption over asset ownership 357–359; promoting tenure neutrality and robust rental markets 363–364; promoting widespread affordability 359–360; property rights, uniform 369; transparent rules 369–370; value of international experience for 355–356
policy development 44–47, 58; construction stage 47, 48–49, 51–53; in Finland (*see* Finland); introduction stage 47–48, 50–51; *laissez-faire* 48; management stage 47, 49, 53; path dependence of housing policies 46–47; public retrenchment or privatization stage 47, 49, 53–54, 56–57; scrutinization of stage model 54–57; tax subsidies 47
political order, in East Asian economies 270
poor people's housing situation 48
population aging: structural change 6–7
population and construction density, close to CBD 65–66
population densities, in East Asian societies 270
Poterba, J. M. 40–41
pre-industrial urbanization 262–263
premium over market value 36
price inflation 279; and rental housing 359; urban housing 307, 309, 311, 321, 325

price-to-income ratio, urban residential housing 291, *292*
priority investments 275
privatizations 5, 21, 273, 275, 277–278
Progressive Wage Model 247
property booms 31–33
property evaluation *see* property valuation/value
property registry, universal 372–373
property rights: duality in 377; and eminent domain 379; uniform 369
property tax 38–40, 41–42, 295–296, 297, 304, 311, 365, 374, 385; -based financing 381; in Singapore 244; in United States 246
property valuation/value 5, 24–25, 30–31, 41–42; central valuation 42; comparable sales prices 33–34; computer-vision method 34–35; enhanced value assessment 36–37; and financial oversight 40–41, 42; hedonic regression vs. human assessor 34; and land development restriction 42n4; national evaluation team 36, 42; and property taxation 39; theory and technology of 33–35; weakened property rights 36, 37
public facilities, provision in Singapore 245
public goods: and land tax 66–67
public housing projects: and opportunity cost of land 365; vs. private-sector projects 365
public parks 367
public–private partnerships (PPPs) 168, 171, 382, 383
public property development institutions, reform of 379–380
public schools, access of tenants to 299
purchasing power parity (PPP) 257
Putnam, R. D. 250n17

racial segregation 231
Rappaport, J. 67
rates of return 276, 280
real estate 257, 278, 379; alternatives, for household retirement savings 385; appraisals 40, 41–42; bubbles, and technology 29–42; as collateral 384; development, and land policy/municipal finance 295; development, competition in 371; dynamic price evaluation for bank (lender) oversight 384–385; overinvestment of China after 1994 in 276–277; property rights *see* property rights; sector policy 289, 300; transactions, and transparency 370
Real Estate Management Law 311
Reardon, S. F. 236, 237
Reckford, Jonathan 232
Redding, S. J. 66, 82
redevelopment of built-up areas 153, 156–157, 161–162, 172
rental housing 46, 308; vs. homeownership 359; market, development 297–300, *298*, 303; and registration 373; in Singapore 247–248; subsidies 363; tenure neutrality and robust markets, promoting 363–364; and young people 362
residential housing: average price index 290, *291*; investment 289, *290*; and land policy/municipal finance 295; prices 292–293, *293*; sales, in square meters 289–290, *290*
residential land supply 330; in Beijing *331*; in Chengdu *333*; in Hangzhou *331*; in Huhehaote *333*; in Taiyuan *332*; urban, evolution of 335–337; Urban Land Supply Plan 336; in Zhengzhou *332*
residential land supply, and housing prices: annual supply of new housing 341–342, **343**, **346–347**; benchmark regression 339–341; descriptive statistics **339**; determinants of urban housing prices 338; determinants of urban residential land supply 337; future housing prices 342, **344**; geographical and man-made land constraints 335; land-use control 334; literature review 334–335; model specifications and variables 338–338; policy implications of study 346; research data 339, **348–349**; tests of alternative channels of influence 341–342, **343–345**; urban residential land price 342, **345**
Residents' Network 248
resource transfer policies, in China 275
resource use: efficiency of 364–365
retirement savings, alternatives to real estate for 385
Rex, J. 209
Robert-Nicoud, F. 230
Roberts, M. S. 90
Rodrik, D. 259
Rognlie, Matthew 7–8, 230
Rossi-Hansberg, E. 63, 67, 74
Rotterdam 71, 74

rural land: conversion, for urban expansion 293, 302; integrating rural and urban land ownership systems 377–378; public rental housing on 299; registration of 373; *see also* agricultural land; farmland preservation policy
rural-urban conversions 383–384

Saez, E. 235–236
Saitama 108
Saiz, A. 74
Scandinavian welfare model 12–13, 45
Schelling, Thomas 231, 250n14
schools: access, in Singapore 246; public, access of tenants to 299; rural 320
sector policy 289
segregation 232; ethnic 239–241, **240**; income 231, 236–239; racial 231
Seongnam 153
Seoul 134, 137, 138–139, 142, 144, 153, 171; Green Belts 93–94, 154; land readjustment scheme 162; Two Million Housing Drive (TMHD) 154–155
Seoul Capital Region (SCR) 143, 150, 154
Sewell, William H. 46
Shanghai 95, 299
Shell-less Snail Movement 177, 180
Shen, X. 302
Shenzhen municipality 18
Shiller, R. J. 32, 40
Shoard, Marion 90–91
simultaneous-equation model 338, 339
Sinai, T. 31
Sinclair, I. 90
Singapore 19–20, 232–233; Additional Buyer Stamp Duty (ABSD) 244; buyer stamp duties 244; Central Provident Fund (CPF) 242; distribution of households by household income 237, **238**; ethnic ghettoes 248; ethnic integration policies 245, **246**, 250n14; ethnic segregation 239–241, **240**; Executive Condominium Scheme 243, 244; expatriates in 248–249; financial repression policies 274; fiscal transfers 244–245; Government Land Sales (GLS) program 243; homeownership rate of 233, *234*; Housing and Development Board (HDB) 233, 236, 240, 242, 243, 244, 245, 247–248, 273; housing estates 362; housing investment in GDP in 266; housing policies in 276, 359, 360, 363; inclusive, land and housing policies for 241–246; inclusive society building 246–249; income inequality and housing wealth distribution 234–236, *235*; income segregation 236–239; intergenerational mobility in 231, 247; Lease Buyback Scheme 247; Ministry of Finance 235, 247; neighborhood diversity in 248; pension system in 242; population aging 247; pro-homeownership policies 363; proportion of households residing in low-, middle-, and high-income neighborhoods 237, **238**; proportion of population residing in low-, middle-, and high-income neighborhoods 238–239, **240**; public housing in 272–273, 361, 363, 365, 366; relationship between average household income and house type 237–238, **239**; segmentation of housing markets 244; Singapore Land Authority 243; and social mobility 247; socio-economic characteristics by ethnic group **240**; speed of development of 261; transfers in 244–245; welfare system 247, 248; work permits in 248–249
Sinitic cultural heritage, of East Asian societies 270
Skinner, G. William 262
small property rights (SPR) housing 289, 302–303
Smith, S. J. 219, 220
social cohesion, and neighborhood diversity 248
social housing policies 357–358, 360, 362
socialist market economy, China as 273, 284n8
socialist welfare housing system 272, 307, 325
social security system, and household savings 297
social value of local public goods and total land rent surplus 65
South Korea: financial repression policies 274; growth transition and 50% level of urbanization in 269; population aging in 7; reconstruction following 1950–1953 Korean War 272; speed of development of 261, 272; underinvestment in housing 276; *see also* Korea
squatting 35, 77, 153, 233
State Council 296, 300, 309, 310, 335

state-owned enterprises (SOEs) 275, 277, 278, 279, 280
Sterk, V. 184
Stiglitz, J. 230
Strange, W. C. 74
streetscapes, and housing structure 361
structural changes, in Chinese housing policy 4–6; climate change and environmental constraints 11; demographic change and population aging 6–7; economic inequality 7; GDP growth rate 10–11; household debt 9–10; income inequality 7; wealth inequality 7–9
structural problems, in housing market 309
structure, housing 360–363
Sturm, D. M. 66, 82
subprime lending and price grown 31–32
subsidies: demand-side 365, 366, 367, 376; direct provision 366; efficiency of 365; homeownership 358, 362; in Hong Kong 209–211, 210, 214–215; industrial land 295; inefficiency in use of 367; in Japan 124, 126–127; and policy development 47; rental 363; supply-side 365–366; in Taiwan 190–191
Sufi, A. 31
Summers, L. H. 40–41
Sun, L. 302
superstar cities 8
supply-side subsidies 365–366
surplus labor 259–260
sustainable growth (economic development) 259, 260, 261, *261*; housing assets during 270; housing investment in GDP during 266–267
Svarer, M. 74
Svensson, Stefan 58n1
Sweden 45
Swyngedouw, E. 90

Taipei 180–181, 186–188, 189, 192, 193
Taiwan 14, 17, 177–180, 274; affordability problem 178–180, 185, 190; affordable housing strategies 189–193; challenges 193–195; corruption cases 199n16; Equalization of Land Rights Act 193; executive suggestions for 197, 198; finance system 196–197; government policies 178–180; growth transition and 50% level of urbanization in 269; homeownership rate 182–183, 185; Housing Act 190; housing market strengthening strategies 191–193; Housing Movement 193–194; housing policies, changes in 189–191; housing prices and housing mobility 184–189; housing problems and causes 180–184; housing protest 177, 180, 194; housing strategy 273; housing subsidy programs 190–191; Immovable Property Gains Tax 193; impact of rapid increases in housing prices 181–183, 190, 197; Integrated Housing and Land Tax 193; intra-city mobility rate (ICMR) 186–189; Land Administration Agent Act 193; Land Value Increment Tax 181, 192, 193; Land Value Tax 181; Luxury House Tax 193; marriage rate 185; migration, influence on housing price 184; migration rate (MR) 186–189; mortgage market 197; mortgage payment-to-income ratio (MIR) 180–181; mortgage subsidies 191; mortgage-to-price ratio (MPR) 192; net migration rate (NMR) 186–189; Overall Housing Policy 180, 190, 191; perspectives 195–197, 198; price-to-income ratio (PIR) 180–181; public housing 190, 199n15, 199nn12–13; Public Housing Construction Loan Act 189–190; Real Estate Brokerage Management Act 193; real estate cycles 177–178, 180; real estate industry management, strengthening of 196; real estate taxation reforms 192–193, 196, 199nn7–8; real estate transaction transparency 193; rental housing market enlargement 195–196; rental yield 199n10; rent-seeking 199n14; rent subsidy 191; residential mortgages loans 194; Selected Goods and Services Tax (Luxury Tax) 192; selective credit controls 191–192, 198n3; social housing units quantity 195; structural change in 194–195, 198; total fertility rate (TFR) and general fertility rate (GFR) 195–196; urban regeneration and redevelopment pace 195
take-off stage (economic development) 259, 260, 261, *261*, 263–265, *264*; of East Asian economies 271, 272; emergence of new Chinese housing system during 274–281; housing investment in GDP during 266, 267
Tang, B. S. 205–206
Tang, Y. 98
Tan, M. 97

Taoyuan 182, 183, 185–186, 189
tax 30; and asset ownership 358; credits, for homeownership 365, 366; housing transaction tax 310–311; land tax 39–40, 66–67; local tax revenue 294, *294*; property tax *see* property tax; -sharing system 294; subsidies 47; in Taiwan *see* Taiwan; transaction-based taxes 39–40, 310–311, 381
tenant-owner 45, 58n1
tenure neutrality, promotion of 363–364
Teulings, C. N. 63, 74
Thompson, A. K. 32
Tilburg 71, 74
Tokyo 108, 111–112, 113, 115, 116, 117, 118, 119, 122, 125, 127; Green Belts 93–94
Tokyo Olympics (2020) 117
total fertility rates (TFRs), decline in 6
total land rent surplus and social value of local public goods 65
total quantity dynamic equilibrium of basic farmland (TQDEBF) 95
trade theory 263
transparency: in property development process 379; of rules 369–370
transportation: access, in Singapore 245; allocation of space for 367; costs, and housing consumption 361; and urban planning 382
Tsai, I. C. 183
Tu, F. 302

undermaintenance problem, infrastructure 383
United Kingdom 108, 123; development rights in 363; Green Belts 89–93, 99, 381–382; housing wealth in 230–231; London Green Belt Act (1938) 89; Restriction on Ribbon Development Act 1935 89; social housing policies in 357; Town and Country Planning Act (1947) 89
United Nations: Habitat III conference 135, 232
United States 29, 108, 123, 263–264; Affordable Housing Movement 360; construction costs, and housing prices 334; cooperative buildings in 375; Dodd-Frank Law (2010) 384; economic development of 263–265; eminent domain 35–36; eminent domain process 378, 379; financial crisis 112, 117; financial oversight 40; free entry and competition in 370; GDP growth rate 10; growth take-off of 264–265; growth transition of 265; housing assets of 258; housing crash (2007–2008) 386–387n2; housing policies of 358; housing regulation in 357; housing vouchers 376; impact of housing location in 362; impact of low poverty neighborhoods on children 237; income segregation in 236; Indian Land Reform Law (2013) 378, 379; Land Assembly Districts 37; local schools 246; Low-Income Housing Tax Credit 366; mean reversion in 384; planning in 370; post-World War II, public housing projects in 360–361; pro-homeownership policies of 363; property booms in 31–32; proportion of households residing in low-, middle-, and high-income neighborhoods 237, **238**; regional specialization in manufacturing and other activities (1860–1987) 264; rental housing in 359; wealth distribution of 236; wealth redistribution in 8
urban and rural land, institutional differentiation between 97
urban density 66–67
Urban Development Investment Corporations (UDICs) 379
urban dowry 90
urban housing challenges 288–289; affordable housing policy and implementation **300**, 300–302; average monthly rent in Beijing *298*; average price-to-income ratio 291, *292*; average residential housing price index 290, *291*; commodity housing markets 292–293, *293*; factors influencing housing demand 296–297; golden period and its aftermath 289–293; impacts of land-based finance on housing markets 293–296; land concession revenue and local tax revenue 294, *294*; population and urbanization rate 296, *296*; real estate stocks **292**; rental market development 297–300; residential housing investment 289, *290*; sales of residential housing in square meters 289–290, *290*; small property rights (SPR) housing 302–303
urban housing market 23, 307–308, 325; average annual salary and housing price in urban areas 309, *310*; average price of commercial housing 311, **312**;

cash and investment upward flow 324; control vs. support policies 308–312; county town (case study) 312, 314–322, *315*, *316*, *317*, **318**, *318*; migration and it's impacts on 322–325, *323*, *326*; regulation policy options **313–314**; structural problems 309
urbanization 62, 282; and housing investment 265–267, *266*; Krugman model 263–265, *264*, 270, 271–272; Lewis model 258, 259–262, *261*; slow pre-industrial, von Thünen model 262–263
Urban Land Institute 232
urban planning reforms 13, 372, 381; alternative infrastructure provision mechanisms 382–383; long-term urban change 381–382; rural-urban conversions 383–384
urban public services, access of tenants to 299
urban sprawl 13, 88, 91, 94, 97–98, 99, 114
urban villages 299; rental housing 299–300
user cost of capital 279

vacancy rates 279–280; urban housing 291, 305n6
Vasoo, S. 248
Vermeulen, W. 63, 74
von Thünen model 258, 262–263

wages: and efficiency of land use and city size 65; wage surplus 64, 65
weakened property rights 36, 37

wealth distribution, inequality in 7–8
Western housing experiences 12–14
Woerkens, C. van 76
Wolf, Martin 10
Wolf, N. 66, 82
Wood, G. 211
World Trade Organization (WTO) 277
World War II (WW2): actual development record since 267–270; East Asian development policies after 271–272; economic growth of East Asian economies after 270; and Japan 268
World Wealth and Income Database (WID) 108
Wu, J. 32, 284n8

Xi Jinping 98, 268, 304, 305n8, 359
Xiong, Wei 31

Yang, J. 97
Yantai City 96
Yip, C. S. 247
Yuan, S.-M. 199n15
Yu, W. G. 262

Zhang, J. 205–206
Zhu Rongji 277
Zipf's law 75
zoning, land-use 368, 369–370; online 374–375; and rural-urban conversions 383
Zucman, G. 236
Zwick, E. 42n1

Taylor & Francis eBooks

www.taylorfrancis.com

A single destination for eBooks from Taylor & Francis with increased functionality and an improved user experience to meet the needs of our customers.

90,000+ eBooks of award-winning academic content in Humanities, Social Science, Science, Technology, Engineering, and Medical written by a global network of editors and authors.

TAYLOR & FRANCIS EBOOKS OFFERS:

- A streamlined experience for our library customers
- A single point of discovery for all of our eBook content
- Improved search and discovery of content at both book and chapter level

REQUEST A FREE TRIAL

support@taylorfrancis.com

Routledge
Taylor & Francis Group

CRC Press
Taylor & Francis Group